THE HARMONY OF ILLUSIONS

THE HARMONY OF ILLUSIONS

INVENTING POST-TRAUMATIC

STRESS DISORDER

Allan Young

PRINCETON UNIVERSITY PRESS   PRINCETON, NEW JERSEY

Library of Congress Cataloging-in-Publication Data

Young, Allan, 1938–
The harmony of illusions : inventing post-traumatic
stress disorder / Allan Young.
p.   cm.
Includes bibliographical references and index.
ISBN 0-691-03352-8 (cloth : alk. paper)
1. Post-traumatic stress disorder—Philosophy.
2. Social epistemology. I. Title.
RC552.P67Y68   1995
616.85′21—dc20   95-16254

This book has been composed in Times Roman

Princeton University Press books are printed
on acid-free paper and meet the guidelines
for permanence and durability of the Committee
on Production Guidelines for Book Longevity
of the Council on Library Resources

Printed in the United States of America by Princeton Academic Press

10 9 8 7 6 5 4 3 2 1

For Roberta

Contents

Acknowledgments ———————————————————————

I OWE a debt to colleagues and friends in the Department of Social Studies of Medicine and the Department of Psychiatry at McGill University: I thank Don Bates, Alberto Cambrosio, Margaret Lock, Faith Wallis, George Weisz, and Laurence Kirmayer for their invaluable advice. I owe a similar debt to Atwood Gaines and Ronnie Frankenberg, who read and commented on earlier drafts. Arthur Kleinman has shown me many kindnesses, both intellectual and professional, while I prepared this book. I am grateful for these favors, but most of all for his sympathetic understanding of my overall project. I am likewise grateful to Ian Hacking for his advice, and I hope that the influence of his ideas on my book will be obvious to my readers. I also want to thank Mark Micale for many helpful suggestions, particularly in connection with the historical sections.

The ethnographic segment of this book would have been impossible without the many favors shown to me by Jack Smith, Glenn Davis, David Lebenthal, and Susan Johnson. I am likewise grateful for the consideration given to me by both the patients and staff whom I met and was permitted to listen to at the Veterans Administration medical unit described in the following pages. It would be difficult to exaggerate my special obligation to Paul Emery. He has been a friend and a mentor. He has been generous with his time and his knowledge and has treated me with a measure of kindness that will always place me in his debt.

I thank William Schlenger, Terrence Keane, and John Fairbank for the valuable information they provided in connection with the National Vietnam Veterans Rehabilitation Study. I also want to express my gratitude to the Veterans Administration Medical System for its material assistance during my field research. McGill University has generously provided me with a faculty research grant to support the preparation of the final manuscript.

I also want to thank Mary Murrell for her expert opinion, patience, and encouragement, and Vicky Wilson-Schwartz for her meticulous attention to my manuscript. The infelicities and obscurities that remain are entirely my own.

The research for this book began in 1986. In the intervening years, I presented drafts of chapters at seminars and conferences in Canada, the United States, the United Kingdom, and Germany. I am grateful to the colleagues who were present on these occasions for their many helpful suggestions. I apologize for failing to express my gratitude to them individually.

My greatest debt is to Naomi Breslau. Without her help and encouragment, I would never have found my way to the subjects and places that are described in this book. Without her guidance and instruction, I would never have understood the things that I now claim to understand.

THE HARMONY OF ILLUSIONS

Introduction _____

As FAR BACK as we know, people have been tormented by memories that filled them with feelings of sadness and remorse, the sense of irreparable loss, and sensations of fright and horror. During the nineteenth century, a new kind of painful memory emerged. It was unlike the memories of earlier times in that it originated in a previously unidentified psychological state, called "traumatic," and was linked to previously unknown kinds of forgetting, called "repression" and "dissociation."

The new memory is best known today in connection with a psychiatric malady, post-traumatic stress disorder (PTSD). PTSD was adopted by the American Psychiatric Association as part of its official nosology in 1980, and it rapidly attracted the attention of clinicians and researchers throughout the Americas, Britain, Australia, Scandinavia, and Israel. Soon afterward, a contributor to the *British Journal of Psychiatry* reported that he had discovered evidence of the syndrome in the *Diary of Samuel Pepys*, in the pages where Pepys describes his mental condition following the Great Fire of London in 1666 (Daly 1983:67). Pepys's self-reported symptoms are said to correspond to the diagnostic features listed in the official nosology—intrusive images of his frightening experience, feelings of detachment and estrangement, survivor's guilt, memory impairment, and so on—and the *Diary* is said to establish that PTSD existed in the past. In the following years, writers uncovered progressively earlier historical evidence of PTSD. The shadow of traumatic memory was spied beneath the surface of this passage from Shakespeare's *King Henry IV, Part One* (Trimble 1985:86):

> Tell me, sweet lord, what is't that takes from thee
> Thy stomach, pleasure and golden sleep?
> Why dost thou bend thine eyes upon the earth,
> And start so often when thou sit'st alone?
> Why hast thou lost the fresh blood in thy cheeks,
> And given my treasures and my rights of thee
> To thick-eyed musing and cursed melancholy?[1]

More recently, evidence of PTSD has been discovered in the *Epic of Gilgamesh*, which takes the disorder and its memory back to the dawn of recorded history (Boehnlein and Kinzie 1992:598; also Parry-Jones and Parry-Jones 1994).

In the following chapters, I argue that none of these writers—neither Pepys, nor Shakespeare, nor the author of *Gilgamesh*—was referring to the thing that we now call the traumatic memory, for this memory was unavailable to them. In everyday usage, the term "memory" has three meanings: the mental *capacity* to retrieve stored information and to perform learned mental operations, such as long division; the semantic, imagistic, or sensory *content* of recollections; and the *location* where these recollections are stored. John Locke and David Hume proposed that memory, in the second and third senses, is intrinsically connected to our conception of "self" and "self-awareness" (Richards 1992:159–161; Warnock 1987:57–60). By connecting self-awareness with the past, memory provides the body with a subject and subjectivity. It is the source of the "I" that initiates the body's purposeful acts and the "me" who experiences its pleasures and vicissitudes and must accept responsibility for its actions. Without memory, the I/me would fail to transcend momentary states of awareness and self-consciousness. It would be a string of experientially unconnected points, a "dehumanizing" condition that is associated with certain neurological pathologies (cf. Parfit 1984:202–217). Our sense of being a person is shaped not simply by our active memories, however; it is also a product of our conceptions of "memory." What occupies me in this book is how certain of these conceptions have changed over time, together with the practices through which memories are retrieved, interpreted, and narrated.

In eighteenth-century Europe, the prevailing conception was that a memory consists of mental images and verbal content: a person sees, says, or otherwise apprehends the things that he remembers.[2] During the next century, the boundaries of memory were expanded to include contents located in acts and bodily conditions (e.g., automatisms, hysterical contractures) as well as words and images. Further, the very fact that these acts and conditions were "memory" was unknown to the person who owned the memory.

This new conception was based on the idea that intensely frightening or disturbing experiences could produce memories that are concealed in automatic behaviors, repetitive acts over which the affected person exercised no conscious control. Without the intervention of an expert, the owner of a "parasitic" memory remained unaware of its content and ignorant that it influenced aspects of his life—an idea that would have been literally unthinkable in the previous century.

The discovery of traumatic memory revised the scope of two core attributes of the Western self, free will and self-knowledge—the capacity to reflect upon and to attempt to put into action one's desires, preferences, and intentions (Dworkin 1988:chaps.3 and 4; Harris 1989:3; Johnson 1993:chap. 6; Ouroussoff 1993:287–295). At the same time, it created a new language of self-deception (Rorty 1985) and justified the emergence of a new class of authorities, the medical experts who would now claim access to memory contents that owners (patients) were hiding from themselves.

Most of what has been written about traumatic memory and PTSD over the last two decades has described these developments from the viewpoint of researchers and clinicians who work on PTSD and are convinced that it is timeless (Trimble 1985; Herman 1992). This view is based on a chronicle of events that runs more or less as follows.

Although there is a "lack of historical and theoretical continuity" in the evolution of psychiatric knowledge of PTSD, the disorder itself has been around from earliest times, hence its appearance in *Gilgamesh* and elsewhere (Gersons and Carlier 1992:742). One of the first physicians to describe the syndrome was John Erichsen, who identified it during the 1860s while examining victims of railway accidents. Erichsen attributed the syndrome, which he called "railway spine," to vaguely defined neurological mechanisms. Charcot, Janet, and Freud subsequently concluded, on the basis of clinical evidence, that the syndrome could also be produced by a psychological trauma. During World War I, the disorder's locus classicus shifted from railway spine (Erichsen) and hysteria (Charcot, Janet, Freud) to the battlefield, where large numbers of soldiers were diagnosed with traumatogenic shell shock. Interest in the syndrome declined after World War I but revived during the 1940s, when Abram Kardiner, an American psychoanalyst who had treated traumatized veterans during the 1920s, codified its criterial features and correctly identified its delayed and chronic forms. Despite Kardiner's achievement, the psychiatric establishment ignored the classification. The diagnosis achieved general acceptance only in 1980, when PTSD was included in the official psychiatric nosology—the third edition of the American Psychiatric Association's *Diagnostic and Statistical Manual of Mental Disorders* (*DSM-III*)—following a political struggle waged by psychiatric workers and activists on behalf of the large number of Vietnam War veterans who were then suffering the undiagnosed psychological effects of war-related trauma.

I will argue that this generally accepted picture of PTSD, and the traumatic memory that underlies it, is mistaken. The disorder is not timeless, nor does it possess an intrinsic unity. Rather, it is glued together by the practices, technologies, and narratives with which it is diagnosed, studied, treated, and represented and by the various interests, institutions, and moral arguments that mobilized these efforts and resources. If, as I am claiming, PTSD is a historical product, does this mean that it is not real? Is this the significance of my book's title? On the contrary, the reality of PTSD is confirmed empirically by its place in people's lives, by their experiences and convictions, and by the personal and collective investments that have been made in it. My job as an ethnographer of PTSD is not to deny its reality but to explain how it and its traumatic memory have been *made* real, to describe the mechanisms through which these phenomena penetrate people's life worlds, acquire facticity, and shape the self-knowledge of patients, clinicians, and researchers. It is not doubt about the reality of PTSD that separates me from the psychiatric insider. It is our divergent ideas about the *origins* of this reality and its universality (the fact that we now find it in many places and times).

This book is divided into three parts, the first of which concerns the origins of the traumatic memory. Until late in the nineteenth century, the term "trauma" was identified with physical injuries. Trauma was extended to include psychogenic ailments through an analogy that connected the newly discovered effects of surgical shock to effects that could be produced via "nervous shock." Through this analogy there emerged what might be called an "affect logic," whose starting point is the experience of *fear*, conceived as a memory, both individual and collective, of traumatic pain. The more familiar version of traumatic memory, the notion of thoughts and images located in the mind, was born in the closing decades of the century, when this analogy and its distinctive affect logic were conjoined with the practices and proofs of clinical hypnosis. These proofs, which culminate in the clinical narratives of Janet and Freud, point to the existence of parallel domains of psychic life—the conscious mind and the subconscious (Janet) or unconscious (Freud)—thereby making it possible to say and show that the mind keeps (traumatic) secrets from itself.

Once born, the traumatic memory attracted relatively little attention. Freud, a keen investigator of traumatic memory during his association with Josef Breuer, set the subject aside once he settled his attention on the infantile origins of psychoneurosis. Janet continued to write about the traumatic memory into the twentieth century, but mainly by recycling his previously published accounts. Chapters on traumatic neuroses continued to be published in psychiatric textbooks, but it was not a subject that commanded much interest or discussion during this period. This neglect of the traumatic memory ended with World War I.

Chapter 2 focuses on the war years, 1914 to 1918, and is organized around the work of W.H.R. Rivers. Rivers had the opportunity to observe a variety of psychiatric casualties—officers as well as other ranks, pilots as well as infantrymen. His articles on the psychogenic origins of war neuroses were widely read, and he was instrumental in establishing a course in psychological medicine for war doctors. Rivers was also a distinguished anthropologist and perhaps the first ethnographer to show an interest in the forms of reasoning underlying medical beliefs and practices. He is depicted in conventional accounts of PTSD as a standard bearer for psychiatric recognition of the traumatic memory (Herman 1992). What is generally ignored in these accounts is that the traumatic memory was a thoroughly heterogeneous thing during this period, and that Rivers's own views—specifically, those concerning traumatic time and the role of suggestion—are at odds with current ideas about PTSD. In retrospect, the most significant event of this period, in terms of its influence on current ideas, was the publication of Freud's brief account of war neuroses in *Beyond the Pleasure Principle* (1920), based on his second-hand knowledge of traumatized Austrian and German soldiers.

The second part of my book, consisting of chapters 3 and 4, describes how the traumatic memory was ultimately transformed into post-traumatic stress disorder. Chapter 3 covers the years following the publication of Freud's account up through the publication of *DSM-III*. Although it was nominally just another edition of an official nosology, *DSM-III* was fundamentally different from the preceding volumes, both in its positivism and in its authority. The adoption of *DSM-III* was part of a sweeping transformation in psychiatric knowledge-making that had begun in the 1950s. These changes profoundly altered clinical practice in the United States and prepared the way for a new science of psychiatry, based on research technologies adopted from medicine (experimentation), epidemiology (biostatistics), and clinical psychology (psychometrics). In the course of these developments, the traumatic memory, up to this point a clinically marginal and heterogeneous phenomenon, was transformed into a standard and obligatory classification, post-traumatic stress disorder.

PTSD is a disease of time. The disorder's distinctive pathology is that it permits the past (memory) to relive itself in the present, in the form of intrusive images and thoughts and in the patient's compulsion to replay old events. The space occupied by PTSD in the *DSM-III* classificatory system depends on this temporal-causal relation: etiological event → symptoms. Without it, PTSD's symptoms are indistinguishable from syndromes that belong to various other classifications. The relation has practical implications also, since it is the basis on which PTSD qualifies as a "service-connected" disability within the Veterans Adminstration Medical System. (A service-connected designation is a precondition for getting access to treatment and compensation.) On the other hand, there are numerous clinical cases that resemble PTSD in every respect except that time runs in the wrong direction, that is, from the present back to the past. The patient fixes upon his etiological memory/event retrospectively, either by intention (to acquire a service-connected diagnosis) or through psychological processes that operate outside of his conscious control (consequent to preexisting depression and anxiety disorders). Critics of the PTSD classification have argued that, in cases of delayed onset and chronic PTSD, time and causation usually take this form, going from present to past. (Most PTSD research is based on this clinical population: more specifically, Vietnam War veterans diagnosed with chronic PTSD.) Further, there is no effective way to distinguish these cases from the cases in which time runs in the opposite (approved) direction. Chapter 4 describes how PTSD knowledge-workers have responded to this criticism by developing technologies that provide the disorder's otherwise invisible pathogenic process with a visible presence.

The third part of the book describes how PTSD is made real through psychiatric practice and psychiatric science. Chapters 5, 6, and 7 are about diagnosis and clinical practice, and they focus on a specialized psychiatric unit of the Veterans Administration Medical System, the pseudonymous National Center for the Treatment of Post-Traumatic Stress Disorder. I conducted field research at the center between 1986 and 1988. My research focused on the diagnosis and assessment unit and the inpatient ward, where I was permitted to attend all therapeutic sessions and staff meetings except for individual psychotherapy (one therapist with one patient). Once a month, I introduced myself to new patients at a ward meeting, where I identified myself as an anthropologist and described the scope of my study. Two years after I completed this research, the Veterans Administration revised its original conception of a national center dedicated to war-related PTSD. Divisions were created for clinical education, behavioral science, clinical neuroscience, clinical evaluation, and women's health, and these functions were distributed among medical centers in various geographic regions. As part of the reorganization, the original site, described in this book, was closed.

Aside from its distinctive Freudian orientation, the center was fairly typical of Veterans Administration units specializing in the treatment of PTSD. As in most of the PTSD units, sessions of psychotherapy were intercalated with sessions teaching techniques of psychological self-management, such as rational thinking and biofeedback. In common with other units too, the clinical ideology identified the patients' disorder with a loss of ontological security that was traced to the veterans' inability to reconcile their traumatic memories of Vietnam (often involving atrocities) with their cognitive schemas, the moral codes, self-concepts, beliefs about human nature, and notions of cosmic justice through which these men attempted to impose a sense of order and meaning on the world.

PTSD patients are assumed to have three main ways of responding to the cognitive dissonance that originates in traumatic experiences. They can attempt to reframe their traumatic memories, making the memory content consistent with their preexisting cognitive schemas. They can attempt to revise the cognitive schemas, making them consonant with their memories. Or they can try to empty the memories of their salience and emotional power, by assimilating them (responses one and two) or erecting defenses against them via denial, efforts at avoiding the stimuli that trigger recollections, generalized emotional numbing, and so on. The psychological process through which dissonant memories are assimilated is believed to consist of phases and cycles: the conscious mind engages the traumatic memory → this encounter generates anxiety → the conscious mind disengages from the memory, through denial, self-dosing with alcohol or drugs, et cetera → the level of anxiety is reduced, the conscious mind reengages the traumatic memory and attempts to process it (via responses one and two) → anxiety increases, and a new cycle begins. Normally, cycling and processing continue until the memory is metabolized, at which point it becomes part of the individual's inactive memory. That is, it is retrievable but no longer intrusive: in effect, it is buried in the past. PTSD is exceptional in this respect, because its traumatic memory generates a high level of anxiety. Consequently, the engagement phase is brief and ineffective, and the memory cannot be buried. It lives on for decades, a source of suffering and socially and psychologically maladaptive behavior (Horowitz 1986:24–27, 41, 86, 94–97).

Thus the sine qua non for an effective treatment program is a clinical milieu in which patients are enabled and obliged to confront their traumatic memories and lengthen the periods during which memory content is processed, and patients are provided with realistic—that is to say, rational—cognitive schemas to replace their old self-defeating ones. The underlying premise is that the patient's traumatic experience has initiated an irreversible transformation. He will never be his prewar self again. The job of the clinical program is to complete the man's transformation or, more realistically, to provide him with tools and insights that will enable him to complete the change. Chapters 6 and 7 describe the efforts undertaken at the center to create this milieu. It is also an account of how patients and many therapists resisted these efforts, and how their resistances were appropriated by the clinical regime.

The historical formation of the traumatic memory in the twentieth century is tied to concurrent changes in the ideas and practices of psychiatric science. This is the subject of the book's concluding chapter. What counts as a reasonable question, a satisfactory answer, a significant difference, an anomalous finding, or even an outcome—the criteria for each of these changed during this period. What did not change was the belief in the solidity of scientific facts and the conviction that psychiatry's facts, being scientific, are essentially timeless.

Ludwik Fleck, writing in *Origins and Genesis of a Scientific Fact* about laboratory research on infectious diseases, traced the perceived timelessness of facts to a "harmony of illusions" that emerges in the course of successful research.

The research worker gropes but everything recedes, and nowhere is there a firm support. Everything seems to be an artificial effect inspired by his own personal will. Every formulation melts away at the next test. He looks for that resistance and thought constraint in the face of which he could feel passive. (Fleck 1979 [1935]:94)

The researcher struggles to distinguish between the phenomena that obey his will, because they are his unintended creation, and the phenomena that arise spontaneously, because they originate in nature, and *resist* his manipulations.

Fleck's point is that the scientist's phenomena are products of his technologies, practices, and preconditioned ways of seeing. Every scientific phenomenon is simultaneously a *techno-phenomenon*. Even if certain resistances index objects whose unity owes nothing to science—the physiological effects of the endogenous opiates described in chapter 8, for example— these objects and resistances can be reached and represented only through technologies and practices. There is no representing without intervening (Fleck 1979:38–39, 94–95, 105–106, 108; Hacking 1985; Latour 1987:180– 207 and 1993:39–43, 49–59, 85–90, 103–109).

The techno-phenomena of psychiatric science are possible because they are preceded by tests and standards that can tell the researcher whether the things that he is examining are comparable (e.g., cases) and whether his findings are worth keeping and circulating (e.g., because they have/not crossed a threshold of statistical significance). Standards and tests are themselves possible because psychiatric researchers have access to systems of measuring and calibrating, for example, the techniques for determining the inter-rater reliability of the psychometric scales described in chapters 3 and 4. Scientific tests, standards, systems of measurement, and methods of calibrating are efficacious (providing criteria of truth and grounds for belief) because they are thought to be neutral and objective. And they acquire this characteristic efficacy as parts of an autonomous and self-vindicating style of scientific reasoning (Hacking 1982, 1992b, 1992c; Collins 1992; Arbib and Hesse 1986: chaps.2 and 3).

To say that traumatic memory and PTSD are constituted through a researcher's techno-phenomena and styles of scientific reasoning does not deny the pain that is suffered by people who are diagnosed or diagnosable with PTSD. Nothing that I have written in this book should be construed as trivializing the acts of violence and the terrible personal losses that stand behind many traumatic memories. The suffering is real; PTSD is real. But can one also say that the facts now attached to PTSD are *true* (timeless) as well as real? Can questions about truth be divorced from the social, cognitive, and technological conditions through which researchers and clinicians come to know their facts and the meaning of facticity (Florence 1994:315– 316)?[3] My answer is no. Does it matter, though? The ethnographer's job is to stick to reality, its sources and genealogies; that should be enough.

Part I

THE ORIGINS OF TRAUMATIC MEMORY

One

Making Traumatic Memory

A CENTURY AGO, a new kind of memory was born, at the intersection of two streams of scientific inquiry: somatic and psychological. The somatic stream dates from the 1860s and the discovery of a previously unidentified kind of assault, called "nervous shock." The psychological stream begins earlier, in the 1790s, and leads to the discovery of a previously unidentified kind of forgetting, called "repression" and "dissociation." By the 1890s, nervous shock and repression/dissociation have been conjoined to produce *the traumatic memory*, the subject of the present study.

This chapter is divided into three parts. The first provides a history of nervous shock and how it evolved into the idea of a memory that is embodied in the neurophysiology of pain and fear rather than in words and images. The second part recounts the history of repression and dissociation and the pathogenic secret underlying them—a memory so awful that its owner is compelled to hide it from himself. The final portion of the chapter describes events that follow the birth of the traumatic memory, up to the dawn of the First World War.

Traumatic Fright

The earliest entry for "traumatic" in the *Oxford English Dictionary* is 1656: "belonging to wounds or the cure of wounds." This definition mirrors the term's Greek root and was the only sense in which the word was used until the nineteenth century, when it was, for the first time, extended to include mental injury. Historical accounts of the traumatic event routinely trace this innovation to the publication of *On Railway and Other Injuries of the Nervous System* (1866) by John Erichsen, a professor of surgery. Erichsen, like other British physicians responsible for diagnosing and assessing injuries and symptoms attributed to railway accidents, divided these patients into three categories. There were the cases that had originated in powerful blows or "shocks" that damaged neural tissue in ways that would be visible to postmortem examination; the cases resulting from shock that originated in only trivial blows or from shaking and jarring and produced damage that was generally invisible; and the cases in which people fabricated their symptoms in order to collect compensation. In all three cases, the symptoms were outwardly the same (Erichsen 1866:10, 113–114).

In his book, Erichsen did not specify the precise mechanism by which "these Jars, Shakes, Shocks, or Concussions of the Spinal Cord directly influence its action" in the second group of cases, but he suggested the following analogy:

> We do not know how it is that when a magnet is struck a heavy blow with a hammer, the magnetic force is jarred, shaken, or concussed out of the horse-shoe. But we do know that it is so, and that the iron has lost its magnetic power. So, if the spine is badly jarred, shaken, or concussed by a blow or a shock of any kind communicated to the body, we find that the nervous force is to a certain extent shaken out of the man. (Erichsen 1866:95)

Like other early writers on traumatic episodes, Erichsen was also an authority on surgery, and his description of the symptoms of nervous shock parallels his account of surgical shock given in *The Science and Art of Surgery*:

> [The effects of shock] consist in a disturbance of the functions of the circulatory, respiratory, and nervous systems, the harmony of action of the great organs being disarranged. On the receipt of a severe injury the sufferer becomes cold, faint, and trembling; the pulse is small and fluttering; there is a great mental depression and disquietude; the disturbed state of mind revealing itself in the countenance, and in the incoherence of speech and thought; the surface becomes covered by a cold sweat; there is nausea, perhaps vomiting, and relaxation of the sphincters. . . . In extreme cases, the depression of power characterizing shock may be so great as to terminate in death. (Erichsen 1859:106)

Erichsen's descriptions of nervous shock and surgical shock are pathognomonic: they identify sets of characteristic symptoms but indicate no causal mechanisms. This is probably not the whole story, though. While it is true that Erichsen does not specify how the organ systems—nervous, circulatory, and respiratory—are connected in cases of shock, knowledgeable physicians of the period would have been able to supply the missing details. What these ideas were likely to have been can be discerned from a popular text on the topic of trauma, Edwin Morris's *Practical Treatise on Shock after Surgical Operations and Injuries* (1867). Like Erichsen, Morris was a surgeon with a special interest in railway accidents. He defines shock as an effect "produced by violent injuries from any cause, or from violent emotions." The idea that the effect of violence inflicted on one part of the body might then be transmitted to other parts and to internal organs presupposes the existence of an anatomical structure that connects all of these parts. Further, the structure must be able to perform its function without leaving behind any obvious postmortem evidence, since shock can occur in the absence of lesions or hemorrhages. The only structure capable of all this is "the nervous system, acting directly upon the great nervous centre, the brain." Once nervous influences on the heart's action are suspended, the flow of blood to the brain, nervous system, and respiratory organs is interrupted, and a negative feedback is created, producing an effect that accompanies surgical shock and, in the form of syncope, many cases of nervous shock: "the brain could not perform its normal part without the blood any more than the heart could beat without the brain" (Morris 1867:9–10, 11, 17, 19).

The account works only if one can find a plausible mechanism that connects the nervous system to the heart. One explanation offered by medical authorities was that traumatic shock has the effect of exciting or irritating nerves to the point of exhaustion. Once exhausted, the nerves cease to excite the heart (or other organs); it slows down and finally ceases activity. This seems to have been Erichsen's position, as illustrated by his magnet and hammer analogy. There was another possibility, however. In 1845, E.F.W. Weber had stimulated a frog's vagus nerves with an electrical current and had observed a decrease in its heartbeat, to the point where the heart was induced to stop. (In humans as in frogs, the paired vagus nerves connect the brain to the heart and other organs located in the thorax and abdomen.) Up to this time, it had been assumed that nerves function only to excite action, but Weber's experiment produced evidence that a nerve could also have a specific inhibitory effect (Smith 1992:80). This finding would put the phenomenon of shock into a new context, since it connected shock with the brain—which was Morris's position, but not Erichsen's. The eventual effect was to situate shock within a self-regulating organ system that integrated the inhibitory (parasympathetic) and excitatory (sympathetic) nervous systems, but it was not until the 1880s, following three decades of experimentation, that vagal inhibition came to be seen in this light (Smith 1992:82). In the meantime, it was possible to see Weber's experiment as confirming rather than challenging received views of the nervous system (e.g., Morris 1867:30).

A satisfactory account of trauma had to explain one more feature, namely the part that *fear* seemed to play in cases of both surgical shock and nervous shock. Morris mentions fearful patients who died before their surgery and links their deaths to the power of their emotion. But how does an emotion, acting alone, produce effects that duplicate the consequences of a serious physical trauma? The puzzle is solved once one accepts that fear is simply an assault, comparable in its action to a physical blow or injury: "shock through the medium of the brain is such as to suspend the faculties of sense and volition, and to act directly upon the heart as a powerful sedative, producing a prostration of the nervous system." (Morris 1867:20–21).

This then is the background knowledge that Erichsen's knowledgeable

readers would have brought to his account. Unlike Morris and other authorities, Erichsen implied that, in railway injuries, traumatic shock occurs exclusively through concussion to the spine. While the initial effects of the
concussion are invisible, detectable organic changes gradually develop.
These changes may take months and are characterized by the "softening
and disorganization" of the cord, ruptures in its membranes, and inflammations—effects similar to the changes found in chronic meningitis and my-
elitis (Erichsen 1866:95, 112–114).

Erichsen describes the typical "uninjured" patient as being initially calm
and asymptomatic. The effects of his injury begin once he gets home. "A
revulsion of feeling takes place. He bursts into tears, becomes unusually
talkative, and is excited. He cannot sleep, or if he does, he wakes up suddenly with a vague sense of alarm. The next day he complains of feeling
shaken or bruised all over." A week later and he cannot exert himself or
resume business. As the (hypothesized) spinal lesions develop, he grows
pallid, careworn, and anxious; his memory suffers, and his thoughts become confused; he is fretful and irritable; his head throbs, and he is giddy;
he suffers from double vision and photophobia; his hearing is either hypersensitive or dulled, and he is troubled by a loud and incessant noise; his
posture alters, and his gait becomes unsteady; the strength of his limbs
diminishes, sometimes to the point of paralysis; he experiences a change or
loss of sensation, such as numbness or tingling; his senses of taste and
smell are likewise affected; his pulse is feeble, unnaturally slow in the earlier stages of his disorder and unnaturally fast in the later stages; he suffers
from local pain and tenderness in the area of the spine (Erichsen 1866:95–
110).

Erichsen's account mentions, in passing, the victim's mental state during railway collisions: "[I]n no ordinary accident can the shock be so great
as in those that occur on Railways. The rapidity of the movement, the momentum of the person injured, the suddenness of its arrest, the helplessness
of the sufferers, and that natural perturbation of mind that must disturb the
bravest, are all circumstances that of necessity increase the severity of the
resulting injury to the nervous system" (Erichsen 1866:9). Once more
Erichsen is vague, failing to identify the mechanisms—are they psychological? physiological?—through which states of helplessness and perturbation (fear, fright, confusion) might intensify the effects of physical injuries.

Erichsen's account was criticized by other physicians for insisting on the
pathoanatomical distinctiveness of railway accidents: "Surely the concussions of the spine, as such, are precisely similar, whether produced by a
tumble off a ladder or a jumble in a railway train which has come to grief"
(Morris 1867:45). The most sustained criticism came from Herbert Page, a
fellow of the Royal College of Surgeons and consulting physician for the

London and Western Railway Company. In 1883, Page had published a monograph on railway accidents that summarized 234 case histories, many of which involved injuries "without apparent mechanical lesion." In this book, he criticizes Erichsen for attempting to explain the entire span of traumatic railway cases by a single mechanism—spinal concussion resulting in lesions—and also for underplaying or misunderstanding the role of mental factors, especially fear and the desire for compensation, in the onset of symptoms. In 1846, Parliament had passed the Campbell Act, which compensated families of persons killed in accidents resulting from the negligence of a second party. In 1864, an amendment to the Act extended its provisions to include victims of railway accidents. In the following year, juries awarded over three hundred thousand pounds to people injured on the railways (Morris 1867:55). According to Page, the British public was fully aware of the provisions of the Campbell Act, and people involved in railway accidents were now unable to think of injuries in isolation from their possible monetary significance. He advises physicians to take care when diagnosing people with trivial or invisible injuries: to consider not only the possibility of conscious fraud but also the possibility that the patient's state of mind might be affected in "wholly unconscious ways" by a desire for compensation:

> [T]he knowledge of compensation . . . tends, almost from the first moment of illness, to colour the course and aspect of the case, with each succeeding day to become part and parcel of the injury in the patient's mind, and unwittingly to affect his feelings towards, and his impressions of, the sufferings he must undergo. . . .
>
> Even in perfectly genuine cases . . . compensation acts as a potent element in retarding convalescence, as evidenced in numberless instances by the speed with which recovery sets in as soon as the settlement of pecuniary claims has been accomplished. (Page 1883:255–256; 261)

Page's account of the physiology of shock is based on a surgical monograph by John Furneaux Jordan (1880). Like Morris, Furneaux Jordan argued that fear plays a determinative role in certain cases of shock, and this explains why surgical shock "is not always proportionate in its intensity to the severity of the wound" (Jordan 1880:1–2, 10, 13). According to Page, the "medical literature abounds in cases where the gravest disturbances of function, and even death . . . have been produced by fright and fright alone," and the same mental element explains why people who suffer broken and mangled limbs in railway accidents arrive in hospital in a state of shock characteristically more severe than that suffered by people with similar injuries from other causes (Page 1883:117).

A typical case of nervous shock involves "S.W.," a man who suffered a broken nose and bruises (relatively minor injuries) in "a very severe and destructive collision."

> [S.W.] lay for several days after the accident in a state of great nervous depression, with feeble and rapid pulse, and inability to eat or sleep. He suffered at the same time much distress from the fact that a friend sitting beside him in the carriage had been killed; and this seemed to prey constantly upon his mind. The bodily injuries progressed rapidly towards recovery.... [But even after two months,] his mental condition showed extreme emotional disturbance. He complained that he had suffered continuously from depression of spirits, as if some great trouble were impending. (Page 1883:151–152)

At this point, S.W. suffered from shortness of breath, his pulse was weak, and he was too shaken and feeble to manage more than a bit of reading. His voice was weak and often inaudible; he cried easily, slept poorly, and was troubled by distressing dreams. His claims for compensation were settled two years later, but his medical problems continued. He remained depressed, tired easily, slept poorly, and still suffered palpitations and other symptoms four years after the accident (Page 1883:151–152).

A smaller number of Page's cases consist of "neuromimesis," instances of "purely functional disorders" imitating neurological disease (Page 1883:198–199). A typical case concerns "R.C.," an army officer who had been pitched back and forth in his carriage during a nighttime collision. R.C. did not recall being injured, but

> the next morning he felt very ill and vomited, and he soon began to suffer from pain across the loins, queer sensations all over the body, nausea, giddiness, and want of sleep. On ... the twelfth day after the accident he fell suddenly and struck his nose against the corner of a table.... His own description was that the "fit came on about three in the afternoon, I fell down and screamed, and then began to cry and sob violently. During it I was unconscious, although I knew that people were around me...." He called this fit an "hysterical attack," and the doctor who saw him immediately afterwards, and who found him more or less unconscious, thought that this was its nature. Six weeks after the accident he complained of pain in the back, loss of memory, inability to apply himself, occasional giddiness, nausea, and want of sleep. (Page 1883:216–217)

According to Page, R.C.'s hysterical seizures originated in syncope, a temporary loss of consciousness that is caused when the heart fails to provide the brain with blood:

> He was reduced to a condition in which he was ready to be alarmed, and when, after the fainting, he became partially conscious upon the floor, he screamed hysterically . . . and [manifested] increased fear. And each subsequent fit began

the same way, by a sensation of syncope . . . which by the alarm it caused him at once determined the screaming and sobbing which were characteristic signs of each attack. With returning strength and cardiac tone the seizures lessened in frequency and severity, until at length they died away. (Page 1883:218)

Page traced R.C.'s symptoms (and, one supposes, S.W.'s) to "morbid changes of the nerve centres which underlie them." These changes were unlike those proposed by Erichsen, since they were not confined to the spinal cord nor could they be compared with the effects of meningitis. Indeed Page disavowed any analogy: "That there are changes is almost certain; what those changes are we do not know. One thing, however, may be said with some degree of confidence, that they differ very materially from the gross pathological changes in structure which we are accustomed to see upon the post-mortem table, or . . . [through] the microscope" (Page 1883:198–199).

Jean-Martin Charcot

Jean-Martin Charcot, the famous director of the Salpêtrière infirmary, proposed perhaps the earliest psychological account of the syndrome that Erichsen had called "railway spine." According to Charcot, patients like S.W. and R.C. were most likely to be suffering from hysteria. He was familiar with the monographs of Erichsen and Page, and like the latter, Charcot believed that intense fright could produce the traumatic syndrome. But unlike Page, he believed that fear produced its effects during a self-induced hypnotic state, when "the *mental spontaneity*, the *will*, or the *judgment*, is more or less suppressed or obscured, and suggestions become easy." Once in this state of autohypnosis, the slightest injury to the arm or leg would be transformed into a paralysis or contracture, that is, neuromimesis. "It is in this way that one so often sees after railway accidents cases of monoplegia, paraplegia, or hemiplegia, simulating organic lesions although they are no other than dynamic or psychic paralyses" (Charcot 1889:335). Charcot reported that he had himself induced similar effects at Salpêtrière when he placed a woman in a hypnotic state and informed her that she had "just had an attack during which she received a blow upon the hip." This attack was described to her in detail, with a prediction that severe pains would follow. After being roused, the woman recalled the attack and suffered the pains that Charcot had described. At the same time, she was "absolutely ignorant of our intervention" (Charcot 1889:334).[1]

Charcot had a special interest in cases like R.C.'s. In previous decades, hysteria had been considered a disorder that only rarely affected men. From the 1870s on, the French medical literature told a different story, and Charcot could now claim that "cases of male hysteria can be met with frequently enough in everyday practice" (Charcot 1889:221; Micale 1990). Male hysteria, according to Charcot, was unlike the female malady in that it presupposed no constitutional or acquired vulnerability:

> One can conceive that it may be possible for a young effeminate man, after excesses, disappointments, profound emotions, to present hysterical phenomena, but that a vigorous artisan, well built, not enervated by high culture, the stoker of an engine for example, . . . should, after the accident of a train, by a collision or running off the rails, become hysterical for the same reason as a woman, is what surpasses our imagination. (Charcot 1889:222)

Charcot complained about the "deeply rooted prejudice" that many physicians held against the term "hysteria" and about their ill-informed resistance to the idea that "railway spine" syndrome might occur "apart from any traumatic lesion." At the same time, he acknowledged that these negative attitudes were understandable given the nature of this disorder. Some cases said to involve invisible "injuries" were certainly fraudulent and difficult to detect. Further, even in certain bona fide cases, symptoms might not appear until long after the traumatic event, and a physician might reasonably raise questions concerning their actual origins. However, while these considerations made differential diagnosis more difficult, they were not good reasons to reject the hysterical character of these cases out of hand (Charcot 1889:221–222).

Charcot was also critical of German physicians like Hermann Oppenheim, who proposed to give railway spine and syndromes with similar origins and symptomatologies a special status, calling them "traumatic neuroses" and distinguishing them from "hysteria." In 1888, Oppenheim had published a monograph, *Die traumatischen Neurosen*, where he argued that these disorders share a common neurological basis. This thesis is repeated in the various editions and translations of his widely read text on nervous and psychiatric diseases: "No problem arises in distinguishing traumatic neuroses from other disorders of the nervous system; the problem is, rather, in determining whether a case is a matter of disease or simulation" (Oppenheim 1894:711). According to Charcot, though, "the pretended 'traumatic neurosis' does not exist as a morbid entity." Its characteristic features—paralyses, contractures, anesthesias, melancholia, alterations in the field of vision, et cetera—are indistinguishable from the symptoms found in other cases of hysteria, and its particular origin (fear, fright) provides, by itself, insufficient grounds for making a differential diagnosis (Charcot 1889:224–225).

Because of his differences with Erichsen, Oppenheim, and others, Charcot seems to be odd man out in this history of traumatic memory. But this is true only in the sense that his career marks the point where two trajectories of science cross: the line that leads from Erichsen through Crile, Cannon, and Pavlov, tracing the evolution of the somatized traumatic memory; and the line that leads from Charcot through Janet and Freud, tracing the evolution of a psychologized traumatic memory.

The Architecture of Fear

By the 1880s, knowledgeable medical men in England, France, and Germany believed that experiences of extreme fear could produce consequences comparable to surgical shock. Over the following half-century, ideas about nervous shock continued to evolve, shaped by animal experiments conducted by two American physicians/physiologists, George W. Crile and Walter B. Cannon, and their Russian counterpart, Ivan Pavlov.

Erichsen connected nervous shock and surgical shock to the same pathoanatomical source (spinal lesions); Page traced them to parallel but different neurological structures; Charcot rejected purely somatic explanations in favor of a psychoneurological account. Nevertheless, they could agree on the nature of "fear." Intense fear—characteristically, fear plus the element of surprise—is an assault equivalent (Erichsen, Page) or analogous (Charcot) to physical violence. The meaning of fear is in its pathogenic effects.

If fear (nervous shock) and injury (surgical shock) produce similar effects, how are they connected to each other? Erichsen and Page concluded that it was through pathoanatomical and/or pathophysiological pathways. Crile and Cannon accepted this proposition (e.g., Crile 1899, 1915:75) but argued that it was only part of the story: there is one more element connecting fear with injury, and it is *pain*. Pain is an experience that the organism strives to avoid ("dumb pain"), but it is also a signal of bodily injury and a portent of mortality. Indeed the (non-human) organism does not fear injury per se but rather the sensation of pain that accompanies injury.[2] Without knowledge of pain, the organism would be free to pursue its own destruction.

So pain has at least two meanings, experiential and evolutionary. Fear, like pain, is no simple thing. Crile's and Cannon's conception of fear, as a complex phenomenon, parallels the account given by Herbert Spencer, years earlier, in *The Principles of Psychology* (1855:594–600). Fear, according to Spencer, is the memory of pain. These memories are acquired ontogenetically through the organism's own experience with pain, and phylogenetically through instinctive (inherited) fears. But fear is more than memory. It is a state that connects memory to desire (goal-oriented arousal). It may seem incongruent that memory, being retrospective, and desire, being prospective, coexist in a single state (fear), but it is not. Behaviors evoked by desire "have been previously presented in experience;

and the representation of them is the same thing as a memory of them" (Spencer 1855:597).

Fear is the expression of a somatic state that connects it to its opposite expression, anger:

[T]he psychical state which we call fear, consists of mental representations of certain painful results; and . . . the one we call anger, consists of mental representations of the actions and impressions which would occur while inflicting some kind of pain upon another. . . . [T]hese passions are partial excitations of those states involved in the reception or infliction of injury. (Spencer 1855:596; cf. Darwin 1965 [1872]:74, 77 and James 1896:415)

What makes Crile's and Cannon's accounts of fear and anger different from Spencer's is that they are narrated in the context of experimental physiology (English 1980).[3] These emotions are represented as phases of physiological mobilization in an internal environment that is perpetually striving to adjust itself, through the actions of the sympathetic nervous system and the endocrines, to changes and challenges in the organism's external environment:

[T]he bodily changes which occur in the intense emotional states—such as fear and fury—occur as results of sympathetic discharges, and are in the highest degree serviceable to the organism in the struggle for existence. . . . Thus are the body's reserves—the stored adrenalin, and the accumulated sugar—called forth for instant service; thus is the blood shifted to nerves and muscles that may have to bear the brunt of struggle; thus is the heart set rapidly beating to speed the circulation; and thus also, are the activities of the digestive organs for the time abolished. (Cannon 1914:275; also Cannon 1929: chaps.12, 14)

Figures 1 and 2 compare the conception of fear proposed by Crile and Cannon with earlier ideas (Page).

It might appear that Cannon has depathologized fear, by transporting its traumatic associations to the field of evolutionary biology and by redefining its neurophysiology as a state of adaptive arousal. But one has to follow Cannon only one step further to see that the opposite is true. Cannon's classic account of how a survival mechanism (mobilization) is transformed into its opposite (pathogenic process) is found in an article titled "'Voodoo' Death," published at the end of his career in the *American Anthropologist*.

In records of anthropologists and others who have lived with primitive people in widely scattered parts of the world is the testimony that when subjected to spells or sorcery or the use of "black magic" men may be brought to death. . . . The phenomenon is so extraordinary and so foreign to the experience of civilized people that it seems incredible. . . .

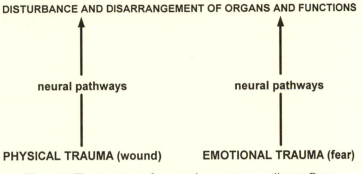

Figure 1. The structure of traumatic events according to Page

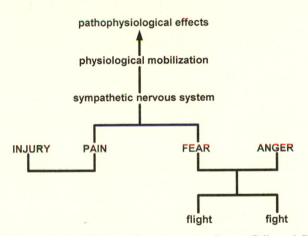

Figure 2. The structure of traumatic events according to Crile and Cannon

[T]he phenomenon is characteristically noted among . . . human beings so primitive, so superstitious, so ignorant that they are bewildered strangers in a hostile world. Instead of knowledge they have a fertile and unrestricted imagination, which fills their environment with all manner of evil spirits capable of affecting their lives disastrously. (Cannon 1942:169, 175)

According to Cannon, episodes of voodoo death move through three phases: the public laying on of a curse, the isolation of the victim by members of his community, and the convergence of the community on the wretched man "in order to subject him to the fateful rite of mourning." Over the course of these proceedings, the victim is filled with "powerless misery." He is bathed in fear and anger, physically primed either to escape or to attack the source of his danger but unable to follow either course. "If these powerful emotions prevail, and the bodily forces are fully mobilized for action, and if this state of extreme perturbation continues in uncontrolled possession of the organism for a considerable period, without the occurrence of action, dire results may ensue" (Cannon 1942:176).

Cannon's ideas concerning the pathophysiology of voodoo death are based on experiments he conducted on decorticated cats earlier in his career. In these experiments, Cannon had severed connections between the cat's cerebral cortex and the rest of its nervous system, producing a state of "excessive activity of the sympathico-adrenal" system that he allowed to continue unabated. Cannon called this state "sham rage" and claimed that it replicates the states of intense fear and anger that occur naturally in whole animals. Several hours of sham rage produced a gradual drop in the animal's blood pressure, to the point where its heart stopped beating:

> This is the condition which during World War I was found to be the reason for the low blood pressure observed in badly wounded men—the blood volume is reduced until it becomes insufficient for the maintenance of an adequate circulation. . . . Thereupon deterioration occurs in the heart, and also in the nerve centers which hold the blood vessels in moderate contraction. A vicious cycle is then established; the low blood pressure damages the very organs which are necessary for the maintenance of an adequate circulation. (Cannon 1942:178)

This is Cannon's solution to the mystery of voodoo death. Victims "die from a true state of shock, in the surgical sense—a shock induced by prolonged and intense emotion." Although the hocus pocus of voodoo death is foreign to civilized people, similar psychophysiological processes occur. Cannon compares voodoo death with cases of soldiers who fall into shock following "wounds . . . so trivial that they could not be reasonably regarded as the cause of the shock state." He tells a story of a soldier who had been rescued after being buried by an exploding shell. The "classic symptoms of shock" were already observable, and treatment proved ineffective. The man died after forty-eight hours, and postmortem examination revealed no gross injury (Cannon 1942:179).

A Self-Perpetuating Disorder

In Cannon's account of voodoo death, fear is inescapable and its traumatizing effects are unremitting. An unending state of arousal leads to exhaustion, a fall in blood pressure, and death. But what happens when exposure to traumatic shock is intermittent rather than continuous? Cannon and Crile refer to this phenomenon as "summation," defined as a process during

which the physical effects of multiple exposures accumulate and eventually lead to the changes associated with voodoo death. At about the same
time, animal experiments conducted in Russia by Ivan Pavlov pointed to
another possibility, viz., that during periods between exposures, the organism returns to a state of homeostasis, but a state different from the status
quo ante (Pavlov 1927). This is a transformation rather than summation. In
the experiments and accounts of Cannon and Crile, a distance had been
preserved between the victim and the source of its pain and fear. Pavlov's
experiments collapsed this space and interiorized the source of pain, he
made it part of the internal environment.

In a classic conditioning experiment, a laboratory animal is exposed to
a source of inescapable pain, such as electric shock delivered through a
metal floor plate. The victim's response is spontaneous, "unconditioned."
It is repeatedly reexposed to this source, and each repetition produces pain
and provokes physiological arousal. At the same time, the victim is aware
of phenomena in its immediate environment—sights, sounds, smells, et
cetera—that co-occur with, but are only incidental to, the source of its pain.
The victim learns to associate these phenomena with the shock, and they
accumulate a mnemonic power: whenever he encounters them, he is forced
to remember and relive his distress and arousal. This is a "conditioned"
response, and its scope is gradually extended, through association (via
analogy and contiguity), to objects and events located outside the original
place of pain. Escape is now impossible, since each reexposure revives the
victim's pathogenic memory and the potency of the conditioned stimuli.

A properly conditioned victim has several ways of responding to its
pathogenic memory. Pavlovians focused on two reactions: some victims
develop strategies and routines that allow them to avoid noxious stimuli
(phobias), and other victims simply give up (learned helplessness). There
is a third possibility; let us call it neo-Pavlovian, since it emerges much
later—most recently in connection with post-traumatic stress disorder (van
der Kolk et al. 1985; Pitman et al. 1990). The basic idea is that victims of
traumatic experiences may seek out circumstances that replicate their traumatogenic events. This conclusion is based on anecdotal and experimental
evidence that suggests that endogenous opioids (endorphins) may be released into a victim's bloodstream during moments of traumatic shock.
This would be an adaptive response to fight-or-flight emergencies, for
endorphins would produce a state in which the individual is undistracted
and undeterred by pain. In cases of post-traumatic stress disorder, the endorphins would have a tranquilizing effect, reducing the feelings of anxiety, depression, and inadequacy that are associated with this syndrome.
Over time, these people would become addicted to their endorphins and to
the memories that release these chemicals. When intervals between exposures grow too long, people can be expected to experience the symptoms of

opiate withdrawal—anxiety, irritability, explosive outbursts, insomnia, emotional lability, hyperalterness—and these symptoms would exacerbate the ongoing distress intrinsic to post-traumatic disorders. Pain would build to the point where the individual is induced to self-dose with his endorphins, by reexposing himself to traumatogenic-like situations.

Trauma and Memory

There is a conspicuous difference between the post-traumatic process described by Erichsen, Page, Charcot, Oppenheim, Crile, and Cannon and the process as it is described in the official psychiatric nosology of our time (*DSM-III, -IIIR, -IV*). The current account places verbal memory at the center of the disorder. (Nonverbal memory, embodied in a patient's phobic symptoms, is also a recognized component of PTSD. However, its diagnostic status as "memory" depends on the patient's being able to connect it to a significant verbal memory. The importance of verbal memory in diagnosing PTSD is discussed in chapter 4.) Erichsen mentions a few cases that involve "forgetting," but he refers to the deterioration of intellectual powers—an erosion of the ability to calculate, an inability to recall the spelling of common words, and so forth—rather than the loss of recollections (Erichsen 1866:71, 98; 1883:75). In his summary of 234 victims of railway accidents, Page mentions several patients who were "stunned" or "dazed" at the time of the collision or derailment and were consequently unable to recall these moments. He mentions four other patients who, long after the accident, suffer amnesia. The first is a woman who claims that the accident has caused total amnesia: she recalls nothing at all about her life previous to the accident. Page says she is a fake, someone out for insurance money. A second woman cannot recall the moment of the collision, and Page explains merely that she "lost her senses from fright." In the other two cases, he simply mentions the loss of memory in passing (Page 1883:308, 334, 348–349). In a chapter on "Shock to the Nervous System," Page includes a short section on "loss of memory." His treatment of this subject is similar to Erichsen's:

> The "loss of memory," of which we have here to speak, is not an inability to recall the events and incidents of past life, but rather an incapacity for sustained thought, and for continued application to the work which may be taken in hand. It is lack of the power of volitional attention, and is a symptom of easily induced fatigue. . . . True loss of memory is very rare. (Page 1883:165)

Erichsen and Page both mention patients who are given to ruminations. In nearly every case, however, the focus of the patients' attention is on their symptoms, their disabilities, and the culpability of the railway company. Page's description of the case of S.W., described above, is an exception. Months after the accident, S.W. is said to be still disturbed by the fact that a friend sitting beside him in the carriage had been killed; he also reports "distressing dreams." Page does not dwell on this point, however, nor does he concern himself with the content of S.W.'s dreams (1883:151–152).

Oppenheim's treatment of memory is essentially the same as Erichsen's and Page's.

For Crile and Cannon, verbal memory is no issue, since their research concentrates on animal experiments. This is Cannon's description of a key experiment:

> When under brief ether anesthesia, the cerebral cortex of a cat is quickly destroyed so that the animal no longer has the benefit of the organs of intelligence, there is a remarkable display of the activities of lower, primary centers of behavior, those of emotional expression. . . . [E]ther anesthesia, if light, may release the expression of rage. In the sham rage of the decorticate cat there is a supreme exhibition of emotional activity. The hairs stand on end, sweat extrudes from the toe pads, the heart rate may rise from about 150 beats per minute to twice that number, the blood pressure is greatly elevated, and the concentration of sugar in the blood soars to five times the normal. [After four hours,] the decorticate remanent of the animal . . . ceases to exist. (Cannon 1942:177)

Charcot is the qualified exception. He describes the case of Le Log——, a man diagnosed with hystero-traumatic paralysis. Le Log—— is said to complain of "a feeling of heaviness, of weight, almost a sensation of *absence of his legs* . . . ," and Charcot connects these symptoms to Le Log——'s conviction that "the wheels of the van which knocked him over 'passed over the body,' as he puts it. Nevertheless, this conviction, which has even appeared to him in his dreams, is absolutely erroneous" (Charcot 1889:386). As one would expect, Charcot traces Le Log——'s symptoms to autosuggestion. But he goes on to say something that is worth pausing over: "[I]n the case of Le Log——, as in others of the same kind, the paralysis was not produced at the very moment of the accident, but . . . only after an interval of several days, after an incubation stage of unconscious mental elaboration" (Charcot 1889:387). A constellation of ideas settles in Le Log——'s mind, like a parasite. Cut off from the rest of the mind and outside of the control of the ego, this fixed idea expresses itself through corresponding motor phenomena:

> The patient . . . does not preserve any recollection, or he preserves it in a vague manner. . . . Questions addressed to him upon this point are attended with no result. He knows nothing or almost nothing. Briefly one can compare the process in question to a sort of reflex action [whose arc passes through] . . . regions of the grey cortex, where the psychical phenomena relative to voluntary movements of

the limbs are situated. By reason of the easy dissociation of the mental unity of the *ego* in cases of this kind, these centres can be set in operation without any other region of the psychic organ being interfered with or forming part of the process. (Charcot 1889:387)

Although there are affinities between Charcot's model and the traumatic memory, it would be a mistake to read too much into his remarks. When Charcot refers to "unconscious or sub-conscious cerebration, mentation or ideation," he is not thinking of the dynamic unconscious elaborated by Freud. Nor does he have a particular interest in the *content* of these nonconscious ideas, except insofar as it serves to explain, in empirical terms, the neurological pattern and location of such hysterical symptoms as paralyses affecting the lower part of the body or loss of feeling affecting one side of the body.

Pavlovian and post-Pavlovian accounts of traumatic events are different from the earlier explanations in precisely this respect. Unlike Erichsen and the others, the neo-Pavlovians memorialize traumatic events. Of course, it is a special form of memory—a pathogenic secret embodied in conditioned responses and withdrawal symptoms rather than in words and images— and, as we will see now, it appears late in the game, preceded by two centuries of writing and experiment on pathogenic secrets.

Pathogenic Secrets

By the end of the nineteenth century, the word "memory" is being used in at least four different ways in the medical literature: it denotes a cognitive skill, such as remembering how to calculate sums; a faculty for storing and retrieving remembrances; a remembrance of a specific past event; and a pathogenic secret.

The traumatic memory is a kind of pathogenic secret (Ellenberger 1993). Such memories are "pathogenic" because they are reputed to cause psychiatric disorders—hysteria in the late nineteenth century, post-traumatic stress disorder in the late twentieth century—and "secret" because they are acts of concealment. Two kinds of concealment are possible. In one, the owner wants to hide the contents of his recollection from other people. In addition, he wants to forget the memory himself or, failing this, he wants to push it to the edges of awareness. This kind of secret memory has a long history in the West. In Latin and Orthodox Christianity, it is the hiding place of sins and transgressions, and it is the object of rituals of confession. In certain varieties of Protestantism also, secrets of this sort are causes of moral suffering, as well as sickness. In the eighteenth century, these secrets were called up and neutralized by a confessional technique known as the "cure of souls" (Ellenberger 1993:343–345). The second kind of conceal-ment involves a memory that the owner is hiding from himself. He knows that he has a secret memory, because he senses its existence, but he is unable to retrieve it; or, what is more common, he does not remember that he has forgotten and has to learn about his memory from someone else, typically a therapist.

Ordinary memories fade. They are eventually confused and conflated with other ordinary memories and are assimilated into webs of remem-brance. They belong to the past. When they penetrate into the present, it is as nostalgia, regret, and a desire for things now gone. In each of these respects, the pathogenic secret is different. Years after its creation, it re-mains unassimilated, a self-renewing presence, perpetually reliving the moment of its origin. An ordinary memory is a trace, the pathogenic secret is a mental "parasite," to use Théodule Ribot's term (1883:108–109).

Ribot was the author of an influential nineteenth-century monograph on forgetting and remembering, *Diseases of Memory: An Essay in the Positive Psychology* (1883). He was a professor at the Collège de France; a psychol-ogist and philosopher whose postivism was shaped by the evolutionary theories of Herbert Spencer and John Hughlings Jackson, by British associ-ationist psychology, and by German experimental neurophysiology (Roth 1989:50; Gasser 1988; Harris 1985:204–205; for Ribot's influence on Freud, see Otis 1993).

Ribot's book is simultaneously an account of memory and the nature of the self and self-awareness (*le moi*). As Ribot represents it, the self is expe-rienced as two interpenetrating states of being.[4] The first of these is the sum of all current states of awareness. Ribot analogizes this sense of self to a field of vision: perception at the center of the field is sharp and detailed but grows less precise toward the periphery. In the ordinary course of events, the impressions and memory traces at the center of consciousness are dis-placed by more recent and more salient impressions. The displaced impres-sions are pushed to lower states of consciousness, at the periphery of awareness. The impressions that remain at the center of consciousness now group themselves around the new impressions, just as they had done be-fore. The emerging center of attraction "becomes the point about which other new associations are formed; and thus a new complexus, a new *Ego* [*le moi*], is formed" (Ribot 1883:107, 108). This loose and fluid association of states of consciousness is equivalent to what one might call the syn-chronic self.

The diachronic self, the "conscious personality" according to Ribot, is known and experienced in its continuity with the past. It is *the self as the subject of its own history*. As such, it is a thing that is formed, nourished, and renewed by memory.[5] Like the synchronic self, it is a protean phenom-enon, ceaselessly passing through phases of growth, degeneration, and reproduction. But self-renewal is possible only because room is continually being made for new memories and new associations. And room is available only because old memories fade, and their ability to evoke the emotions with which they were first associated weakens, until they are often no more than memories of memories. In other words, Ribot's thesis is that forgetting is both normal and necessary: "To live is to acquire and lose; life consists of dissolution as well as assimilation. Forgetfulness is dissolution. . . . Without the total obliteration of an immense number of states of consciousness, and the momentary repression of many more, recollection would be impossible. Forgetfulness, except in certain cases, is not a disease of memory, but a condition of health and life" (Ribot 1883:61).

In Ribot's scheme, the self and self-awareness occupy an existential space that is bounded, on one side, by *hypermnesia*, remembering too much, a "condition in which past acts, feelings, or ideas are brought vividly to the mind, which in its natural condition has wholly lost the remembrance of these" (Tuke 1892:602; also Ribot 1883:174; Roth 1989:57–59; Luria 1968). On the other side, the boundary of self-awareness is *amnesia*, remembering too little. Ribot describes four categories of amnesia: a congenital form associated with diminished cognitive capacities, a temporary form marked by sudden onset and sudden remission, a progressive form marked by the structured dissolution of memory, and a form that he calls "periodic amnesia" (Ribot 1883:79). In the first three categories, amnesia is equivalent to a loss of memory that results in either the pathological impoverishment or decomposition of the self. In the fourth category, amnesia is an act of concealment and it results in the *pathological transformation* of the self.

Cases of periodic amnesia are divided into two types. The "developed form" consists of alternating conscious personalities: each personality is sustained by its own store of memories, and neither personality is able to supplant the other. Ribot calls this state "double consciousness." In addition to this developed type of periodic amnesia, there is an "undeveloped form" that Ribot associates with the "victims of somnambulism, natural or induced" (Ribot 1883:102–103).

Somnambulism corresponds more or less to what we would now call a hypnotic or dissociative state. Investigations of this phenomenon go back to the 1780s and predate the discovery of alternating personalities. Nineteenth-century writers were familiar with three kinds of somnambulists. The first kind were adepts, people who put themselves into a somnambulistic state, usually to perform feats of clairvoyance. Then there were people who fell into the state involuntarily and whose susceptibility to it was usually regarded as a sign of sickness or mental weakness. Finally, there were people who were induced by others to fall into the somnambulistic state, usually through techniques of "animal magnetism," in which the magnetizer (hypnotist) touched the subject's body or passed his hands over it, often while holding a magnet. In one respect the three categories of somnambulism were the same: once returned to consciousness, the somnambulist was generally unable to recall the acts that he or she had performed or experienced in the alternate state. There were exceptions, consisting of people who *could* remember what happened to them during the somnambulistic state, but only when they went back into such a state. (In the waking state, they remembered nothing about the somnambulistic events.) One of the earliest accounts of this phenomenon involves a servant girl who suffered periodic somnambulistic "paroxysms" (episodes). At first, she responded to nothing that was said to her during these episodes.

> In her subsequent paroxysms, she began to understand what was said to her, and to answer with a considerable degree of consistency. . . . She also became capable of following her usual employments during the paroxysm . . . her eyes remaining shut the whole time. The remarkable circumstance was now discovered, that, during the paroxysm, she had a distinct recollection of what took place in former paroxysms, though she had no remembrance of it during the intervals.

During one of the episodes, the girl was "abused in the most brutal and treacherous manner"—raped by a young man who had been let into the house by another servant woman: "On awakening she had no consciousness whatever of the outrage; but in a subsequent paroxysm, *some days afterwards, it recurred to her recollection, and then she related to her mother all the revolting particulars*" (Colquhoun 1836: 1:311–313, cited in Gauld 1992:260; see also Hacking 1991a:136, for a genealogy of this case).

This is a case in which a secret has been implanted in the somnambulist's mind accidentally. Although this phenomenon was a commonplace by the end of the century, it was still an unusual occurrence in the 1830s, when Colquhoun described it. At this time, nearly all accounts of memories being implanted during somnambulistic states involved what we would call posthypnotic suggestion, a very early example of which is recorded in the 1780s. M. de Mouillesaux put a young woman into a somnambulistic state as part of an experiment. While she was in this state, he instructed her to go, at nine o'clock the next morning, to a certain house that she had never entered before. Precautions were taken to ensure that she would not learn about the suggestion when she awoke. At the appointed hour, de Mouillesaux observed her to pass and repass in front of the house several times. She seemed preoccupied and eventually moved on. In a short while, and "to our great surprise," she returned and entered the house, where de Mouillesaux was waiting. The woman was greatly embarrassed and "told

us that ever since she had got up she had the idea of taking this step, that . . .
an irresistible impulse had made her brush aside all opposition, and that she
then found herself much relieved" (de Mouillesaux 1789:70–72, cited and
translated in Gauld 1992:58–59).

Experiments involving posthypnotic suggestion continued to be performed over the next hundred years. In 1883, for example, a competition
developed among physicians associated with the Nancy school of psychiatry to determine who could produce the longest delay between the moment
of posthypnotic suggestion and its execution. Bernheim produced a delay
lasting sixty-three days; Beaunis topped this with an interval of nearly six
months; and Liégeois succeeded in doubling this. To the physicians and
psychologists who were familiar with these experiments and others like
them, the demonstrations were proof that concealed memories could endure over very long periods and that, at the end of these intervals, the memories could produce "the right effect at the right time or at the right signal"
(Gauld 1992:454–455).

Throughout the nineteenth century, posthypnotic suggestion was employed as a technique for implanting secret memories and secret motives.
This concept—the idea of a secret idea and secret determination to act the
idea out—provided the prototype of the pathogenic secret that has passed
into twentieth-century psychiatry. Historically, the pathogenic secret's first
home is in periodic amnesia and double consciousness. By the time of
Janet's *L'automatisme psychologique* (1889), the pathogenic secret is no
longer restricted to these rare conditions but is deployed over the entire
span of hysterical disorders.

Pierre Janet

Among Janet's subjects, there was a middle-aged woman called Léonie
whose somnambulistic experiences are said to have started when she was
three years old. When Janet placed Léonie under hypnosis, she revealed a
second personality, whom Janet labeled "Léonie 2." Léonie 2 refused to
identify herself with Léonie 1 but acknowledged her existence. In several
experiments with Léonie, Janet employed a technique he called "distraction," a practice he had used previously with other subjects. As he describes it, the technique is simple: the subject is engaged in talk with a
second party; while she is absorbed in conservation, Janet stands behind
her and speaks in a low voice, asking questions or giving directions. Janet's
questions are answered through automatic writing: "[I] ask a question:
'What is your age? In what city are we now? . . . etc.' Léonie's hand moves,
it writes an answer on the paper, and her conversation continues without
interruption." In other experiments employing distraction, Léonie was di-

rected to perform certain acts, such as lifting her arm after Janet clapped his hands a designated number of times. Léonie performed as Janet had instructed. All the while, Léonie 1's conversation continued. When Léonie 1 was interrogated by Janet afterward, she was unaware of either his commands or Léonie's compliance. On the other hand, when Léonie 2 was called into consciousness during hypnosis, she remembered both the commands and Léonie's actions (Janet 1889:243–244).

The results of these experiments with distraction followed a pattern that other authors had observed with regard to posthypnotic suggestion. "[T]he subject, on awakening, executed *the post-hypnotic suggestions* without knowing who had given them to her, but *retrieved this memory on entering a new somnambulistic state*" (Janet 1889:330). The evidence provided by the technique of distraction went beyond this, however, by pointing to the existence of *simultaneous* rather than alternating consciousnesses (Janet 1889:323). Indeed, events permitted Janet to go a step further. In a later session with Léonie, hypnosis produced a third personality, labeled Léonie 3. While neither Léonie 1 nor Léonie 2 had access to Léonie 3's memories, Léonie 3 had access to the memories of both Léonie 1 and Léonie 2. This likewise suggested to Janet that, under certain circumstances, streams of consciousness and stores of memories might coexist *simultaneously*, within a single nervous system. It was an idea anticipated by Hippolyte Taine in *De l'intelligence*: "the coexistence, at the same instant and in the same individual, of two thoughts [*pensées*], two wills, two distinctive actions—the one conscious, the other unconscious and attributed to some invisible existence" (Taine 1870:1:16, cited in Janet 1889:244).[6]

Janet's conception of the subconscious should not be confused with our contemporary ideas. For Janet, the split between the conscious and subconscious is pathological. He associated it with behaviors called "psychological automatisms," which he divides into two categories. In the case of *total automatisms*, the entire body is placed outside the control of the conscious personality. These states include alternating personality, hysterical somnambulism, fugue (a state in which a person loses awareness of his identity and often departs to some other place), and catalepsy (a comalike state in which patients are usually insensible to stimulation and pain, and their limbs are in a condition of "waxy" rigidity, remaining fixed in whatever position they are placed by a second party). *Partial automatisms*, Janet writes, affect only parts of the body. They include common hysterical symptoms—for instance, paralyses, anesthesias, and contractures affecting half of the body (hemi-, para-) or a single appendage—and behaviors such as automatic writing and water-divining (where the movements of the hand are outside the subject's conscious control) and automatic speaking (where the vocal apparatus is beyond conscious control).

According to Janet, partial automatisms often originate in traumatic experiences (see Leys 1994). Janet's interest in traumatic events dates back to his early work, and cases are discussed in *L'automatisme psychologique* (1889:160, 208, 211, 439). In these early studies, Janet is concerned entirely with traumatic memories that his patients can retrieve but want to conceal from other people, usually out of shame. The patients are generally women, and the typical traumatic events involve seduction, rape, incest, and the birth of children out of wedlock. In each of the cases, somatic symptoms have attached themselves to a painful and intrusive memory (hypermnesia). The somatic symptoms fall into two classes: "stigmata," which are typical of hysteria and "have remained the same since the Middle Ages," and "accidental" symptoms, which are connected symbolically to their traumatic origins, so that the "thought of the accident determines the nature of the symptoms" (Janet 1901:358). Janet illustrates how accidental symptoms might form with a reference to certain of his female patients, who are afflicted with a paraplegia (paralysis of the legs) characterized by contraction of the abductor muscles. Janet calls these muscles the "guardians of virginity"—they provide a defense against vaginal penetration—and suggests that these contractures have been "brought about by the memory of rape or by that of sexual relationships with a husband who had become odious" (Janet 1925:592).

Up to this point, Janet had focused on traumatic events that his patients remembered but wanted to hide. By 1898, however, he had concluded that pathogenic secrets might also include memories that are concealed from the conscious personality (amnesia):

> [M]any of the most important traumatic memories might be imperfectly known by the subject. . . . It was necessary, therefore, to institute a search for hidden memories . . . which the patient preserved in his mind without being aware of them. The gestures, the attitudes, the intonations of the patient, would lead us to suspect the existence of such submerged memories. Sometimes we had to look for them when the patient was in a special mental condition; sometimes lost memories would crop up in the somnambulist state, in automatic writing, in dreams. (Janet 1925:594–595)

Here then are the two kinds of pathogenic secrets: one based on pathological remembering, the other on pathological forgetting. Janet calls them "subconscious fixed ideas." The remarkable capacity of fixed ideas to endure over long periods and to produce behaviors (automatisms) over which the conscious personality has no control, is explained by their isolation: "They grow, they install themselves in the field of thought like a parasite, and the subject cannot check their development by any effort on his part, because they are ignored, because they exist by themselves in a second field of thought detached from the first" (1901:267).

Janet was not the first writer to argue that autonomous groups of ideas might be fixed in a person's mind by traumatic experiences. Charcot had made a similar claim about ideas—such as memories of fictitious accidents—that are implanted in the mind by suggestion:

> "The suggested idea or group of ideas," said M. Charcot, very justly, "find themselves in their isolation sheltered against the *control* of that great collection of personal ideas, a long time accumulated and organized which constitute consciousness properly so called, the Ego." . . . We ask permission to preserve [Charcot's] striking metaphor: Suggestions, with their automatic and independent development, are real parasites in thought. (Charcot quoted in Janet 1901:267, also 280–281)

But the parallel between ideas implanted by suggestion and ideas fixed by traumatic experience is not perfect. Suggestion is an intentional act, in which a second party slips ideas into the subject's subconscious. Memories of traumatic experiences have a patently different genesis—there is no intentional act, no second party—and Janet must explain why these memories enter the subconscious as fixed ideas (and pathogenic secrets) rather than moving into the conscious mind as ordinary memories.

To answer this question, he returns to Ribot's ideas about the relation of memory to self-consciousness. Memory, Janet writes, is an action: "[E]ssentially, it *is the action of telling a story.* . . . The teller must not only know how to do it, but must also know how to associate the happening with the other events of his life, how to put it in its place in that life history which for each of us is an essential element of his personality" (Janet 1925:662; also Janet 1889:324). The traumatic memory becomes a fixed idea rather than an ordinary memory because the individual is unable to assimilate its meaning and now finds himself in a situation that is "too complicated and too difficult." He attempts to adapt to it, he fails and makes new efforts at adaptation. He goes on and on, one futile attempt after another, until he has reduced himself to a state of mental exhaustion. The unassimilated memory endures as a split-off element of consciousness (Janet 1925:663; also 671, 679). Janet identifies the patient's malady with the fixed idea, rather than its symptomatic manifestations. Symptoms and attacks may be intermittent, but the malady is continuous: a woman "has" hysteria not simply when she experiences attacks but also during the intervals between attacks (Janet 1889:345).

The role of therapy is to help the patient discover his fixed idea and bring it into consciousness. Then he must recite it and re-recite it, until the recital and its memory are independent of the event and the emotions that they memorialize. Once this point has been reached, patient and therapist move to the final step: they find a resting place for the memory—in Janet's words, as a chapter in the patient's life history (Janet 1925:662, 666).[7]

Sigmund Freud

Freud's interest in traumatic events was limited to two periods: the years between 1892 and 1896, when he (working together with Josef Breuer) examined the causes of hysterical attacks, and the years following World War I, when, in the aftermath of that conflict, he turned his attention, very briefly, to the etiology of the war neuroses.

In his work of 1892–1896, Freud traces hysterical attacks to three origins: a traumatic event, whose exemplar is the railway accident; an accumulation ("summation") of lesser frights, mortifications, and disappointments; and incidents too slight to be called a trauma. Events belonging to the third group are capable of producing attacks only in people with hysterical dispositions. Attacks resulting from the other categories of events require no predisposing factor (diathesis), although one may be present (Freud 1966b [1892]:152–153; Breuer and Freud 1955 [1893]:12–13).

In traumatogenic hysterical disorders, "the determining process continues to operate in some way; or other for years." Like Janet, Freud rejected the explanation—offered by Erichsen, Page, Oppenheim, and others—that the "determining process" is a direct result of forces produced during traumatic events. Rather, the pathogenic agency is invested in the patient's *memory* of the trauma. *"Hysterics suffer mainly from reminiscences"*—a position which, as Breuer and Freud indicate, had been anticipated in the work of Moritz Benedikt and others (Breuer and Freud 1955 [1893]:7–8, 210; Ellenberger 1993:352–354).

Traumatic experiences are always charged with high emotion. Reactions to them—ranging from tears to acts of revenge—have the effect of discharging the attached affect. When this happens, memories of the events become ordinary recollections and are accessible to the conscious mind. A reaction discharge is not always possible, however. Perhaps the traumatic event involves circumstances that the individual wants to conceal, or an inconsolable loss, such as the death of a loved one; or perhaps the event occurs in circumstances that inhibit the individual from reacting in an overt and effective way; or the individual's unusual mental condition at the time of the event—a state of paralysing fear or autohypnosis, for example—blocks a proper reaction discharge (Breuer and Freud 1955 [1893]:10–11). Whatever the cause, the failure to discharge the attached affect produces a situation in which the nervous system must manage a sudden surge of excitation. This is a problem because the nervous system "endeavours" to keep the sum of excitation constant (Freud's "constancy principle"), and its ability to accomplish this is a precondition of health: "If we start out from this theorem, . . . we find that the psychical experiences forming the content of

hysterical attacks have a characteristic in common. They are all of them
impressions which have failed to find adequate discharge" (Freud 1966b
[1892]:153–154).

Unabreacted (undischarged) memories are said to enter a "second consciousness," where they become secrets, either isolated from the conscious
personality or available to it only in "a highly summary form" (Freud
1966b [1992]:153). Because the memories are painful and unmanageable,
the conscious personality wishes to banish them from awareness (suppression). Once the banished memories have entered the second consciousness,
the conscious personality gradually loses access to them, as an act of self-
defense. The memory becomes "a foreign body which long after its entry
. . . is still at work" (Breuer and Freud 1955 [1893]:6). Work with hypnotized patients suggested that these secret memories might persist for long
periods. Twenty-five-year-old memories were found to be "astonishingly
intact" and to possess "all the affective strength of new experiences"
(Breuer and Freud 1955 [1893]:9–10).

In addition to this discovery, Breuer and Freud made two further claims:
first, that they had success in treating these patients and, second, that the
putative cures vindicated their theory of the etiology of hysteria. When
traumatic memories were revived together with the attached affect—
whether through hypnosis, narcosis, or persuasion—they lost their pathogenic power:

> [E]ach individual hysterical symptom immediately and permanently disap-
> peared when we had succeeded in bringing clearly to light the memory of the
> event by which it was provoked and in arousing its accompanying affect, and
> when the patient had described that event in the greatest possible detail and had
> put the affect into words. Recollection without affect almost invariably produces
> no result. (Breuer and Freud 1955 [1893]:6)

At this point in time, Freud's and Janet's ideas about traumatic memories and pathogenic secrets were quite similar (Erdelyi 1990:6–14). According to Freud, the most important difference between their views was
that Janet "attributed to hysterical patients a constitutional incapacity for
holding together the contents of their minds." For Breuer and Freud, "dis-
positional hysteria" did certainly exist, but hysterical attacks could also be
found among "people of the clearest intellect, strongest will, greatest character and highest critical power" (Freud 1955d [1923]:237; Breuer
1894:232; cf. Janet 1925:602, 610–611, 622, 657–658).

The development of Freud's ideas on the traumatic origins of neurosis
following the publication of *Studies on Hysteria* (1895) is too well known
to require extensive treatment here, and I will review it only briefly.

By 1896, Freud had made a sharp separation between two kinds of neurotic formations: *psychoneurosis*, which originates in experiences in early childhood, and *actual neurosis*, which is produced by events that occur later in life (Freud 1962 [1896]:195–196). While both kinds of neuroses are said to originate in traumatic experiences, cases of psychoneurosis are invariably traced to sexual traumas (Freud's "seduction theory"): "If we have the perseverance to press on with the analysis into early childhood, as far back as a human memory is capable of reaching, we invariably bring the patient to reproduce experiences which . . . must be regarded as the aetiology of his neurosis for which we have been looking. These *infantile* experiences are . . . *sexual* in content" (Freud 1962 [1896]:202). The alternative diagnosis, actual neurosis, is reserved for cases that can be traced to significant traumas. Even here, Freud writes, one must exercise caution in drawing any conclusions, since the memory of an event that occurred later in life may screen an infantile trauma that is the actual origin of the patient's malady. By this stage of his career, Freud had very little interest in the actual neuroses, and they now disappear from his writings.

Between 1896 and 1914, Freud abandoned the seduction theory and, with it, his ideas concerning the traumatic origins of the psychoneuroses: "Influenced by Charcot's view of the traumatic origin of hysteria, one was readily inclined to accept as true and aetiologically significant the statements made by patients in which they ascribed their symptoms to passive sexual experiences in the first years of childhood—to put it bluntly, to seduction" (Freud 1958 [1914]:17; see also Israëls and Schatzman 1993; Grünbaum 1993:345–347). Freud now settled on his permanent position: (1) psychoneuroses originate in the dynamic interplay of instinctual impulses that are present at infancy, and (2) the sexual fantasies that derive from these dynamics operate with the force of real experiences (Strachey 1962).

The abandonment of traumatic origins is accompanied by a shift in technique: from abreaction to free association, and then to a form of psychoanalysis

> in which the analyst gives up the attempt to bring a particular moment or problem into focus. He contents himself with studying whatever is present . . . on the surface of the patient's mind, and he employs the art of interpretation mainly for the purpose of [provoking and] recognizing the *resistances* which appear there, and making them conscious to the patient. . . . [W]hen these [resistances] have been got the better of, the patient often relates the forgotten situations and connections without any difficulty. (Freud 1958 [1914]:147; my emphasis)

For a short period at the close of World War I, Freud returned to the subject of traumatic etiologies, specifically in connection with the traumatic neuroses of war (Freud 1955a [1919], 1955b [1920], 1955c [1920]). I will examine this aspect of his work toward the end of the next chapter.

The Traumatic Memory, 1914

The traumatic memory emerges at the end of the nineteenth century, at the intersection of two evolving fields of medical knowledge: knowledge of how trauma affects the nervous system and, through it, the rest of the body, and knowledge of how pathogenic secrets impact on the mental life of their owners.

There is no "turning point" in the history of the traumatic memory. One looks in vain for a key discovery, or a paradigmatic experiment, or even a prophetic figure (cf. van der Kolk and van der Hart 1989). What one finds is that the traumatic memory comes together in two anonymous developments.

The first of these is the *medicalization of the past*. In the years leading up to World War I, a small number of medical men acquired the technical and rhetorical means to demonstrate three claims to the satisfaction of their audiences: that traumatic neuroses are produced by memories of events rather than by the events themselves; that the memories are pathogenic secrets, merging concealed ideas with concealed urges; and that medical men have privileged access to these secrets and their meanings.

The medicalizing process is captured in two biological images. There is the image of the traumatic memory as a kind of *parasite*, an idea that recurs in the work of Ribot, Charcot, Janet, Freud, and the neo-Pavlovians. And there is the image of the traumatic memory as *mimesis*, a memory that is inscribed simultaneously *in the mind*, as interior images and words, and *on the body*, where it is disguised in perverse postures, sensations, and absences (catalepsies, anesthesias, etc.).

The second development is the *normalization of pathology*. Pathology is denied its uniqueness and separateness and is now either a *loss* or *displacement* of normal functions, followed by the release of lower-level normal functions (this point will be elaborated in chapter 2, in connection with Hughling Jackson's influence on W.H.R. Rivers), or an *exaggeration* or *overextension* of normal functions, resulting in a disequilibrium or depletion of functions and vital energies.[8]

The most significant effect of normalization is the idea, much favored by nineteenth-century positivists, that pathological states are a window onto normal states:

> In studying the psychology of the individual, sleep, madness, delirium, somnambulism, hallucination offer a far more favorable field of experience than the normal state. Phenomena, which in the normal state are almost effaced because of their tenuousness, appear more palpable in extraordinary crises because they are exaggerated. The physicist does not study galvanism in the weak quantities found

in nature, but increases it, by means of experimentation, in order to study it more easily, although the laws studied in that exaggerated state are identical to those in the natural state. (Renan 1890:184, cited in Canguilhem 1991:44–45)

This process, connecting normal and pathological states, is played out twice in the history of the traumatic memory:

First, there is the migration of the "second consciousness," the hide-out of pathogenic secrets. It starts as "alternating amnesia," a sympton specific to a rare disorder, double personality. By the 1890s, it is transformed into a parallel but pathological stream of consciousness (Janet) and then into the unconscious (Freud), where it is transmuted into a universal (normal) part of the mind.

The normalization process plays out a second time when the embodiment of pathology is transformed from physical lesions into physiological and quasi-physiological functions organized for self-defense and self-regulation. Thus Janet's "fixed idea" is a pathological entity; at the same time, it is a device for protecting endangered bodies from total nervous exhaustion (by isolating indigestible memories). Likewise Freud: neurotic symptom formation is both pathological and inherently defensive. In neuroses originating in traumatic events, symptoms function to maintain the constancy of nervous energy, a precondition of health, according to Freud. A similar normalizing principle operates throughout the work of Cannon, Crile, Selye (1950), and the neo-Pavlovians. Pathology is transmuted into a "disease of adaptation," as the notion of the lesion is subordinated to that of the disturbance of functions:

> To speak [as Cannon does] of the wisdom of the body leads one to understand that the living body is in a permanent state of controlled equilibrium, of disequilibrium which is resisted as soon as it begins, of stability maintained against disturbing influences originating without: it means, in short, that organic life is an order of precarious and threatened functions which are constantly reestablished by a system of regulations. (Canguilhem 1991:260; also 272)

By 1914, the traumatic memory exists in the sense that (1) a number of medical men working in Europe and North America are familiar with the idea that the memory of an experience can produce syndromes resembling hysterical and neurological disorders, and (2) a much smaller segment of medical men has incorporated this idea into their own clinical practices.

Medical interest in the traumatic memory at this point in time was not intense. Developments in experimental and clinical science—especially with regard to the pathophysiology of epilepsy, syphilis, and the toxic psychoses—inclined many physicians to wonder whether hysteria (locus of the traumatic memory) was, after all, a discrete disease entity or merely a residual category, a wastebasket for otherwise unclassifiable cases (Micale

1993:504–515). Moreover, the trauma research of Crile and Cannon, while it captured the attention of large numbers of physicians and researchers, focused on experiments with animals and subcortical processes rather than memory. Even within Freud's own circle, the traumatic memory had lost any compelling clinical or intellectual interest. Significant nonmedical influences were also missing: the railway accidents that had sparked the initial interest in traumatic experiences were now relatively infrequent, and injuries were now managed through established medico-legal routines.

With the start of World War I in August 1914, the situation changes. By Christmas, the British regular army had been wiped out, and the French and German armies together had suffered a million casualties. The next spring, at Ypres, 60,000 casualties were inflicted on the British. Several months later, a similar number of British soldiers were lost at Loos, and in June 1916, another 60,000 were destroyed or wounded on just the first day of the Battle of the Somme:

> I see men arising and walking forward; and I go forward with them, in a glassy delirium wherein some seem to pause, with bowed heads, and sink carefully to their knees, and roll slowly over, and lie still. Others roll and roll, and scream and grip my legs in uttermost fear. . . .
>
> And I go on with aching feet, . . . and my wave melts away, and the second wave comes up, and also melts away, and then the third wave merges into the first and second, and after a while the fourth blunders into the remnants of the others, and we begin to run forward to catch up with the barrage, gasping and sweating. (Henry Williamson, quoted in Fussell 1975:29–30)

A year later, another battle was fought near Ypres and the British suffered 160,000 more casualties. In July, the British attacked at Ypres once again:

> This time the artillery was relied on to prepare the ground for the attack. . . . The bombardment churned up the ground; rain fell and turned the dirt to mud. In the mud the British assaulted until the attack finally attenuated three and a half months later. Price: 370,000 British dead and wounded and sick and frozen to death. Thousands literally drowned in the mud. (Fussell 1975:16)

The following March, the Germans counterattacked at the Somme and inflicted 300,000 casualties on the British. Between these battles, there were thousands of other deaths, mutilations, and heart-stopping shocks.

A half-century after the publication of Erichsen's first book on railway accidents, physicians serving in the Royal Army Medical Corps, like their counterparts in the other combatant armies, were witnesses to an epidemic of traumatic paralyses, contractures, anesthesias, and aboulias. It was as if a hundred colossal railway smashups were taking place every day, for four years. By war's end, 80,000 cases of shell shock had been treated in RAMC medical units, and 30,000 troops diagnosed with nervous trauma had been evacuated to British hospitals. After the war, 200,000 ex-servicemen received pensions for nervous disorders (Stone 1988:249).

The following chapter is an account of this epidemic and the traumatic memory's place in it. I examine events mainly from the perspective of British physicians and focus on the ideas and experiences of one man, W.H.R. Rivers. Rivers deserves this attention because he had much to say about the causes of the war neuroses and was thoroughly familiar with contemporary thinking about traumatic memory. He is an important figure for another reason also, namely, his place in current historical accounts of post-traumatic disorders, where he is singled out, not only for his humane practices and advocacy of psychotherapy, but as a bridge that connects the classical age of the traumatic memory, culminating in Janet and Freud, to the modern period (Herman 1992; Leed 1979; Showalter 1987; Stone 1988).

Two

World War I

By daylight each mind was a sort of aquarium for
the psychopath to study. In the daytime, sitting
in a sunny room, a man could discuss his psychoneurotic symptoms with his doctor, who
could diagnose phobias and conflicts and formu-
late them in scientific terminology. Significant
dreams could be noted down, and Rivers could
try to remove repressions. But by night each man
was back in his doomed sector of a horror-
stricken Front Line, where the panic and stam-
pede of some ghastly experience was reenacted
among the livid faces of the dead. No doctor
could save him then, when he became the lonely
victim of his dream disasters and delusions.

(*Siegfried Sassoon,* Sherston's Progress)

THE "RIVERS" who is mentioned in this passage is W.H.R. Rivers, a temporary captain in the Royal Army Medical Corps (RAMC) serving as a psychiatrist at the Craiglockhart Military Hospital, near Edinburgh. Over the
preceding two decades, Rivers had established an international reputation
as an ethnographer and a pioneer researcher on nerve regeneration. It is
Rivers the anthropologist who is best remembered today—Rivers the
member of the Cambridge Expedition to the Torres Straits (1898), the orig-
inator of the "genealogical method" of investigating kin relations and ter-
minologies (1900), and the author of a classic ethnography of South Asia,
The Todas (1906).[1]

In November 1915, Rivers traveled to London to deliver the FitzPatrick
Lectures to the Royal College of Physicians; his subject was "Medicine,
Magic, and Religion." He had two messages to convey in these lectures.
The first was that in every society, even the most primitive, people possess
beliefs and practices that are identifiably "medical." That is, they possess
beliefs and practices that allow them to control, or to believe that they

control, the natural phenomena that reduce their physical and social functioning; and they use terms equivalent to our word "disease" to distinguish these phenomena as pathological. Rivers's second message was that the medical beliefs and practices of primitive people are not ragbags of superstition and hocus-pocus, but, like our own, therapeutic systems. Moreover, the forms of reasoning that underlie these systems are characteristically rational: "Such positive knowledge as we possess concerning the psychological processes underlying the blend of medicine and magic leads us into no mystical dawn of the human mind, but introduces us to concepts and beliefs of the same order as those which direct our own social activities" (Rivers 1916:65).

The FitzPatrick Lectures coincide with Rivers's departure from Cambridge to take a temporary position as a civilian physician at the Maghull Military Hospital, part of a network of hospitals treating soldiers invalided with "war neuroses." At two points in the lectures, Rivers compares the varieties of medicine practiced in primitive and Western societies, making observations and reaching conclusions that can be seen, with the benefit of hindsight, as prologue to the clinical experiences and psychiatric writing that would occupy him from Maghull onward.

The Power of Suggestion

Rivers's first point is that while Western medicine is generally rational—in the sense that its etiological and physiological beliefs are coherent and consistent, and its diagnostic and therapeutic practices are consonant with these beliefs—Western medicine is not *uniquely* rational. Rivers draws the attention of his audience to certain tribal societies of Melanesia:

> [T]hese practise an art of medicine which is in some respects *more rational than our own* in that its modes of diagnosis and treatment follow more directly from their ideas concerning the causation of disease. [There is] a *logical consistency* which it may take us long to emulate in our pursuit of a medicine founded upon the sciences of physiology and psychology. (Rivers 1916:122–123; my emphasis)

We can assume that Rivers intended this comparison to be taken literally. Unfortunately, he does not elaborate this idea, and the audience is never told why Western medicine has failed to reach a higher level of logical consistency. Nor do they learn whether certain branches of Western medicine, say psychiatry, might be more problematic in this respect than others.

In claiming that the medicine practiced in primitive societies is rational, Rivers is taking a position in opposition to the views of his famous contemporary, the French philosopher Lucien Lévy-Bruhl (Rivers 1918:123). In *Les fonctions mentale dans les sociétiés inférieures* (1905), Lévy-Bruhl associates primitive societies with a distinctive "mentality" that he describes as "prelogical" (Cazeneuve 1972:1–23; Tambiah 1990: chap. 5 discusses the implications of this debate for the anthropology of science and medicine). Prelogic does not antecede logical reasoning, nor does it underlie logic in the way that the unconscious might be said to underlie the conscious: "It is not *antilogical*; it is not *alogical* either. By designating it 'prelogical' I wish to state that it does not bind itself down, as our thought does, to avoiding contradiction" (Lévy-Bruhl 1985 [1905]:20–27, 361, 379–86).

Rivers counters Lévy-Bruhl's ideas with the intellectualist argument favored by British anthropologists at this time (and afterward): although one might question the premises on which primitive people base their conclusions—ideas about witchcraft, sorcery, malevolent ghosts, et cetera—the process by which these people reason from premises to conclusions is no different from our own.

Rivers's second comparison between medical systems primitive and Western seems, at first glance, to undermine the intellectualist argument. It is at this point that he explains the part played by suggestion—a "process by which one mind acts upon another unwittingly"—in the production and cure of disease among such people as the Melanesians.

> There can be no question that such processes . . . are efficacious. Men who have offended one whom they believe to have magical powers sicken, and even die, as the direct result of their belief and if the process has not gone too far they will recover if they can be convinced that the spell has been removed. . . .

> Doubtless, with this real factor of suggestion there is mixed up much deception. . . . At the same time there is reason to believe that . . . [the dissimulator, typically a healer,] is not wholly a deceiver, but in some measure shares the general belief in his powers. (Rivers 1916:122)

During this period, the psychiatric literature distinguished two forms of suggestion, according to their respective sources: heterosuggestion (often, simply "suggestion"), in which an idea or influence is implanted by someone else, and autosuggestion, in which the target and source of suggestion are the same person. Thus the Melanesians mentioned in Rivers's quotation are made sick by either hetero- or autosuggestion and are cured by heterosuggestion. Rivers describes suggestion and "intelligence"—by which he usually means reason—as opposed principles. This seems to recall Lévy-Bruhl's distinction between prelogic and reason, but there are fundamental differences. For one thing, Rivers wants to emphasize that suggestion also plays an important part in Western medicine. Indeed, he feels that the Melanesians may have something to teach the British in this regard: "Not only will the study of people of rude culture help us to estimate aright the part taken by fraud and deception in certain forms of the medical art of the civilised world, but, what is far more important, it will help us also to understand better the place taken by suggestion both in the production and the treatment of disease" (Rivers 1916:122).

Ideas about pathogenic and therapeutic suggestion have a long genealogy in Western medicine. The term "imagination" (*imaginatio*) was used by Renaissance physicians and philosophers in ways that included the idea of suggestion. Montaigne, in his *Essays* (1581), cites imagination/suggestion as a cause of physical and emotional disease, and Ludovico Muratori employs the term, in *On the Power of Human Imagination* (1745), in a way that anticipates the psychiatric accounts given by Bernheim, Charcot, and their contemporaries (Ellenberger 1970:111–112, 149, 151). By Rivers's time, suggestion was a very common form of psychiatric intervention, often used in combination with hypnosis, electrotherapy, and "education."

The Ghost of John Hughlings Jackson

Rivers's own ideas about suggestion are highly distinctive and essential to his understanding of primitive culture and the war neuroses. They are described in a set of essays that are collected in his most famous work on psychiatry, *Instinct and the Unconscious* (1922). To understand his position, one needs to go back to 1903 and the neurological experiment that he conducted with Henry Head (Rivers and Head 1908).

The experiment began with surgery performed on Head's forearm. The radial and external cutaneous nerves of the left arm were severed and then sutured. Over the next four years, the men charted the return of sensation to the skin, as the severed nerves regained full function. Rivers and Head describe the regeneration process as passing through two stages. The first stage is characterized by "protopathic" sensitivity. Head's sense of feeling had a nonlocalizable and all-or-nothing quality: he was unable to discriminate, for example, between stimuli at twenty degrees centigrade and those at zero degrees, yet his pain threshold was so radically reduced that even pricks with a needle induced high levels of discomfort. Over time, Head acquired the ability to localize his sensations and discriminate them as to quality and intensity. The researchers called this capacity "epicritic" sensibility.

According to Rivers and Head, protopathic and epicritic sensibility represent separate systems of nerve fibers that coexist within a neurological hierarchy. The epicritic develops later than the protopathic but does not evolve from it. Rather, it overlays the protopathic, restricting and modulating its all-or-nothing quality. The additional claim is that the neural systems responsible for protopathic and epicritic sensibilities extend beyond the peripheral nervous system and are rooted in the structure of the brain itself: "the inner layer of the cerebral cortex, the first of its parts to develop in the human species, governed individuals' protopathic responses, whereas the outer layer governed higher functions. . . . [Thus a] weakened human organism would regress toward the protopathic level" (Kucklick 1991:159).

These ideas are prefigured in the work of John Hughlings Jackson (1835–1911), the most famous British neurologist of this period. According to Hughlings Jackson, the nervous system is composed of functional levels laid down at different evolutionary moments. The archaic functions are highly organized (we would say "hard-wired"), simple, and automatic and are overlaid by more recently acquired functions that are less organized, more complex, and more voluntary. Thus the development of levels within the nervous system parallels the evolutionary history of life forms, as described earlier by Herbert Spencer.[2] Neurological diseases, such as aphasia, and most kinds of insanity occur when higher-level functions become inoperative or impaired and lower-level functions escape from their restraining effects. Hughlings Jackson called this a process of "dissolution"—a term that he borrowed from Spencer—and described it as the "opposite of evolution" (Jackson 1931a [1880–1881]:318). The release of lower-level functions occurs temporarily in healthy people during sleep and dreaming. "The deeper the dissolution [whether in sleep or disease], the shallower the level of evolution remaining" (Jackson 1931b [1884]:49–51; also Clark 1983; Engelhardt 1975; Starr 1992:215–221).

Hughlings Jackson describes himself as being a "materialist" regarding the origins of mental states. He believed that mental states originate in the structures and processes of the nervous system, and that alterations in the nervous system effect changes in mental states. At the same time, he writes, "I do not concern myself with mental states . . . except indirectly in seeking their anatomical substrata. I do not concern myself about the mode of connection between mind and matter. It is enough to assume a *parallelism*" (Jackson 1931a [1875]:52; my emphasis). Rivers is also this kind of materialist. In *Instinct and the Unconscious*, he writes that he wants "to provide a foundation for a biological theory of the psycho-neuroses," but he settles down to a variety of parallelism in which mental activity is loosely connected to neurological activity through what are, in effect, vaguely defined bridge principles, viz., the protopathic and epicritic (Rivers 1920:119).

Now we can return to Rivers's account of suggestion. He writes that suggestion comprises three elements: *mimesis*, the action element, consisting of "unwitting imitation"; *sympathy*, the affect element, consisting of reciprocal and spontaneous responses between individuals; and *intuition*, the cognitive element, equivalent to the direct apprehension of what is going on in someone else's mind. He discounts the variety of suggestion that takes place in the consulting room: clinician and patient are conscious participants and, under these conditions, the process is "specialized and artificial." Rivers also ignores autosuggestion, even though it is often mentioned by army doctors as an important source of hysterical conversions (Rivers 1920:92–93).

Rivers has two main interests in suggestion: its place in the evolutionary history of humankind and its role in the etiology and epidemiology of the psychoneuroses. Because his account of the medical context is grounded in his ideas about evolution and the nervous system, we begin there.

According to Rivers, the survival of animal species depends on their ability, as individuals and as groups, to respond to danger. The responses are rooted in a survival instinct that provides three options: fighting, fleeing, or immobility. Rivers places humanity in the category of gregarious animals, but adds that the world's peoples vary in this respect: "the Melanesian is distinctly more gregarious than the average European." Among gregarious animals, the initial response to danger is immobility. This makes sense from an evolutionary perspective, since it is more efficient for such animals to avoid attracting the attention of predators than to attempt to flee or fight. But immobility is an efficient mechanism only if every member of the group remains still. Movement by even one individual makes the group visible. The evolution of a capacity for suggestion solves this problem, making it possible for "all members of the group [to share] a mental content so similar that all act with complete harmony towards some common end." This collective effect is itself made possible by two concurrent processes: the process that disposes individuals to immobility (via mimesis, sympathy, and intuition) and a process of *suppression*, through which other instinctive tendencies (notably flight) are restrained (Rivers 1920:94–98).

A similar chain of events reshaped sleep, another all-or-nothing instinct that, through suggestion, has been made "capable of gradation in a high degree." The survival of gregarious animals requires them to "respond in sleep, not merely to sounds or movements which threaten danger, but also to the sounds or movements of the other members of the group" (= the process of mimesis, sympathy, and intuition). But the individual must also restrain himself from awaking to unimportant stimuli (= the process of suppression). Without this capacity, he would lose "the recuperation which is the special function of sleep" (Rivers 1920:116, 117). During sleep, "[suggestion] acts progressively upon successive levels of mental activity, first putting out of action the experience and modes of mental functioning which have been recently acquired. The deeper the sleep, the greater the number of such levels put out of action and the lower and earlier the levels

which are left to manifest their special modes of activity in the dream"
(Rivers 1923:92)

In a nutshell, suggestion is a mental process that is born in fear and that
"belongs to an instinct which is concerned with collective as opposed to
individual needs" (Rivers 1920:99). In the course of human evolution, it is
followed by the emergence of intelligence and reason—and, as we shall
see, the mechanism that accompanies them, repression: "The presence in
Man of both suggestion and intelligence shows that the early protopathic
forms of instinctive behavior were modified in two directions, one leading
towards intelligence and the other towards suggestion and intuition" (Rivers 1920:99).

Suggestion is not supplanted by reason, nor do they represent separate
"mentalities." Rather, suggestion and reason coexist as faculties within the
same mind/brain.

According to Rivers, the human mind includes an unconscious. He uses
this notion in two senses. There is what might be called the *simple unconscious*—mental operations conducted outside of awareness—and there is
the *dynamic unconscious*, which resembles Freud's conception. Rivers had
made a systematic study of Freud's writings; he lectured on Freud's psychology of the unconscious (Rivers 1917); he refers to Freud at length in
his own work (Rivers 1918, 1920); and he is credited with making Freud's
ideas and techniques familiar and credible to medical audiences. Rivers
was not a "Freudian" in the strict sense of the word, however. In his writing
and his clinical practice, he did not hesitate to simplify, transmute, or reject
key propositions of psychoanalysis (e.g., Rivers 1917). Further, many of
the notions that one might assume he had borrowed from Freud—the release of the dreaming mind from the control of the conscious mind, for
instance—reflect, instead, their common debt to the work of John Hughlings Jackson and their shared attraction to "a style of explanation that may
be labeled 'evolutionary naturalism.' For any mental, social, or generally
'higher' human achievement, the default assumption (as we say) was that
the explanation should describe how it evolved or could have evolved from
earlier, unproblematically natural phenomena, paradigmatically, from conditions shared with animals" (Kitcher 1992:66–67).

Rivers describes the dynamic unconscious as a storehouse of (1) memories that conflict with the prevailing constituents of consciousness and that
produce pain and discomfort when they enter consciousness and (2) desires, instinctive tendencies, and affective elements. He bases his belief in
the existence of the unconscious on two kinds of evidence. First, certain
memories are not immediately accessible to the conscious mind but present
themselves only during sleep, hypnosis, free association, and physical
states such as fevers—in other words, when control by the conscious mind
is weakened. Second, there are certain mental phenomena that, according

to Rivers, are meaningless unless one supposes the operation of an unconscious. These include the content of dreams (to be understood through the mechanisms described in *Freud's Interpretation of Dreams*); the morbid thoughts, images, and "dreads" that intrude into one's sleeping and waking states; characteristic "solutions" (adaptations) to mental conflicts, such as suppression, hypochondria, phobic avoidance, compulsive ritual, paranoia, and the use of alcohol; and sudden personality changes (Rivers 1920:9, 14–15, 36, 38, 139–147).

Diagnosing the War Neuroses

Rivers takes this theory of the mind and mental mechanisms to Maghull and Craiglockhart Military Hospitals. To understand what he discovers when he arrives there and how his perceptions fit with those of other army doctors, we must take a long detour, through the maze of technical terms that were then employed for making diagnoses, choosing interventions, interpreting outcomes, and communicating with colleagues and patients.

For most of the First World War, the Royal Army Medical Corps (RAMC) divided the war neuroses into four diagnoses: shell shock, hysteria, neurasthenia, and disordered action of the heart.

Shell shock is the war's emblematic psychiatric disorder. The initial and most restrictive definition associated the disorder with three features: exposure to an etiological event sufficient to cause neurological damage, symptoms indicating the loss or impairment of functions of the central nervous system, and the presumption of organic changes that would be sufficient to connect such events to such symptoms.

As the name "shell shock" suggests, the typical etiological event involved exposure to forces generated by high explosives. Mere proximity to an explosion was said to be sufficient, since the effect of coming into contact with the shock waves would be equivalent to violently striking a man's head and spine with a solid body. According to an account published in *The Lancet*, when a soldier is struck by these shock waves,

> the vibration is transmitted through the bony structures to the cerebro-spinal fluid and thence to the brain and spinal cord, causing a molecular disturbance of the delicate colloidal structures of the neurones. . . . [Compression is followed by an equal decompression,] causing the liberation of bubbles of gas in the blood and tissues leading to embolism. [The combined] forces of compression and decompression act in producing vascular disturbances in the central nervous system, causing arterio-capillary anaemia and venous congestion. (Mott 1917:614)

In addition to injuries produced by shock waves, explosions were said to cause concussions by dislodging sandbags from parapets and by throwing men against the walls of their dugouts and other solid objects. Shell shock symptoms would be still more severe if the victim, lying unconscious and half-buried, inhaled the noxious gases produced by these explosions (Mott 1916:441–448, 545).

Microscopic hemorrhages and other vascular changes had been observed in the brains of two British soldiers who had died as a result of explosions and were without visible external injuries (Mott 1917). The first man had been in a dugout when a shell exploded ten feet away. Following this, he suffered tremors, general depression, and periods of crying. The next day, he was unable to talk "or do anything" and refused to answer questions. That evening, he entered "a state of acute mania, shouting 'Keep them back, keep them back.' He was quite uncontrollable and . . . impossible to examine. He was quieted with morphine and chloroform, and got better and slept all night. . . . Next morning he woke up apparently well, and suddenly died." The second soldier was in an ammunition shed when it was hit by a shell. He became unconscious at once and died soon afterward (Mott 1917:612–613).

In addition to the postmortem evidence supplied by these two cases, physicians in the French army reported that, "if a lumbar puncture was performed . . . within a few hours of the onset of symptoms, the cerebro-spinal fluid was generally under increased pressure and contained albumin, blood and slight excess of lymphocytes" (Mott 1917b; Hurst 1941:112; Mott 1919:709 also refers to experiments on animals conducted in France and the United States).

There was, however, no effective technique for demonstrating organic effects in shell shock survivors. If cerebrospinal fluid was examined more than forty-eight hours after the injury, "abnormalities were no longer present. The cerebro-spinal fluid was consequently almost always normal when the lumbar puncture was performed at base hospital" (Hurst 1941:112).

Diagnosis of *hysteria* was based on symptomatology—the partial or complete loss of control over sensory, perceptual, or motor functions—rather than etiology. The most common symptoms reported among British soldiers included paralyses, contractures, muscle rigidity; gait disorders involving the limbs, extremities, and spine; seizures, tremors, spasms, tics, and uncontrollable blinking; the radical narrowing of the field of vision and blindness; localized numbness and loss of sensitivity; localized pain, and hypersensitivity; mutism, aphonia, and stammering; deafness; fugue states, Ganser twilight state,[3] amnesia, mental confusion, and extreme suggestibility; a persistent sensation of unpleasant smells and tastes or, alternatively, the loss of sensibility in these faculties; various cardiovascular symptoms; enuresis; gastrointestinal symptoms, including vomiting, indigestion, and diarrhea (Adrian and Yealland 1917; Eder 1917:20–47; Myers 1940:28–29).

At first glance, this looks like a ragbag of unrelated symptoms. To the physician, however, they have something in common: each can be produced by lesions and other abnormalities in the nervous system. To learn whether or not a patient's symptoms had actually been produced in this way, a physician would have to rely on careful diagnosis and testing.

The *neurasthenia* diagnosis originates in the writing of George Beard (1880, 1881). As a war neurosis, it was believed to result from exposure to prolonged periods of intense mental and physical strain. Onset could be gradual or sudden, and the precipitating events were often the same as those said to cause shell shock. To be more exact, neurasthenia would develop in someone who had been exposed to a series of these events. The cumulative effect of the exposure would be to reduce his ability to withstand such events in the future. Once his threshold of resistance had dropped below the critical level, the next stressful event would precipitate the syndrome. As in shell shock and hysteria, the somatic medium for this process is neurological—"nerve exhaustion," according to Beard and the army doctors.

Here again, a putative neurological etiology connects a set of symptoms that would otherwise appear to be unrelated: anxiety, depression, emotional lability and irritability; difficulties sleeping, concentrating, and remembering; chronic fatigue and easy exhaustion; headache; loss of appetite; obsessive interest in somatic states and symptoms; loss of self-confidence (Turner 1916:1073; Myers 1940:27).

Disordered action of the heart (DAH) included cardiac sensations and indications that are precipitated or exacerbated by physical effort or psychological stress and that could be caused by an organic abnormality. The symptoms that are mentioned most often by army doctors include palpitation, unusually high or unusually low pulse rate, arrhythmic pulse, anginal or precordial pain, high or low blood pressure, shortness of breath, fainting attacks, giddiness, severe weakness or unsteadiness, sweating, fatigue and easy exhaustion, irritability, sleeplessness, headache, problems concentrating, and (less often) nausea and vomiting. In contrast to hysteria and neurasthenia, DAH often originated in activities and events that took place far from the front line, and it was not uncommon for soldiers to be diagnosed with this disorder while still undergoing training. The syndrome was known by various names, including DaCosta's syndrome, effort syndrome, irritable heart, and valvular disease of the heart (in the British forces), and soldier's heart and neurocirculatory asthenia (in the American forces) (Lewis 1917, 1920; Mott 1919).

In 1917, a fifth classification, *not yet diagnosed (nervous)* (NYD [N]) was added to this list. This was intended to be an interim diagnosis that would serve until men were reexamined in specialized hospital units located further down the line. In practice, though, it was often the only diagnosis that men received for symptoms associated with war neuroses (Wittkower and Spillane 1940b:31; Merskey 1991:258).

The war neuroses—shell shock, hysteria, neurasthenia, DAH, and NYD (N)—shared two features. First, the practical effect of their symptoms was to eliminate or severely reduce a man's effectiveness as a frontline soldier. In this sense, they were equivalent to a debilitating wound or injury. Second, they were characterized by the RAMC as being "functional" rather than "organic" disorders, a conventional medical distinction that dates back to the nineteenth century. Authors of standard medical texts of the period, such as Gowers (1903) and Oppenheim (1911), cautioned against the common error of equating "functional" with "psychological." (For a more recent commentary on the distinction, see Marsden 1986.) The approved usage, which was followed by the army doctors writing in journals such as *The Lancet* and the *British Medical Journal*, employs "functional" as a label for symptoms and syndromes that deviate from normal functioning and *might* have biological origins. As this term was applied to war neuroses, it covered syndromes that were presumed to originate in invisible (submicroscopic and unidentified) neurological lesions and abnormalities (Bury 1918:297); syndromes presumed to originate in visible (identifiable) neurological conditions that are undetectable under ordinary circumstances, such as the conditions described a moment ago in connection with shell shock; syndromes presumed to originate in unidentified biochemical changes that interfere with normal neurological processes (Gowers 1903:591); and syndromes presumed to originate in psychogenic factors, including psychic conflicts (Eder 1916).

When psychogenic factors are posited, they are usually described as operating together with physical causes.

In some instances, psychological causes (fright) and physical causes (typically shock waves) are described as having *concurrent* effects (see pages 59–60 for a detailed account). Psychological factors are also described as operating *consequent* to physical causes (organic injuries). According to one army doctor, the hysterical conditions most commonly found among the British troops were paralyses and contractures. In the great majority of these cases, these conditions were "grafted" onto wounds and injuries to these same body parts:

> The immobility and spasm may arise as a voluntary or reflex response to pain, or they may be due to localized tetanus or the application of splints or bandages, the abnormal posture assumed and the immobility and spasm being perpetuated by auto-suggestion after the primary cause has disappeared, to which very often is added the hetero-suggestion involved in treatment by electricity and massage when this is not really required.

A similar pattern was said to follow injuries to the central nervous system. As the invisible neurological lesions healed, a man's organic symptoms would be gradually replaced and imitated by hysterical symptoms (Hurst 1941:8–9).

Finally, psychological factors are also said to produce effects through the *mediation* of undetectable physical changes: "Even in a passing emotion such as terror or anger, some change must occur in the brain—we know not whether it is merely a disturbance in the supply of blood to certain parts or whether it is a chemical change affecting for the time being the atoms and molecules of which the nerve tissue is composed" (Bury 1896:189; also Bury 1918:97–98).

Fredrick Mott, a distinguished physician and pathologist, and by 1916 a colonel in the RAMC, reported that the majority of war neurosis casualties occur among soldiers with "an inborn timidity or neuropathic disposition, or an inborn germinal or acquired neuropathic or psychopathic taint," that is, conditions that would naturally make these men vulnerable to "the terrifying effects of shell fire and the stress of trench warfare" (Mott 1916:331).[4] In 1918, J. M. Wolfsohn, an American army doctor under Mott's supervision at Denmark Hill Military Hospital (London), analyzed the family and personal histories of one hundred nonofficer patients diagnosed with war neuroses. Three-quarters of the patients were found to have positive [pathognomonic] family or personal histories: "insanity occurred in the family in 34 per cent., epilepsy in 30 per cent., stigmata of degeneration, e.g. onychophagia [nail-biting], adherent ear lobes, high arched palate, etc., in 10 per cent. . . . [A] history of phobias of various kinds, insomnia, superstitions, "nerves," or hysterics is present in 64 per cent., and irritability of temper in 36 per cent" (Wolfsohn 1918:178, 180). Half of the patients had alcoholic parents or grandparents, and a third had lineal relations who were total abstainers (regarded as equally significant by Wolfsohn and others). When he compared the one hundred neurotic men with one hundred nonneurotic soldiers hospitalized at Denmark Hill for war wounds, Wolfsohn found striking differences. There were no histories of insanity or epilepsy among the nonneurotic patients; the rate of familial alcoholism was half that of the neurotic patients; and so on (Wolfsohn 1918:178).

Wolfsohn's report includes histories of two men who are described as being typical of the neurotic and nonneurotic patients. First, the neurotic:

Diagnosis: Shell-shock, constitutional psychopath.
[A] high-explosive shell burst near by, badly dazing the patient. He had amnesia for several days following, terrifying dreams, much general tremor, and had occipital and frontal headache. He remains in one place very sad and morose. Physical manifestations of fear; no somatic injury. *Family history*: Father used alcohol

to excess, one sister insane and two sisters nervous. *Personal history*: Never did well at school; could not learn; bit his nails until 14 years of age. He was always of a timorous nature, nervous and fearful. He smoked excessively and is a staunch teetotaller. He had many attacks of "melancholia" and always liked to be alone; much depression and asocial; masturbation. (Wolfsohn 1918:179)

And now, a typical control patient:

Diagnosis: Compound fracture of the tibia and fibula.
In 1915 he was wounded by shell fire and returned to active service in two months. In August 1917, he was struck by a piece of shrapnel which fractured his tibia and fibula, after which he crawled 900 yards for help. On examination he had no tremors: memory and concentration were good; no fears, no insomnia; only occasional dreams; very cheerful; no signs of nervousness. *Family history*: Negative throughout [i.e., no pathology]. *Personal history*: Negative throughout. He was always cheerful, experienced no fears, and drank moderately of liquor. He was never previously nervous or timorous. (Wolfsohn 1918:179)

Two years earlier, David Eder, an army doctor serving in Malta, had analyzed a comparable sample of men invalided from the Gallipoli campaign. Wolfsohn had found evidence of significant hereditary and constitutional factors affecting three-quarters of his patients, but Eder detected these factors in only one-third of his soldiers (Eder 1916:144). It is Wolfsohn's figures, rather than Eder's, that are cited in the medical literature, however. Further, it seems that, like Wolfsohn, most army doctors were inclined to believe that flawed heredity and constitution have a determining effect in the great majority of cases of war neuroses (Adrian and Yealland 1917:868; Mott 1918a:127; Smith 1916:813, 815; see Pick 1989:chaps. 6 and 7 for the nineteenth-century background to these degenerationist theories).

Functional Disorders

While army doctors could agree that the war neuroses were functional disorders, there was no consensus on what "functional" signified in the case of specific disorders, particularly with regards to the effects of psychological factors. On the other hand, it was unnecessary to know the etiology of these disorders in order to diagnose them or treat them, and questions about the causes of the neuroses interested only a minority of physicians. Most army doctors found their tacit knowledge, consisting of vague ideas about organic and hereditarian causes, sufficient for their clinical needs (Culpin 1931:15). Therapies were simultaneously pragmatic and eclectic—relying on time-proven combinations of suggestion, hypnosis, electrotherapy, sedation, reeducation, rest, and dietary regimen—and were not specific to any etiology (Adrian and Yealland 1917). Psychologically oriented doctors like Rivers were an exception in this respect.

In practice, calling the war neuroses "functional disorders" intersected invisible mechanisms with ignorance. This is how one authoritative text characterized neurasthenia:

> Books have been written about it, and it has been divided into numerous classes. . . . It is convenient to be able to designate the condition by one word . . . , but there is no more justification for regarding neurasthenia as a . . . disease due to a definite morbid process, or even as an affection characterized by a well-marked group of symptoms, than there is for adopting a similar course with re-gard to "debility" among general diseases. (Gowers 1903:1045)

This is not just any kind of ignorance, though. Rather, it is a knowing ignorance, based on a diagnostician's familiarity with the families of dis-eases that the war neuroses imitate and resemble: the neurological and cardiac disorders that medical science and its technologies had success-fully connected to visible somatic causes, such as localizable neurological damage.

Let us take hysteria as our example. Hysteria was diagnosed by compar-ing a patient's symptoms with the neurological symptoms that they most closely resembled and noting the key differences: the distribution of hys-terical anesthesia is typically regional (for instance, matching the area of hand and wrist that would be covered by a glove) and does not correspond to the anatomical nerve pattern (Eder 1917:123); in cases of hysterical par-aplegia, patients generally exert control over the anal sphincter, a function that would be impossible in cases of paraplegia resulting from neural in-jury; and so on (Jones and Llwellyn 1917:124–125; Head 1922:827–828). However, the pathognomonic boundary between hysterical disorders and the diseases they imitate is not continuous. That is, available diagnostic tests and criteria were (and are) applicable to only some hysterical symp-toms and did not (nor do they now) identify people whose symptoms corre-spond perfectly with organic etiologies. In these cases, it is the patient's anomalous recovery that would confirm the disorder's hysterical character (Hurst 1941:10; Marsden 1986).

Hysteria and Malingering

Like the other three war neuroses, hysteria is defined by things which it is not. That is, hysteria is *not* the combination of neurological problems that it resembles. Nor is hysteria the same thing as "malingering"—the willful imitation of disease symptoms—even though hysteria and malingering are capable of producing the same sets of symptoms (Jones and Llwellyn 1917:115–243; Collie 1917:141–202): "The dividing line between malingering and functional neurosis may be a very fine one and many 'shell-shocks' are of hysterical nature. . . . Hysteria was called 'La grande Simulatrice' by Charcot . . . [and] simulation is common to both malingered 'shell-shock' and to much genuine 'shell shock'" (War Office Committee 1922:140).

To learn how to distinguish hysteria (and the other war neuroses) from malingering, army doctors could consult books that instructed them in the appropriate detection techniques. The books are written for a general medical audience and direct their readers to the various motives that men and women have for malingering. High on the list is the pursuit of insurance money through "compensation neuroses": simulations of syndromes that match traumatic hysteria and traumatic neurasthenia (and also match the neurological diseases that these functional disorders simulate). The typical case is a railway accident that the patient blames for causing symptoms—paralysis, severe chronic pain, tremors, et cetera—that make it impossible for him to resume his normal occupation, activities, and pleasures (Page 1883:226–253; Erichsen 1883:92–95). Military malingerers are a favorite subject of this literature: army doctors are warned to be alert for the dodges of malingerers, especially "in regiments hastily recruited under circumstances unfavourable to progressive and complete discipline" (Jones and Llwellyn 1917:15; also Collie 1917:371–381).

The message was not wasted on the RAMC. As the war continued, and the carnage at the front escalated to levels never before imagined, the old professional army was destroyed. It was replaced by a volunteer army that was, in its turn, superseded by a conscript army. From stage to stage, the problem posed by malingering—its threat to manpower and morale—grew more salient in the minds of military doctors and leaders (Stone 1988:253–254, 258; Leese 1989: chap. 2).[5] Shortly after the war, a War Office committee issued a report, based on testimony by RAMC physicians, identifying three main varieties of malingering and the relative frequency of each:

> (1) *True malingering*, meaning the action of one who deliberately attempted imposition in pretending to be suffering from "shell-shock," was of rare occurrence. . . .

> (2) *Partial malingering*, exaggeration of symptoms or prolongation of a condition no longer remaining was far from uncommon and frequently arose from a desire to avoid service or for a continuation of pension. Such a form of malingering was found most difficult to deal with even by specialists owing to the doubt which often existed in their minds as to the degree of intention present.

(3) *Quasi-malingering*, skrimshanking, skulking. In this group there were included those who with little or no pretence decamped from the battle as opportunity arose, pleading "shell-shock" as the excuse for their evasion. Their numbers were great. For the most part they made but feeble if any attempt at deception and ultimately by persuasion or command returned to duty. (War Office Commission 1922:141)

Detection books on malingering are full of portentous maxims of doubtful utility. For example, doctors are instructed that symptoms of hysteria and malingering are not quite identical: hysteria is a "replica" of a disease, while malingering is a mere "parody." Since patients with hysteria deceive themselves and malingerers seek to deceive others, a doctor can expect hysterics to revel in physical examinations and malingerers to loathe them. Doctors are further informed that it is only through exhaustive and unremitting study of the maladies of the nervous system that they can learn to distinguish between the two conditions (Jones and Llwellyn 1917:122, 123, 127). But these books also describe practical ways of telling hysteria from malingering. In cases of monoplegia, for example, the examining physician should raise the patient's arm and then release it. If it is a case of hysteria, there is a moment of hesitation after which the arm falls slowly, without the deadness that characterizes real neurological injury. The malingerer, on the other hand, is said to allow his arm to drop like a stone (Jones and Llwellyn 1917:119). Similar tests are included for symptoms of blindness, deafness, mutism, and other conditions. The strategic application of electrotherapy is also frequently mentioned:

A military surgeon informs me that, "as a remedial measure for exaggerators and malingerers, a very strong faradic current is the most useful of all remedies. When a case is without doubt diagnosed to be malingering, there need be no compunction about the forcible application of this remedy in the case of a soldier in time of war. . . . The malingerer may stand one or two applications, but he cannot stand the prospect of a daily repetition, and so he quickly gets well." (Collie 1917:9)

However, readers are warned that these tests are useful only up to a point, since the neuroses are characterized by highly variable responses. Patients with genuine hysterical monoplegia sometimes allow their arms to simply drop; painful faradization sometimes produces remissions in genuine hysterics as well as malingerers; and so on. Further, the clever malingerer, the man who knows what he will be up against in the examining room, has his own bag of tricks and can be hard to catch out despite the doctor's tests and ruses (Jones and Llwellyn 1917:83–84, 120–121; cf. War Office Committee 1922:141–144).

During this era, hysteria, as a diagnostic entity, occupies the space opened up between bona fide neurological disease and malingering. It is bounded, on one side, by the technologies and procedures that are transforming neurology into a medical science and, eventually, a clinical specialty. Its boundary on the other side is erected by a repertoire of tests and ruses that, under the right conditions, are supposed to distinguish it from malingering. In other words, hysteria's unity as a syndrome is essentially extrinsic. The same thing can be said of shell shock and disordered action of the heart. Neurasthenia is rather different. It starts the war as a shadowy diagnosis, a congeries of poorly defined symptoms configured around vague ideas about nerve exhaustion. Over time, however, it acquires a degree of positivity that the other war neuroses never develop—a point I shall elaborate presently.

Officers and Other Ranks

Most of what we now know about the procedures for diagnosing the war neuroses is based on three sources: accounts published by army doctors in medical journals and books; the very fragmentary epidemiological data published by the RAMC and the Ministry of Pensions; and Peter Leese's recent monograph on shell shock, in which he traces a sample of patients through hospital records and doctors' notes (Leese 1989). While the actual practices used by army doctors for classifying casualties are not always clear from these sources, one point seems certain. The four diagnostic classifications did not provide a basis for a coherent system of inclusions and exclusions: men with apparently similar symptoms often ended up with different diagnoses.

The "shell shock" diagnosis was initially reserved for conditions that follow a concussive shock delivered by high explosives to the head or spine. By 1916, Army doctors were arguing that the disorder's defining event actually consists of two components, a "commotional shock" (*commotio cerebri*) delivered by the explosive force and an accompanying "emotional shock": "Psychic trauma . . . plays a very considerable part in the production of symptoms of shell shock; in many cases a patient will tell you that he can picture in his mind's eye the shell coming in and can recollect the sound of the explosion, and even recall its terrifying effects, causing death and destruction of comrades" (Mott, quoted in Royal Soc. of Med. 1916:xi).

According to Mott, it would be extremely difficult for a physician to differentiate between the two kinds of shock, since "both might be attended by a state of unconsciousness followed by hysterical and neurasthenic

symptoms" (British Med. Assoc. 1919:709). Another army doctor, Charles
Myers, went a step further, arguing that, under certain conditions, emotional shock alone might be enough to produce these effects. A soldier who
had been weakened through fatigue, poor food, disease, and lack of sleep—
that is, by conditions endemic to life at the front—would be vulnerable to
shocks resulting from sudden fright or scenes of horror, and such experiences could precipitate symptoms of shell shock (Myers 1940:26; Royal
Soc. of Med. 1916:xl). Differential diagnosis was made even more complicated by two widely accepted beliefs: that the onset of symptoms in traumatic neuroses can be delayed—"days, weeks, and even months may pass
[following the etiological event] before they develop"—and that cases can
be accompanied by symptomatic amnesias that block recollection of the
etiological event (Oppenheim 1911:1164; also Kraepelin 1902:372).
Given these various possibilities, army doctors had sufficient grounds for
giving a symptomatic patient the diagnosis of "shell shock" on the mere
presumption of an appropriate etiological event.

By 1916, the link between shell shock and concussion was broken. Although the classification continued to be used, it had lost any semblance of
specificity and was simply a synonym for "war neuroses" (Ross 1941:140).
E. G. Fearnsides analyzed the medical records of a sample of seventy soldiers diagnosed with "shell shock." In one-third of the cases, onset of
symptoms had been acute and followed either the explosion of a shell in
close proximity or burial in a dugout, a sequence of events and symptoms
that was consistent with the restricted definition of "shell shock." In another third, "symptoms came on acutely . . . [but] at a time when the men
were going to trenches from their billets, or when shells were arriving spas-
modically and occasionally"; that is, symptoms followed frightening
events that did *not* subject the men to concussive forces. In the final third,
symptoms were reported to have come on gradually, and the soldiers were
unable to recall any significant (etiological) events; that is, the clinical picture was closer to "neurasthenia" than to the restricted definition of "shell
shock" (Royal Soc. of Med. 1916:xl–xli). Hurst adds a fourth category of
patients, recalling that some army doctors would label soldiers "as suffering from shell-shock, even if, as occasionally happened, they had never
been out of England" (Hurst 1941:11).

Disordered action of the heart (DAH) followed the opposite trajectory:
it came to signify less, not more, than it originally had. Although the classification was used throughout the war and large numbers of soldiers were
given this diagnosis (Wittkower and Spillane 1940a:267), influential physicians, both civilian and military, argued that DAH was not a discrete
disorder. Rather, it simply (mis)labeled two other conditions: undetected
organic heart problems and neurasthenia. In the latter case, "DAH" merely
represented the neurasthenic's obsessional interest in his somatic states.

Cardiac events are a favorite target, because they are both ominous (and therefore worth worrying about) and perpetually available. An "exciting disturbance accidentally affecting his cardiac apparatus makes a very great impression upon a patient, and he develops a phobia concerning his heart." This is possible because the "heart—or, what comes to the same thing, its neural mechanism—reacts to all emotional manifestation" (Dejerine and Gauckler 1915:92–94). In many soldiers diagnosed with DAH, Mott writes, the response becomes part of a self-perpetuating cycle: anxiety (exacerbated by exertion or fatigue) induces the heart's disordered action, which captures the man's attention; the cardiac irregularities vindicate his concern and renew his anxiety, and thus a new round is initiated. It "would be better if this term [DAH] were abolished and simple neurasthenia adopted in its place" (British Med. Assoc. 1919:710; also Myers 1940:64; Wittkower and Spillane 1940b:19).

Following the Battle of the Somme (1916), the RAMC introduced measures to rationalize the assessment and management of casualties. "Advanced neurological centres" were established, each a short distance from the front line. (The RAMC used the term "neurological" to label disorders that we would call "psychiatric." Similarly, physicians given special training in diagnosing and treating war neuroses were titled "neurological specialists.") Patients were brought to the special centers from casualty clearing stations and then sorted into five wards, on the basis of their symptoms, injuries, and special needs. The neuroses were represented by two wards, one for neurasthenia and the other for hysteria (Hargreaves, Wittkower, and Wilson 1940:172). (The remaining wards were for "simple exhaustion," confusional states and psychoses, and miscellaneous medical and neurological disorders.) According to the RAMC's plan, patients who required extended or specialized attention would be moved from the advanced neurological centers to hospitals in rear areas or Britain. In practice, the majority of neurological patients evacuated to British hospitals never passed through an advanced center (Hurst 1941:11).

Today, we are unable to reconstruct the reasoning through which army doctors made their diagnoses. We do not know what proportion of the patients treated in neurasthenia wards were given the diagnosis "neurasthenia," or "DAH," or "not yet diagnosed (nervous)." We do not know how physicians decided which men would be sent to a "neurasthenia" ward rather than a "simple exhaustion" ward. Nor do we know, in detail, how the diagnostic process played out once men reached Britain. One gets the impression, reenforced by data collected in Leese's monograph, that diagnostic practices were not uniform but varied from one medical unit to another and, often, from one doctor to another. Medical data on officers present a special problem, since these hospital records and clinical notes describe symptoms in characteristically vague and nonspecific terms, and their pathognomonic significance remains opaque (Leese 1989:285, 242–245).

What *is* clear from the medical literature is that (1) very large numbers of men of other ranks (nonofficers) were diagnosed with "hysteria" or "neurasthenia," and (2) proportionate numbers of officers suffered functional neurological disorders. (There is ambiguous evidence to suggest that the number of officers who developed war neuroses may have been disproportionately large [Salmon 1917:29].) The distribution of diagnoses among officers was radically different, however. Many officers were diagnosed with "neurasthenia," but very few of them were diagnosed with "hysteria"—less than 1.5 percent of all officers diagnosed with nervous disorders, according to one survey (Wittkower and Spillane 1940a:266). Here again, the historical record does not permit a systematic analysis, and we are obliged to piece together an explanation with only very fragmentary information. Three points can be made with some degree of confidence, however.

Army doctors were treating officers for symptoms that were considered pathognomonic of hysteria among other ranks, for example, functional paralyses. Yealland (1918) describes four such cases; Culpin (1940:45), two more; Brown (1919:834), two at Craiglockhart; and Rivers, another at the same hospital (see below). Needless to say, this is hardly a scientific sample. My point is simply that a substantial fraction of the cases of officers' war neuroses that are described in the medical literature involve hysteria-like symptoms. Further, when officers exhibited hysteria-like symptoms, they were treated with the same medical procedures that doctors used for treating "hysteria" in other ranks.

Why were officers infrequently diagnosed with "hysteria," even when they exhibited the disorder's familiar symptoms? The answer is that symptoms would become signifiers of hysteria only if they were found to occur in an appropriate context. "A single hysterical symptom must not be regarded as more than a pointer to the general condition" (Culpin 1940:53; also Gowers 1903:1049 and Oppenheim 1911:1131). It could be assumed that soldiers developed symptoms of hysteria because they were weak, mentally and morally. This weakness represented an inherent inferiority whose emblematic expression was the "neuropathic personality"—the condition that Wolfsohn had discovered in the majority of his patients. "Neurasthenia" might also point to a predisposing weakness, but there was a second possibility as well. The disorder could originate in a weakness imposed from the *outside*, a weakness that was the result of a mechanical process, in which a man's physical defenses were gradually worn down through repeated exposure to the traumas and hardships of the front line.

Neurasthenia could be weakness without stigma, and the doctors who wrote about the war neuroses in the medical journals—men like Mott,

Hurst, Brown, and Rivers—promoted this view. On the other hand, it is clear from anecdotes and memoirs, such as Sassoon's, that officers, as well as their families and friends, often regarded this diagnosis as a source of shame. Doctors recognized the moral ambiguity attached to "neurasthenia" and routinely diagnosed the affected officers as suffering from "exhaustion," a somatic term, so long as these men returned to their units immediately following brief treatment (often simply sleep, hot food, and encouragement) at an advanced neurological center (Hargreaves et al. 1940:169–170).[6]

Clinical perceptions of hysteria were shaped by the widely accepted belief that it is the product of suggestion and suggestibility. This was the context—one which presupposes weakness of character and will—that allowed doctors to translate symptoms into signifiers. In the case of the typical officer, such a context was improbable (in the physician's view) because of what could be assumed about the patient's breeding and education and the process of selection and monitoring that produced frontline officers. It was this assumption, rather than the presence or absence of any particular symptom, that argued against hysteria. Thus Culpin recalls an officer patient he treated for a functional paralysis of the arm:

> Even so long after the event he was able to analyze his mental processes and recognize [that] there had been in his mind the thought of becoming . . . honourably removed from a situation in which he had reached his limits of endurance. Thus, in spite of an energetic and self-driving temperament that was the opposite to all common conceptions of the hysteric, his symptom was hysterical and the result of suggestion. Such a diagnosis would have been undoubtedly true, *but the truth in it was negligible.* (Culpin 1940:37; my emphasis)

In this system of medical reckoning, it is often the things that go without saying that are, in the end, the most telling.

The Return of the Protopathic

In *Instinct and the Unconscious*, Rivers proposes to explain the epidemiology of hysteria and neurasthenia among British soldiers (Rivers 1920:132, 207–209). Like many army doctors, he believed that these disorders originate in the soldier's reaction to his fear of being killed or maimed. When one's life is threatened, the instinctive response is to flee, fight, or freeze. But flight and fight were generally unavailable options: desertion was extremely dangerous, if not altogether impossible (Babington 1983), and opportunities for attacking the enemy were sporadic. According to Rivers, hysteria is a neurologically primitive response to these circumstances, a "collapse into sickness" that goes through three steps: (1) the flight and fight impulses are suppressed, (2) the neurological apparatus switches to the only available response, immobilization, and (3) the response is embodied in the disorder's characteristic symptoms: paralyses, contractures, anesthesias, et cetera (Rivers 1920:97–98, 102, 130).

When man's ancestors roamed the ancient plains and forests, the freeze response functioned to protect the *group* from shared dangers (predators), by giving it collective invisibility.In the era of modern warfare, this phylogenetic legacy acquires a new significance. Take the example of mutism, which, Rivers writes, is "[o]ne of the most frequent features of the hysteria of warfare":

> [Crying out is] a mode of reaction, by which the individual warns the rest of the group of the danger from which he is himself reacting instinctively by means of flight. . . . If, however, a group of animals should adopt the reaction to danger by means of immobility, the cry would be totally out of place. . . . If, therefore, hysteria be primarily a variant of the instinct of immobility, it is natural that one of its earliest, if not its earliest, need should be the suppression of the cry or other sound which tends to occur in response to danger. I suggest, therefore, that the mutism of war-hysteria is primarily connected with the collective aspect of the instinct of immobility. (Rivers 1920:133)

The function of the instinctive response has now shifted, to protect the *individual*, by removing him from the group (Rivers 1919:89; 1920:135, 148–152). At the same time, immobilization works against collective interests and values, depleting the group's defenses and jeopardizing its collective mission, spreading through its ranks like a contagion that is transmitted by suggestion (Rivers 1920:131).

Neurasthenia—or, as Rivers preferred, "anxiety neurosis"—has no archaic narrative, since the possibility of the disorder develops relatively late in phylogenetic time: only after the brain had evolved a cortical layer and a site of self-consciousness. As with hysteria, symptoms of anxiety neurosis represent the soldier's attempt to deal with his fear. Unlike hysteria, which begins with an act of suppression located in the unconscious, anxiety neurosis originates in "repression"—the soldier's *conscious* attempt to "thrust out of his memory some part of his mental content . . . [making it] inaccessible to manifest consciousness." Repression is not necessarily a bad thing on the front line: it benefits the individual, by allowing him to manage the conflict between duty and the survival instinct (up to a point, at least), and it benefits the collectivity, by providing it with an overtly calm and self-confident leader, a man who represses not merely his feelings of fear but also his (natural) inclination to express them outwardly (Rivers 1920:186).

In other words, the ordinary soldier's hysteria subordinates the interests and values of the group to the survival of the individual, while anxiety neurosis (neurasthenia) reverses the formula, since it is a self-sacrifice in the interest of the collective welfare (Rivers 1920:123, 208, 213).

According to Rivers, the goal of military training has traditionally been to teach the ordinary soldier to repress his feelings and expressions of fear, so that he might "act calmly . . . in the presence of events . . . calculated to arouse disturbing emotions" (Rivers 1920:186, 218). The Regular Army succeeded in achieving this goal because its enlistments were long-term, and "training in repression" had to be spread over many years. Presumably, hysteria would have been comparatively rare among the men of Britain's Regular Army, but we cannot know because it was decimated in the initial stage of the war. According to Rivers, the soldiers who made up the front-line troops from 1916 on were different from the regulars in precisely this respect: training in repression had to be conducted only over relatively short periods, months rather than years. At the same time, the new soldiers were the target of practices that intensified their natural suggestibility: notably, close-order drill and routines aimed at habituating men to automatic obedience (Rivers 1920:90–100, 211). This historical change, from an army of regulars trained in repression to an army of conscripts and volunteers trained simply to obey orders and fire weapons, was, according to Rivers, the "most important feature of the present war in its relation to the production of neurosis" (Rivers 1920:213).

An officer's background and breeding would have protected him from the conditions that promote hysteria but would have simultaneously exposed him to the conditions responsible for anxiety neurosis: "Fear and its expression are especially abhorrent to the moral standards of the public schools at which the majority of officers have been educated. The games and contests which make up so large a part of the school curriculum are all directed to enable the boy to meet without manifestation of fear any occasion likely to call forth this emotion" (Rivers 1920:209). While the ordinary soldier was, by nature and training, highly suggestible and had few scruples about either acknowledging or expressing his fear, the mental life of the officer was "more complex and varied"; he possessed "greater intelligence and broader education" and he was expected to be the source of suggestion, not its target (Rivers 1920:123–124, 132).

> Anyone having much to do with those who have taken part in the fighting of the war must have been struck by the extraordinary manner in which an officer, perhaps only just fresh from school, has come to stand in a relation to his men more nearly resembling that of father and son than any other relationship. It seems clear that different battalions show the incidence of neurosis in very different degrees, and this is probably due more than anything else to the nature of the relations between officers and men by which the private soldier acquires towards his officer sentiments of duty and trust. (Rivers 1920:217)

Rivers's conclusion is that anxiety neurosis is the price that an officer pays for repressing fear and its expression under these conditions. Over time, exhaustion, illness, and repeated psychological shocks rob him of the strength that he needs for resisting "the motives arising out of the instinct of self-preservation." Psychological symptoms, such as intrusive thoughts and images, begin to appear, and the officer tries to adapt to them in the same way that he adapted to fear, through willpower and repression and through sublimation, the channeling of instinctive energies (survival impulses) into socially useful ends (tending to the welfare and readiness of the troops). A cycle of affect (anxiety) and action (sublimation) is created, and physical exhaustion is accelerated. When psychobiological resources fall below the level needed for normal functioning, a "breakdown" occurs: "If the nature and causation of anxiety-neurosis were more fully understood, it would be possible to intervene at an earlier stage . . . , for the victims of this form of neurosis suffer through excess zeal and too heavy a sense of responsibility and are likely to be the most valuable officers" (Rivers 1920:225).

How utterly different Rivers's neurasthenic officer is from the common soldier. While the officer continues to be burdened with anxiety and depression even after he has been evacuated to Britain, the typical hysteric is said to be "relatively or positively happy" once his somatized symptoms have removed him from the "conflict between instinct and duty." According to Rivers, this pattern is consistent with another odd epidemiological fact, namely, that cases of hysteria are rarely reported among prisoners of war or soldiers with serious wounds. Because these men are safely hors de combat, hysterical symptoms would be superfluous (Rivers 1920:128–129).

Like many army doctors, Rivers believed that hysterical symptoms become fixed through a combination of suggestion and suggestibility, and that the most effective way to remove the symptoms is through countersuggestion. If hysteria can be called a "suggestion-neurosis," he writes, then neurasthenia might be called a "repression-neurosis" (Rivers 1920:124, 130, 223; cf. 135). Rivers is quite clear that the soldier is not simply repressing his awareness of his fear and his desire to give it outward expression; he is also repressing a traumatic memory. Why would Rivers want to make this connection, between the onset of symptoms and a discrete memory/event, when he knew that some patients had been subjected to a stream of terrifying events prior to the onset of symptoms and that some patients claimed to have no memory of traumatizing events? Rivers does not address this question directly. Given what he does say, however—in the case studies included in *Instinct and the Unconscious*, for instance—it is reasonable to assume that he was swayed by four considerations.

First, some patients at Craiglockhart spontaneously reported that they were troubled by discrete traumatic memories. Further, these reports were consistent with published case studies narrated by other army doctors.

The focus on discrete memories was also consonant with Rivers's interpretation of psychoanalytic ideas. Unlike Rivers, Freud now concluded that his patients' traumatic memories originated in imagined, rather than real, events. But this was less important to Rivers than were Freud's conclusions regarding the therapeutic significance of the memories. On this point, both men agreed: the path to a cure led through the traumatic memory.

Other army doctors, including men not inclined to give psychological explanations, reported that the onset of "shell shock" symptoms was sometimes precipitated by an event that was not markedly different from the events previously experienced by the patient.

Finally, Rivers believed that these memories are repressed because they are distressful. Therefore the inability of a correctly diagnosed patient to recollect such a memory would be evidence of its successful repression.

According to Rivers, the widespread failure to understand this last point is an important obstacle to treating anxiety neurosis. The conventional medical wisdom had it that patients should be advised to further repress their traumatic memories: "Put it out of your mind," "Try not to think of it." This is also common sense, but it is mistaken and countertherapeutic (Rivers 1917:914):

> Instead of advising repression and assisting it by drugs, suggestion, or hypnotism, we should lead the patient to resolutely face the situation provided by his painful experience. We should point out to him that such experience . . . can never be thrust wholly out of his life. . . . His experience should be talked over in all its bearings. Its good side should be emphasised, for it is characteristic of the painful experience of warfare that it usually has a good or even a noble side, which in his condition of misery the patient does not see at all, or greatly underestimates. (Rivers 1917:914)

Remaking Rivers

Rivers was not a typical RAMC physician. Relatively few doctors were familiar with Freud's work, and fewer still wrote original accounts of the origins and epidemiology of the war neuroses. Beyond these obvious distinctions, it is a matter of some dispute as to how Rivers's ideas and clinical practices articulated with those of other army doctors. Perhaps the most frequently cited attempt to position Rivers in the medical community of his day is Eric Leed's *No Man's Land* (1979).

According to Leed, the RAMC established two therapeutic regimes—

that is, "two different techniques of domination, rooted in two different conceptual frameworks and visions of human nature"—for treating and controlling patients diagnosed with war neuroses. One regime was *analytic*, based on the assumption that the mind is "a mechanism of opposed parts that processed—often below the level of consciousness—the needs of the individual and accommodated these needs to the imperatives of reality." Within this regime, the doctor's job was to promote the successful processing of mental conflicts. The second regime was *disciplinary*, based on "principles and techniques derived from animal training." Pain was inflicted on the patient with electricity; commands were shouted at him; he was placed in isolation and on a restricted diet. Then, after he had a taste of this treatment, he was told that he could free himself only by abandoning his symptoms. The function of the regime was "to dramatize and clarify moral issues involved in the conflict between public duty and the private intentions of the patient," and it promoted this end by collapsing the "distinction between legitimate neurosis and malingering" (Leed 1979:170, 171, 173).

Leed's scheme has been adopted by Elaine Showalter in a chapter on male hysteria in *The Female Malady* (1987). For both Leed and Showalter, the emblematic figure of the analytic approach is Rivers, who is juxtaposed in their accounts with Lewis Yealland, representing the disciplinary regime. Yealland was resident medical officer at the National Hospital for the Paralysed and Epileptic, in London. He remained a civilian physician throughout the war and treated soldiers who had been referred to the National Hospital for special care. His book, *Hysterical Disorders of Warfare* (1918), is a collection of case studies describing his therapeutic encounters with British and Belgian soldiers.

Leed represents the disciplinary regime through Yealland's account of a soldier identified as "Case A1" (Yealland 1918:7–15; Leed 1979:174–175). The story of Case A1 is also excerpted in Showalter's book, where this man's treatment at the National Hospital is contrasted with Sassoon's therapy at Craiglockhart, a place where "the patient and doctor were friends; the therapy was kindly and gentle; the hospital was luxurious; the most advanced Freudian ideas came into play" (Showalter 1985:176–178). Yealland and the hapless Case A1 next appear in Pat Barker's *Regeneration* (1991), a fictionalized account of the Rivers-Sassoon encounter. The book's denouement replays Yealland's narrative, only this time Rivers is the horrified eyewitness to the treatment of Case A1 (now named "Private Callan") (Barker 1991:chaps. 20 and 21).

In each of these narrations, the author has located Rivers's place among the British army doctors by juxtaposing him with Yealland and, then, by identifying Yealland with Case A1. In order to follow this chain of representations back to Rivers, we must start with Case A1. Yealland

writes that A1 was a twenty-four-year-old private who had already participated in nine major engagements on the western front. While serving in
Salonika, he had been overcome by heat and fell unconscious for five
hours. On waking, he "shook all over" and could no longer speak. Before
being sent to the National Hospital, A1 had been treated for nine months,
without success, by a variety of methods, including electricity, hypnotism,
and "hot plates" applied to the back of his mouth. What follows are pas-
sages excerpted from Yealland's account of his four-hour encounter with
A1:

> In the evening he was taken to the electrical room, the blinds drawn, the lights
> turned out, and the doors leading into the room were locked and the keys re-
> moved. The only light perceptible was that from the resistance bulbs of the bat-
> tery. Placing the pad electrode on the lumbar spines and attaching the long pha-
> ryngeal electrode, I said to him, "You will not leave this room until you are
> talking as well as you ever did; no, not before." The mouth was kept open by
> means of a tongue depressor; a strong faradic current was applied to the posterior
> wall of the pharynx, and with this stimulus he jumped backwards, detaching the
> wires from the battery. . . .
>
> [I repeated that] "the doors are locked and the keys are in my pocket. You will
> leave me when you are cured, remember, not before." . . . This evidently made an
> impression on him, for he pointed to the electrical apparatus and then to his
> throat. "No," I said, "the time for more electrical treatment has not come. . . .
> Suggestions are not wanted from you; they are not needed. When the time comes
> for more electricity you will be given it, whether you wish it or not." . . .
>
> [Later] "You are now ready for the next stage of the treatment, which consists of
> the administration of strong shocks to the outside of the neck; these will be trans-
> mitted to your 'voice box' and you will soon say anything you wish in a whisper."
> . . . It was not long before he began to whisper the vowels with hesitation. . . . He
> then made an attempt to leave the room, and I said firmly to him, "You will leave
> the room when you are speaking—speaking normally." . . .
>
> I then applied shock after shock to the posterior wall of the pharynx, commanding
> him each time to say "ah," and in a few minutes he repeated "ah" in expira-
> tion. . . . When he was able to repeat the days of the week, months of the year, and
> numbers, he became very pleased and was again quite ready to leave me. I said,
> "Remember, there is no way out, except by the return of the proper voice and the
> door. You have one key, I have the other; when you talk properly I shall open the
> door so that you can go back to bed." With a smile he stammered, "I believe you
> have both keys—go and finish me up."

There are two obvious contrasts between the therapy of Yealland and Rivers. While Yealland uses electricity and inflicts physical suffering and an-

guish on A1, Rivers employs a painless talking cure. The second difference lies in their contrasting attitudes toward the past. During his session with A1, Yealland refers to the past only twice, when he reminds A1 that he has survived frontline ordeals worse than faradic electricity and when he mentions the unsuccessful treatments that A1 previously received for his problem. The connection between past and present is homiletic rather than dynamic: what you have endured before you can endure again; what was done incorrectly the first time will now be set right. A similar attitude is reflected in Yealland's narration of A1's emotions. A1 is described as depressed; at one point, he is said to cry; one is given the impression that he was frightened. In the vignette, these emotions are represented as reactions to A1's immediate situation: his fear of the pain to come and (it would seem) his attempts to evade pain by stimulating Yealland's pity. And that is it: the emotions lead back to nothing else, their meaning is on the surface. A similar theme runs through Yealland's description of the electrical room. The locked door, the closed blinds, the naked electric bulb connected to the battery that powers the faradic device, the ominous language ("You will leave when you are cured . . . , not before")—Yealland planned all this with great care. His intention was to create a very specific environment (see Adrian and Yealland 1917), to confine the arena of action to this room, these two men, and this moment in time. For the next four hours, the consciousness of A1 would be closed off to anything else.[7]

Another notable feature of this vignette is Yealland's attitude toward the mind and its processes. It would be incorrect to describe Yealland as unpsychological, since his therapy is based on ideas about suggestion. But his psychology consists mainly of his conviction that, in clinical encounters of this sort, it is the stronger, more determined intelligence that will dominate the weaker one. Yealland is interested in wills rather than minds. This explains his attitude toward emotions and also his preoccupation with things somatic: the precise placement of the electrodes, the excited pursuit of spasms and tremors over A1's body. Yealland was familiar with the major theories about the war neuroses, and one supposes he would have had no problem explaining his procedures in psychological terms. Given the very high cure rate that Yealland claims to have had with his techniques, it would seem that he had no incentive for delving beneath the surface of a patient's symptoms.

In *Sherston's Progress*, Sassoon describes Rivers as "a wonderful man," who "helped and understood me more than anyone I had ever known" (Sassoon 1936:122, 124, 128). Despite the words of gratitude that Yealland puts into A1's mouth at the end of the session—"Why did they not send me to you nine months ago?"—it is hard to imagine that Yealland inspired in his patients the kind of affection that Sassoon felt for Rivers; and if Craiglockhart was not quite the luxurious place pictured by Showalter—

Sassoon described it as a "dilapidated hydro"—its sunny rooms and therapeutic conversations were a world apart from the electrical room at the
National Hospital.

Leed and Showalter are therefore correct in noting the important differences between the clinical practices at the National Hospital and Craiglockhart. It is their next claim—that the differences signify a split into two
technologies of domination, each attached to a distinctive vision of human
nature—that needs to be questioned. While the targets of these technologies are never made clear, it seems reasonable to suppose that Leed and
Showalter have either or both of the following things in mind: (1) The
regimes impact on the consciousness of army doctors, so that doctors acquire or reaffirm different visions of human nature, depending on the therapeutic regime to which they have been posted, *or* each physician acquires
two different visions of human nature, based on his knowledge or observations of the two regimes. (2) The regimes are apparatuses that control the
minds and bodies of soldiers, by appropriating their (nascent) antiwar sentiments and memories into clinical narratives, that is, making them symptomatic of mental disorder (the analytic regime) and/or persuading soldiers
that life at the front is less painful than it would be in a hospital (the disciplinary regime).

Neither argument is persuasive. First of all, medical units treating war
neuroses combined heterogeneous procedures and practices (including be-
nign neglect) and did not divide up neatly into disciplinary and analytic
regimes. Yealland and Rivers were exceptional in precisely this respect:
each man concentrated on a narrow range of practices, Rivers because he
was given the freedom to follow his intellectual inclinations, Yealland because he was recognized as a specialist and was sent a distinctive subpopu-
lation of patients. Second, the two visions of human nature do not preclude
one another: they can coexist as aspects or (in Rivers's model) levels of a
single nature. Perhaps the most striking difference between Rivers and
Yealland in this regard is the narrowness of Yealland's vision and the
broadness of Rivers's.

The prevailing wisdom, shared by both Yealland and Rivers, was that
cases of hysteria were effectively treated through combinations of countersuggestion (facilitated, as needed, by electricity, hypnosis, dietary regimen,
isolation, etc.) and massage, baths, and physiotherapy (to restore motility
and muscle tone in the case of paralyses and contractures) (Oppenheim
1911:1095; Hurst 1918:21; Mott 1919:130).

> In most cases nothing more was needed than a little reassurance, encouragement,
> and waking suggestion. On one occasion six such cases were admitted to my
> wards at one time. I put them in a row of beds in one cheerful ward, where the
> amusing aspect of the row of mutes was fully appreciated. The following morn-

ing one of the mutes woke up speaking, to his own astonishment and seeming
delight. During the same day all the other five found their voices. (McDougall
1926:238; see also Adrian and Yealland 1917:870)

By 1917, hysteria was regarded as relatively easy to treat and high rates of
remission were reported (Wittkower and Spillane 1940b:31). The difficult
patient was the hysteric "who had shrunk into himself and developed fixed,
circumscribed and obvious objective difficulties, and was usually best dealt
with by a *force majeur*, sent in as a stimulus from the external environment" (Royal Soc. of Med. 1919:438). It was this small minority of patients who were referred to the National Hospital for further diagnosis and
treatment. Although Leed makes heavy weather over Yealland's use of
electricity and conjures up specters of Pavlovian conditioning, electricity
was an established technique for treating functional disorders long before
the war. An electrical room had been opened at Guy's Hospital in 1863,
and three years later a similar one was established at the National Hospital.
In the first decades of its use, doctors speculated on the possible physiological effects that electricity might have on these disorders. By Yealland's
time, however, the use of electricity was largely pragmatic and, in the opinion of many physicians, it was simply a tool in the service of clinical countersuggestion and an instrument for detecting malingerers (Beveridge and
Renvoize 1988:157–160). Yealland makes it clear that he shares the latter
view, when, for example, he advises army doctors that "a strong electric
stimulus will produce a sensation and motion in a limb which is supposed
to be anaesthetic and paralysed, and this in itself will be enough to convince the patient that he is on the road to recovery." (See Royal Soc. of
Med. 1915:ii and British Med. Assoc. 1919:709 for similar comments by
Mott.) For electrotherapy to successfully promote countersuggestion,
Yealland continues, the patient must be convinced that the doctor understands his case and is able to cure him. Therefore the "best attitude to adopt
is one of mild boredom bred of perfect familiarity with the patient's disorder." Discuss the case with the patient as briefly as possible and concentrate
on learning what treatment he has previously received, since clinical procedures that failed in the past are an important source of iatrogenic heterosuggestion (they have the effect of further fixing the patient's symptoms). This
explains why patients not yet treated with electricity can be cured with a
mild current, but those previously treated without success require a strong
and painful current, applied over hours rather than minutes (Adrian and
Yealland 1917:869–871). Case A1 stands out in *Hysterical Disorders of
Warfare* in precisely this respect (i.e., chronicity, duration, and intensity of
electrotherapy): he seems to have been an exceptional patient, even among
Yealland's collection of exceptional patients.

Memory, Abreaction, and Suggestion

During his lifetime, Jean-Martin Charcot's account of hysteria was widely accepted. Using his patients at Salpêtrière to illustrate his thesis, Charcot portrayed hysterical attacks as moving, with mechanical regularity, through clearly delineated phases, each characterized by stereotyped and often grotesque contortions and postures (Goldstein 1987:324, 326–327). Knowledge of the sequence and the stigmata constituted a "positive law," and this, he claimed, pointed to an underlying pathophysiology. By 1914, however, Charcot's thesis had been displaced by rival accounts. The most popular of these among British army doctors was one promoted by Joseph Babinski, who had been Charcot's student and protégé. Babinski claimed that cases of hysteria are products of suggestion and suggestibility. As a diagnostic entity, this disorder is simply the set of symptoms that can be removed by countersuggestion (Babinski and Froment 1918:28, 46; also Hurst 1918:21 and Mott 1918:127). The phases of hysteria that Charcot put on exhibit in the course of his clinical lectures did not arise spontaneously from the patients' conditions but were the products of external influences, viz., Charcot's own unintended suggestion, mimicry among patients (hysteria patients and epileptics were often kept in the same wards), and the intentional and unintentional manipulation of patients by Charcot's staff.

Once Babinski's thesis is accepted—that suggestion is the cause and cure of hysteria—treatment becomes a contest of wills. In essence, this was Yealland's position: therapy pits the patient's determination to hold on to his fixed ideas against the doctor's will to remove them by countersuggestion.

There were other ways to interpret and treat hysteria, and a minority of army doctors preferred some form of abreactive therapy. As practiced by Breuer and Freud, abreaction consisted of bringing patients back to their repressed traumatic experiences, dragging the memories and attached affect up to consciousness, and then releasing these emotions when the memory is verbalized in speech (see chapter 1). However, the Breuer-Freud theory was not adopted by all army doctors employing abreaction. According to William Brown, for example, emotions aroused during traumatic experiences produced motor symptoms by disrupting communication between neurological centers. Abreactive therapy would consist of calling up the traumatic memory in order *to produce* high levels of affect: "[R]einstatement of intense emotion acted physically in overcoming synaptic resistances in specific parts of the nervous system, and so put the nervous system into normal working order again. The effect is more efficient than that of, e.g., an electric current since it is selective and occurs only in just

those parts of the system concerned with the production of symptoms"
(Brown 1919:835).

For Yealland and like-minded physicians, this all boiled down to the
same thing, namely, suggestion. If Brown's technique turned out to be superior, it was simply because patients found it more convincing (Adrian
and Yealland 1917:869). Abreactive therapies differed one from another
only in their particular myths (premises) and rituals (techniques). Claims
about repressed memories and neurological processes notwithstanding,
abreactive therapy operates, like any other successful cure, through a *generic* mechanism, namely, countersuggestion. If patients and their doctors
believe otherwise, and if their beliefs result in a remission of symptoms,
then this merely underlines the power of suggestion.

This frame of reference leaves no space for pathogenic secrets. Yealland
cautions against calling a patient's attention to a supposed trauma, especially if it involves a violent event, such as an explosion. The fact that the
physician shows interest in the event can have the effect of further fixing
the symptoms, by persuading the patient that they have a physical cause.
Yealland's advice to physicians, therefore, is to question the patient "as
briefly as possible," concentrate his attention on the events now taking
place in the clinic, and remember that the only part of the past really worth
investigating is the man's previous course of treatment (Adrian and Yealland 1917:869–870).

Practitioners of the Breuer-Freud type of abreaction saw things differently, of course. For them, the traumatic memory is not a prop in the theater
of countersuggestion but a system of connected images, associations, and
emotions—simultaneously a psychological and a neurological reality. At
the same time, one must not suppose that in abreactive therapy the physician takes the same interest in the patient's memory as he would if he were
practicing psychoanalysis or similar techniques. In abreaction, events must
be remembered and reexperienced, but the doctor's interest is in the mechanical bond connecting memory to stored-up energy or, in Brown's version, to energy that the recalled memory might arouse. The patient's memory and the event on which it is based might be regarded as real, but their
specific nature and significance are, in themselves, irrelevant to the clinical
process. The therapist simply listens to the patient's memory; he has no
motive for helping him to interpret it.

Autognosis

The clinical attitude just described contrasts with the technique called "autognosis" (self-knowledge), used by army doctors for treating cases of neurasthenia and anxiety neurosis. The goal of autognosis was said to be the

reintegration of the patient's mind through a process in which he and his physician scrutinized the traumatic past and its relation to the present: "The method to be employed is that of long persuasive talks with the patient . . . in the course of which one enters into his past mental conflicts and worries, explains fully the origin of his present symptoms, and helps him to see both the past and the present experiences in their right proportions" (Brown 1919:836; also Brown 1920:30–31). Where the proponents of autognosis differed among themselves was in the value they placed on abreactive therapy. McDougall argued that the premises of abreaction ran counter to clinical experience: "If living through a scene of horror produces a psychoneurotic disorder, why should the living through it a second time cure or tend to cure it? On the face of it we might expect that the disorder would be accentuated by the repetition of the emotional experience" (McDougall 1920a:24).

There was no standard explanation for what happened during autognosis. McDougall held that these disorders originate in the dissociation of the memory of the etiological event, which is experienced as a symptomatic amnesia. At the traumatic moment of dissociation, the corresponding synaptic junctions are disconnected from the higher cortical levels of the brain but remain connected with the fear center in the basal ganglia. A closed circuit is created, within which the movement of energy is confined. In cases of neurasthenia, this energy is experienced as distressing emotion. In cases of hysteria, the energy finds an outlet in motor symptoms. For example, a tremor of the right hand develops from a violent spasmodic effort by the same hand at the time of the event, such as repeatedly operating the breech mechanism of a rifle. Symptoms cease when the dissociation of the synaptic junctions is overcome. That is, in the act of realizing the *true relation* between his present circumstances and the terrifying incident, the patient restores connections between the circuit and "other cortical dispositions." Emotional energy is no longer confined to one narrow system and now takes a more normal course, spreading to many points in the brain (McDougall 1920a:28–29).

Like other accounts of autognosis, McDougall's explanation transforms the meaning of memory. In abreactive therapy, meaning is invested in a mechanical function: memory is simply a vehicle for dredging emotion to the surface. In autognosis, memory acquires a double meaning: a neurological meaning (it restores synaptic connections) and a meaning in the patient's inner life. I have been able to find no detailed account of treatment by autognosis. How were these clinical sessions organized? Did therapists employ specific clinical strategies and techniques? We do not know, but one gets the impression, mainly from Rivers, that the process was rather casual and low-keyed, at least in cases where abreaction was not also a goal. What is clear is that doctors believed that three categories of patients are inappropriate for the treatment.

The first category consists of patients who are diagnosed with hysteria and who would be served better by other procedures, such as countersuggestion or abreaction.

In the second category are patients who cannot be induced to retrieve or disclose events suitable for building clinical narratives. Rivers describes such a case, involving an officer who, while serving at the front, fell into unconsciousness for a brief period. When he awoke, he discovered that he had lost function and sensation in both legs. He was then sent on home leave and instructed to avoid both reading newspapers and talking to people about the war (i.e., he was, according to Rivers, encouraged to repress his memories and feelings). At the end of his leave, he appeared before an army medical board but broke down when he was asked questions about the war. He was sent to a hydro and treated unsuccessfully with electricity, baths, and massage. Then he was sent to Craiglockhart. Upon arrival, his legs were nearly useless, and he was anxious and emaciated; when he tried to sleep, his mind was crowded with distressing thoughts about the war; and when he fell asleep, he dreamt of disturbing war scenes. Rivers suggested to him that his initial unconsciousness (in France) may have resulted from an explosion. The patient replied that he had no memory of such an event, and during his course of treatment, he disclosed nothing concerning possible etiological events. Rivers advised him "to give up the practice of repression, to read the papers, talk occasionally about the war, and gradually accustom himself to thinking of, and hearing about, war experience." The man complied in "a half-hearted way," but from then on he improved: he slept better; war scenes occurred less frequently in his dreams; and he read the news without distress. At the end of his treatment, he informed Rivers that he did not believe "that his improvement was connected with this ability to face thoughts of war"; he would have recovered on his own, he said, if his home leave had been extended (Rivers 1920:193–194).

The final category of patients unsuitable for autognosis consists of those who have experienced traumatic events that are simply too horrible to be integrated into normal mental life. Rivers narrates a case that began when the officer was thrown down by an explosion.

> [H]is face struck the distended abdomen of a German several days dead, the impact of his fall rupturing the swollen corpse. Before he lost consciousness the patient clearly realised . . . that the substance which filled his mouth and produced the most horrible sensations of taste and smell was derived from the decomposed entrails of an enemy. When he came to himself he vomited profusely, and was much shaken, but "carried on" for several days, vomiting frequently, and haunted by persistent images of taste and smell. (Rivers 1920:192)

The man was sent to Craiglockhart but did not improve. His only relief came during a trip into the countryside, far from things that might remind him of the war. His memory was so horrible that he found it hard to give up repression. Rivers was also reluctant to attempt to lift the repression, since the traumatic event was "so wholly free from any redeeming feature . . . that it is difficult or impossible to find an aspect which makes its contemplation endurable." Over the weeks, the event appeared less frequently in his dreams and in less terrible forms, but it still recurred. Finally, Rivers suggested, and the patient agreed, that "he should leave the Army and seek the conditions which had previously given him relief" (Rivers 1920:192–193).

Sigmund Freud, Continued

At a high level of abstraction, the ideas underlying autognosis (Rivers) and psychoanalysis (Freud) are similar: war neuroses are rooted in conflicts between divergent drives; symptoms are the individual's attempt to defend himself against these conflicts; the etiological events and mental conflicts are incorporated into the patient's dream content; the dreams provide the therapist with invaluable diagnostic and therapeutic information; conflicts can be resolved and symptoms removed through self-knowledge (cf. Eder 1916:267).

When one moves to a less abstract level, however, it is clear that there are fundamental differences between Rivers's ideas and Freud's:

1. Freud traced the psychoneuroses to *unconscious* mental processes, while Rivers connected anxiety neurosis to important *conscious* processes: the patient's "witting" attempt to put memories and emotions out of awareness.

2. According to Rivers, Freud believed that "the essential cause of every psycho-neurosis [lay] in some disturbance of the sexual function" that originated in the early years of life (Rivers 1920:3). Like nearly all the army doctors interested in psychological factors, Rivers rejected this claim (Eder [1917: chap. 5] is an exception), arguing that the vast majority of cases involving "the psycho-neuroses of war . . . [are] explicable as the result of disturbance of another instinct, one even more fundamental than sex—the instinct of self preservation. . . . The awakening of the danger-instincts by warfare produces forms of psycho-neurosis far simpler than those of civil life" (Rivers 1920:4–5).

3. In *The Interpretation of Dreams*, Freud argued that dreams are wish fulfillments, efforts at obtaining pleasure (Freud 1953a [1900]). Rivers argued that this idea was wholly inconsistent with his observations at Craiglockhart (Rivers 1920:37). Rivers's opinion, shared, it seems, by many army doctors (e.g., Mott 1918b), was that the dreams of a patient suffering from anxiety neurosis function to perpetuate high levels of anxiety rather than to gratify desires.

In these comments, Rivers was addressing himself to Freud's ideas circa 1914. However, the war had produced important changes in Freud's own thinking. While Freud did not treat any patients diagnosed with war neurosis, he was familiar with the clinical work of members of his circle who served as military doctors in the German and Austro-Hungarian armies: Ernst Simmel, Karl Abraham, and Sándor Ferenczi. The Fifth Psychoanalytical Congress, held in Budapest in September 1918, included a symposium on the war neuroses. According to Freud, "official representatives from the highest quarters" of government were present and seemed favorably disposed toward creating centers to study "these puzzling disorders and the therapeutic effect exercised on them by psycho-analysis." Nothing came of this, however, since the war ended two months later and "interest in the war neuroses gave place to other concerns" (Freud 1955a [1919]:207). Thus ended Freud's practical interest in the war neuroses. Chronic and delayed-onset cases did not concern him, since he believed that, "when war conditions ceased to operate, the greater number of the neurotic disturbances brought about by the war simultaneously vanished" (Freud 1955a [1919]:207; 1955c [1920]:215). Nevertheless, he had several occasions to write or speak about these disorders in the years immediately after the war. In 1919, he wrote an introduction to a book based on the Budapest symposium; in 1920, he served as an expert witness in an official Austrian government inquiry into the inhumane use of electricity in the treatment of war neuroses; and, in the same year, he devoted a short section of *Beyond the Pleasure Principle* to the war neuroses. I shall here summarize his ideas about the war neuroses.

War neurosis is a subtype of the traumatic neuroses. Its onset is a product of fright and conflict. "Fright" is the name given to the reaction of someone who encounters a danger for which he is unprepared. Where there is fright, there is *no* anxiety, since anxiety signals the anticipation of danger and obviates the element of surprise. When the fright is great—a common situation among frontline soldiers—the neural systems that are responsible for receiving incoming stimuli "are not in a good position for binding the inflowing amounts of excitation." The result is an extensive breach in the shield protecting "the organ of the mind" against excessive stimuli (Freud 1955c [1920]:12, 13, 31).

This flood of unbound stimuli is the source of the physical symptoms of the war neuroses: contractures, paralyses, et cetera. This etiology explains why men with physical injuries do not develop war neuroses, since stimuli are bound to the injury in these cases (Freud 1955c [1920]:33).

At first glance, Freud's etiological account seems to reinstate the old theory that "regards the essence of the shock as being the direct damage to the molecular structure or even to the histological structure of the elements of the nervous system." But the similarities are superficial, since the terms

"excitation," "binding," "protective shield," et cetera do not imply mechanical violence, which is the hallmark of the old theory (Freud 1955c
[1920]:31).

In Freud's view, the somatic symptoms of the war neuroses are a defensive formation, as in all neuroses. The need for a defense is promoted by a
distinctive mental conflict:

> The conflict is between the soldier's old peaceful ego and his new warlike one,
> and it becomes acute as soon as the peace-ego realizes what danger it runs of
> losing its life owing to the rashness of its newly formed, parasitic double. It
> would be equally true to say that the old ego is protecting itself from a mortal
> danger by taking flight into a traumatic neurosis. . . . Thus the precondition of the
> war neuroses . . . would seem to be a national [conscript] army; there would no
> possibility of their arising in an army of professional soldiers or mercenaries.
> (Freud 1955a [1919]:209)

Here again, Freud appears to recall an old theory, in that the formation
of a "parasitic double" echos ideas about double consciousness. But once
more the similarities are superficial. Freud is not referring to alternating
consciousness but rather to a *split* in the patient's conscious personality, in
which the two sides are simultaneously present and capable of internal
monologues of the following sort: "What am I (new ego) getting myself
(my physical survival) into?" and "What am I (old ego) becoming (new
ego)?" The warrior ego represents the penetration into the psyche of a double danger: it simultaneously threatens the survival of the old ego and the
man's physical existence. It is the fear of this *internal* enemy that distinguishes the war neuroses from other ("pure") traumatic neuroses (Freud
1955a [1919]:210).

War neurosis and psychoneurosis are both based on fears of internal
enemies. In psychoneurosis, the disorder originates in a conflict between
the ego and the sexual instincts which it repudiates: "the enemy from which
the ego is defending itself is actually the libido." This seems to be a fundamental difference between traumatic neuroses and psychoneuroses. According to Freud, critics have fastened on this difference as a way of undermining the claims of psychoanalysis to have discovered a unitary (sexual)
etiology for neurotic disorders: "They have been guilty here of a slight
confusion. If the investigation of the war neuroses (and a very superficial
one at that) has *not shown* that the sexual theory of the neuroses is *correct*,
that is something very different from its *showing* that theory is *incorrect*"
(Freud 1955a [1919]:208). The concept of a "narcissistic libido" might
eventually explain the etiology of the war neuroses, Freud adds. If it does,
these neuroses will be brought within a unitary theoretical framework
(Freud 1955a [1919]:209–210; 1955c [1920]:33).

There is some evidence to suggest that Freud believed that the internal conflict in war neuroses can originate in *moral* danger as well as *mortal* danger. In his testimony to the Austrian government commission, Freud proposed that

> the immediate cause of all war neuroses was an unconscious inclination in the soldier to withdraw from the demands, dangerous or outrageous to his feelings, made upon him by active service. Fear of losing his own life, opposition to the command to kill other people, rebellion against the ruthless suppression of his own personality by his superiors—these were the most important sources on which the inclination to escape from war was nourished. (Freud 1955b [1920]:212–213)

In this passage, Freud compares *dangerous* demands with *outrageous* demands, and the fear of losing *one's own life* with the fear of taking *someone else's life*. The first half of each pair signifies the pathogenic power of fear, a point that Freud, like other writers of the time, makes repeatedly. The second half allows another possibility, namely that a man might be traumatized by the violence that he inflicts on others, and that a soldier can be *both the victim and perpetrator* of his traumatic violence. With this observation, a place is opened for traumatic guilt alongside traumatic fear.

Comparison between the symptoms of war neuroses and psychoneuroses poses another problem, this time in connection with the claim that dreams are a form of wish fulfillment. Army doctors found Freud's position difficult to reconcile with their clinical observations. The dreams of war neuroses patients were accompanied by high levels of anxiety, and it was inconceivable that their terrors fulfilled the wishes of the tormented dreamers. In *Beyond the Pleasure Principle*, Freud accepts this criticism and revises his claim.

> [D]reams occurring in traumatic neuroses have the characteristic of repeatedly bringing the patient back into the situation of his accident, a situation from which he wakes up in another fright. This astonishes people far too little. They think the fact that the traumatic experience is constantly forcing itself upon the patient even in his sleep is proof of the strength of that experience: the patient is, as one might say, fixated to his trauma. (Freud 1955c [1920]:13)

If the patient is fixated to his trauma, why does he try so hard to *avoid* thinking about it when he is awake? Freud's answer is that the dreams originate in a *compulsion to repeat*, the patient's unconscious urge to return to the situation in which his pathogenic trauma occurred. The compulsion explains why the dreams recur regularly and why they are invariably accompanied by anxiety. Anxiety is a signal of danger and it is the absence of anxiety at the time of the trauma that explains the pathogenic fright. Therefore, dream anxiety is instrumental. It is an attempt to anticipate (retrospectively) the danger that precipitated the trauma (Freud 1955c [1920]:13, 23, 32).

In this way, Freud finds a home for the pathogenic secret in the patient's anxiety dreams. Because he himself treated no cases of war neurosis, he cannot discuss the place of pathogenic secrets in his own clinical practices, but he does cite, with approval, Ernst Simmel's methods. Simmel was senior physician at a German army hospital specializing in the treatment of war neurosis, and he practiced a form of abreactive therapy modeled after Freud's technique:

> As is known, Breuer and Freud induced patients to remember under hypnosis the original trauma, the conscious memory of which was not accessible to the patient, and to abreact the emotion attached to it. After the patient awakened, the symptom disappeared. . . . Simmel developed a particularly efficient and quick-acting technique; on the basis of previous inquiries, he led the hypnotized patient right into the middle of the traumatic situation, let him experience again with all the details which his conscious memory had lost and also established associative connections with the past. (Freud 1955a [1919]:321)

Freud had discarded this technique in 1905, on the grounds that many patients are not able to be hypnotized and the therapeutic effects of this treatment are not permanent (Grünbaum 1993:347). Now, fifteen years later, testifying before an inquiry into the therapeutic abuse of men suffering from war neuroses, he appears to accept Simmel's claim to have successfully treated war neuroses through hyponosis.

Rivers's Legacy

Rivers died in 1922. In an obituary in *The Lancet*, his close friend and colleague, Grafton Elliot Smith, wrote that

> [Rivers's] psychological work in the military hospitals at Maghull, Craiglockhart, and Hampstead (Royal Air Force) was of momentous importance. His authority was of material assistance in gaining recognition of the value of psychotherapy; and his scientific investigation of the problems of mental disorder not only threw a light upon the mode of causation and the rational treatment of these troublesome conditions, but it also helped to rescue this important branch of medicine from the extremists on the two sides. (Smith 1922:1222)

Regarding Rivers's "momentous importance" for psychological medicine (psychiatry), Elliot Smith proved to be a better friend than a prophet. In the decades following his death, Rivers's name has for the most part been associated with his enduring contribution to the development of social anthropology, his collaboration with Henry Head in their famous nerve experiment, and his friendship with Siegfried Sassoon. Rivers's reputation as a war doctor has been revived twice. At the beginning of the Second World War, he was remembered fleetingly, mainly by colleagues who had returned to service in the RAMC (Miller 1940). The current attention that is given to Rivers the doctor coincides with the entry of post-traumatic stress disorder into the official psychiatric nosology and renewed interest in psychological trauma associated with childhood sexual abuse.

The story of Dr. Rivers and his evil opposite, Lewis Yealland, is retold most recently in Judith Herman's popular book, *Trauma and Recovery*:

> The reality of psychological trauma was forced upon public consciousness once again by the catastrophe of the First World War. . . .

> The most prominent proponent of the traditionalist view was the British psychiatrist Lewis Yealland . . . [who] advocated a treatment strategy based on shaming, threats, and punishment. . . .

> Progressive medical authorities argued, on the contrary, that combat neurosis was a bona fide psychiatric condition. . . . They advocated humane treatment based upon psychoanalytic principles. The champion of this more liberal view was W.H.R. Rivers. (Herman 1992:21)

Set aside the dubious sketch of Yealland (in reality, he was a neurologist, not a psychiatrist; his prominence as a war doctor is retrospective; and there was no unitary "traditionalist view," in Herman's sense). It is Herman's picture of Rivers and his brand of medicine that attracts our attention. "Progressive," "liberal," "humane," and "psychoanalytic" are a single piece, and Rivers is a standard-bearer for these views. This picture of Rivers is questionable for at least two reasons.

Rivers's *theoretical* views of the war neuroses are more indebted to Hughlings Jackson—and standing behind him, Herbert Spencer—than to Sigmund Freud (who had his own debt to Hughlings Jackson). Further, although the views of both Rivers and Freud are anchored in evolutionary biology, it is the experimental physiology of Crile and Cannon that speaks to Rivers and not Freud's deliberately vague notions of cathexis and excitation (Kitcher 1992:73–74).[8] Whether this makes Rivers more or less "progressive," I am unsure. On the other hand, his tendency to associate the epicritic principle with officers and the protopathic with other ranks would seem to cast a shadow on his "liberal" credentials.

Perhaps it is Rivers's *clinical* interest in traumatic memory that makes him appear progressive to progressive people. Once again, the real Dr. Rivers is a disappointment. Shortly before his death, Rivers testified before a War Office committee on shell shock, where he was asked to comment on the "mental wounds" caused by emotional shock:

[Rivers] said he should be inclined to put it this way, that when the man began to have a number of disturbances of different kinds, such as loss of sleep, etc., he either consciously or more or less unconsciously looked for an explanation, and this tended to centre around some particular experience, in many cases a comparatively trivial experience. . . .

The reason why he objected to the term [shell shock] was that so far as he could see the main factor had been stress, and the shock in most cases was merely the last straw. Any disturbance might have produced the same result. (War Office Committee 1922:56)

Put into other words, Rivers is observing that, in most cases, it is not the traumatic memory that produces the physical and emotional symptoms of the war neuroses (anxiety disorder) but rather the reverse: the symptoms account for the memory. He is *not* saying that the soldiers create their memories. His point is that the recalled memory is usually not the effective cause of the syndrome but the patient's way of explaining it. Underlying Rivers's statement to the War Office there is an understanding of the self that is closer to Ribot than to Freud.

It is this conception of the self that informs Rivers's FitzPatrick Lectures (1915), where he advises his audience that "the study of people of rude culture . . . will help us also to understand better the place taken by suggestion both in the production and the treatment of disease" (Rivers 1916:122). Likewise, five years later, speaking before the War Office Committee, Rivers describes the traumatic memory as a form of autosuggestion. At the same time, it would be incorrect to say that Rivers rejects the notion of the pathogenic secret. It is simply that he has no compelling reason, in his clinical work, to distinguish between the effects of suggestion and the effects of pathogenic secrets, since his theory of the war neuroses can support either conclusion. And this explains why "autognosis" seems vague and open-ended compared with analogous psychoanalytic practices.

For Freud, on the other hand, the ability to distinguish between the memory that is self-narrative and the memory that "acts as a foreign body" is critical. Freud requires a way to rule out the effects of suggestion: he needed it in the 1890s, when he and Breuer were explaining the traumatic origins of hysteria, and he needs it in the postwar period, having moved on to psychoanalysis and psychoneurosis. Freud finds a solution in the idea of *resistance*. His clinical interest in resistance is simultaneously epistemological and therapeutic:

In psycho-analysis the suggestive influence . . . is inevitably exercised by the physician. . . . Any danger of falsifying the products of a patient's memory by suggestion can be avoided by prudent handling of the technique; but in general *the arousing of resistances is a guarantee against the misleading effects of suggestive influence.* (Freud 1955d [1923]:251; my emphasis).

[T]he aim of the treatment is to remove the patient's resistances and to pass his repressions in review and thus to bring about the most far-reaching unification and strengthening of his ego, to enable him to save the mental energy which he is expending upon internal conflicts. (Freud 1955d [1923]:251).

It is precisely the epistemological element, reflected in the first passage, that Rivers has left out of his clinical work. He believes that the "repression" (Freud's "suppression") of traumatic memories is an obstacle to achieving a cure and, for this reason, he wants to overcome his patients' resistance to dragging their memories up and verbalizing them. But this is not Freud's idea of "resistance." There is nothing psychodynamic about it, and the subsequent remission of symptoms says nothing decisive to Rivers about the etiology of the disorder or the possible role of suggestion in its cause and cure.

Given that my interpretation is correct, how have Herman and other writers succeeded in misplacing Rivers in the genealogy of the traumatic memory? The answer is that they give us a very selective reading of Rivers's work, overlooking its complexity and originality while, at the same time, praising him for his "wide-ranging intellect" (Herman 1992:21). They view Rivers down the wrong end of the telescope, so that the encounter with Sassoon at Craiglockhart stands for the whole or his theory and practice. What we get is not Rivers at all but a different creature, Rivers-Sassoon. Rivers did not survive long enough to comment on the long-term after-effects of the traumatic memory, nor was he much interested in its short-term effects during the years between the armistice and his death (1922). Sassoon, on the other hand, survived long after the war (he died in 1967) and, according to Herman, "was condemned to relive it for the rest of his life" (Herman 1992:23). By suggesting that Sassoon was "condemned to relive" the war, Herman transposes his poetry and prose writings from the realm of literary imagination to the realm of psychiatry (cf. Fussell 1975:90–105, 207, 216). The effect is to graft Sassoon's sense of the past (now medicalized) onto Rivers's clinical vision.

Conclusion

One of the ironies of the World War is that developments in certain fields of medical science were accelerated by the carnage of the killing fields. Thomas Salmon (1917), a senior American army doctor, predicted that the conflict would provide medical scientists with thousands of natural experiments of the sort that they would rarely encounter in peacetime or, more likely, would ordinarily encounter only in experimentation involving animals. Salmon's prediction proved correct. In the course of the war, army doctors examined and treated tens of millions of traumatic injuries, including unusual and interesting injuries to the central nervous system. G. W. Crile and W. B. Cannon, mentioned in chapter 1, served as army doctors in France, and the latter conducted important field research on wounded soldiers, in a collaborative effort with British doctors, to discover the pathophysiology of traumatic shock (Benison et al. 1991).

Medical interest in disorders involving psychogenic trauma followed a different path during the war. Once again, physicians were presented with unprecedented numbers of cases, but the similarity between the two kinds of trauma ends here. In the case of the psychogenic traumas, there was no accumulation of knowledge, development of new treatments, or revision of established theories to parallel the changes that occurred in biological medicine. Even Rivers, who provided perhaps the most ambitious theoretical account of the war neuroses, was looking back to the nineteenth century, conjuring up the ghosts of Spencer, Hughlings Jackson, and Ribot.

During the postwar years, psychiatric interest in the traumatic neuroses declined. These disorders were now mainly associated with "shell shock" and were more likely to occupy the attention of compensation boards and forensic experts than physicians. Interest in the traumatic neuroses eventually revived but only decades afterward. Indeed, "revival" is too weak a term for what eventually came to pass three conflicts later, in the aftermath of the Vietnam War. Part 2 of this book describes the transformation of the traumatic neuroses and traumatic memory during these years.

Part II

THE TRANSFORMATION OF TRAUMATIC MEMORY

Three

The *DSM-III* Revolution

A FLURRY of publications on traumatic neuroses followed the armistice in 1918. Over the next two decades, however, these disorders attracted little attention, until in 1941, just prior to American entry into the Second World War, a monograph titled *The Traumatic Neuroses of War* was published under the auspices of the National Research Council, a private American foundation (Kardiner 1941; Kardiner and Siegel 1947). This book, by Abram Kardiner, is the first systematic account of the symptomatology and psychodynamics of the war neuroses published in the United States. It is now routinely cited as a landmark in the history of the post-traumatic disorders (e.g., Herman 1992:23–26, 28), and it is a source of the symptom list for post-traumatic stress disorder in the current psychiatric nosology.

Abram Kardiner had been briefly psychoanalyzed by Freud in the early 1920s. Once back in the United States, he worked as a psychiatrist in a Veterans Hospital from 1922 to 1925. (Like W.H.R. Rivers, Kardiner was attracted to anthropology; he played an important part in the "culture and personality" school that flourished in American cultural anthropology from the 1930s to the 1950s.) Kardiner's starting point in *The Traumatic Neuroses of War* is Freud's argument that traumatic events coincide with breaches in the barrier that protects the brain against external stimuli. Freud described the symptom formation that follows these events as *defensive*; it functions to preserve the ego. (See the discussion of Freud in chapter 2.) Kardiner, on the other hand, claimed that the symptomatic reaction is a form of *adaptation*. It is an effort to eliminate or control painful and anxiety-inducing changes that have been produced by the trauma in the organism's external and internal environments. The kind of adaptation that occurs in a particular case will depend on the individual's psychological resources and his relations to his primary social group (Kardiner 1941:141). This is called an "environmental" or "reactive" view of psychiatric problems, and it was widely accepted in American psychiatry (and anthropology) during this period.

In Kardiner's account, traumatic events create levels of excitation that the organism is incapable of mastering, and a severe blow is dealt to the total ego organization. The individual experiences this as a sudden loss of effective control over his environment:

> The activities involved in successful adaptation to the external environment become blocked in their usual outlets. . . . These activities are consummated in some form of *aggression*. This *aggression* is expressed in every function of the sensory-motor apparatus and its adjuncts, the central and autonomic nervous systems. . . . As a result of the trauma, that portion of the ego which normally helps the individual to carry out automatically certain organized aggressive functions of perception and activity on the basis of innumerable successes in the past is either destroyed or inhibited. (Kardiner 1941:116–117; also 79, 84)

In *Beyond the Pleasure Principle*, Freud stresses the importance of the repetition compulsion in traumatic neurosis. This is his way of explaining the recurrent anxiety dreams that accompany the neurosis. Kardiner puts some distance between himself and Freud, by claiming that the idea of the repetition compulsion—a defensive maneuver by the ego to restore mastery—hides the fact that the ego has been significantly altered by the traumatic experience. It is a shrunken version of the pretraumatic ego, and its activities "are no more repetitive than are those of a prisoner in a cell containing nothing but a chair." Whenever some bit of action is initiated, the same blockage (incapacity) is encountered, and failure is the inevitable result (Kardiner 1941:189).

Thus the traumatic event occasions the emergence of a new ego, caught in an external environment that has been transformed into a hostile place, at least as the individual perceives it. The ego is now thrown into a perpetual struggle to regain mastery and responds in various ways: by adapting itself to a level compatible with its altered resources (making fewer demands on the world, reestablishing infantile relations with the world); by obliterating the problematic portion of the world (introjecting it into the offending body part, which is then paralyzed or anesthetized); or by periodically obliterating the entire world in order to reestablish amicable relations with it (through syncopal attacks and loss of consciousness, which symbolize the patient's death and rebirth) (Kardiner 1941:82, 117).

Kardiner writes that traumatic neuroses are accompanied by constant features: irritability, symptomatic of a lowered threshold of stimulation and a readiness for fright; explosive and unpremeditated aggressive reactions that often alternate with reactions of pathological tenderness; contraction in the general level of functioning, including diminished intellectual ability; a loss of interest in the world; and a characteristic dream life (Kardiner 1941:86–100, 204). It is in these dreams that the individual encounters his traumatic experience:

Instead of the condensation and the compactness of action in the dreams of the psychoneurotic, we have here a process of dilution and retardation, *like the picture of a normal piece of action slowed down by the motion-picture camera, the film's being cut off before the action is completed.* The images are redundant and perseverative. Indeed amazing is to hear a patient whose illness is nine years old state that night after night . . . he has dreams which take him back to the war scenes or which consist of feeble transformations of those battle scenes. (Kardiner 1941:88–89; my emphasis)

Kardiner emphasizes the uniformity and transparency of the dream content, its "stereotypy," its "monotonous regularity," and its "feeble transformations." The book includes descriptions of many dreams, including an inventory of sixteen dreams by a single patient, collected eight years after his battlefield trauma. Here are five of them, as described by Kardiner:

(1) I was at a party and a fight started. Someone began shooting and shot me dead right through the head. I woke up frightened.

(2) I was on the subway station. Someone pushed me off, and I was thrown on the tracks. A train came along and ground me up.

(3) I was in a garden somewhere, and there were large roses, larger than myself. I climbed up a ladder to smell them, when a large bee stung me on the back of an ear, and I woke up with a sharp pain which lasted about half an hour.

(4) I was a keeper of a lot of birds in a great big place.

(5) I was taken sick on the street with a spell. When I woke up you [Kardiner] were there to give me some medicine, and you told me to take it. (Kardiner 1941:89–90)

The most common motif is self-annihilation (the dreamer's violent death in dreams 1 and 2); other recurrent themes involve frustration (dream 3?) and aggression (dreams 4 and 5?). Of course, the meaning of the patients' dreams—based on the idea that the dreams replay traumatic events—is not as transparent as Kardiner at first suggests. In practice, their signifying power derives not only from their content but also from Kardiner's interpretations of the patient's concurrent symptoms ("action syndrome"), such as symbolically resonant motor dysfunctions.

The list of symptoms given in *The Traumatic Neuroses of War* is based on clinical observations made in the 1920s. The case studies begin with traumatic events that occurred in 1917–1918. The most often cited monograph about American psychiatric casualties during World War II is *War Neuroses* (1945), by Roy Grinker and John Spiegel. The text was written as an official report in 1943, following the Tunisian campaign, in which an untested American army conducted its first offensive operations against Nazi forces. Although the Americans eventually prevailed, they suffered defeats, high casualties, and low morale, and large numbers of troops developed psychiatric symptoms. Like Kardiner, Grinker and Spiegel adopt an "environmental" orientation, but their book is essentially descriptive and focuses on treatment priorities and modalities rather than pathodynamics. Psychiatric cases are divided into clinical syndromes, based on symptoms rather than etiology: free-floating anxiety states, somatic regressions, psychosomatic visceral disturbances, conversion states, exhaustion states, and psychoses. Each syndrome is said to represent a method by which "the personality attempts to deal with overwhelming anxiety." There is no separate category for traumatic neuroses, and the traumatic memory is found in the free-floating anxiety category (Grinker and Spiegel 1945:2).

The policy of the American Army Medical Corps during World War II was to treat psychiatric casualties close to the front line, where they were provided with hot food, a chance to rest and sleep, a shower, and a dose of reassurance and persuasion. The typical treatment lasted a couple of days, after which the soldier was sent back to his unit. When psychiatric problems were more severe, soldiers were sent to the base area, where they received alternative forms of treatment: abreactive therapy, group and personal psychotherapy, drug-induced continuous sleep therapy, convulsive shock therapy, and occupational therapy. Abreaction usually employed narcosynthesis:

> [This] causes the patient to re-experience the intense emotions which were originally associated with the traumatic battle experiences and which have been perpetuated in various stages of repression up to the moment of treatment. . . . The ego, freed from the impact of the immense forces of the repressed emotions . . . gathers new strength, and restores its contact between the powerful emotional drives and the world of reality, both past and present. (Grinker and Spiegel 1945:78, also 80)

The technique and theory of abreaction are essentially unchanged from the previous war, except that sodium pentothal—"truth serum"—has replaced hypnosis and ether.

Interest in the war neuroses rapidly faded once the war was over. In 1955, the Veterans Administration published a follow-up study of men diagnosed with war neuroses. The report presents its epidemiological data through a mélange of classifications. At some points, data are presented for four diagnostic categories: psychoneurosis, personality or behavior disorder, psychotic reaction (schizophrenic, alcoholic, or post-traumatic), and "other" disorders. Elsewhere in the report, comparisons are made between four different categories: neurotic traits, suggestive neurosis, overt neurosis, and pathological personality. At another point, data are presented for a list of symptom states: anxiety, depression, nightmares, insomnia, alcoholism, headache, hysteria, phobias, obsessions or compulsions, irritability, difficulty concentrating, restlessness, psychogenic somatic complaints, hypochondriacal reaction, psychotic symptoms, overt symptoms of behavior disorder (Brill and Beebe 1955:139–146, 167–176). No attempt is made to match the diagnostic categories with the list of symptoms within the report, nor is there an obvious way to compare the data sets in the report with the "symptom-complexes" (syndromes) described by Grinker and Spiegel in *War Neuroses* (Grinker and Spiegel 1945:2; cf. Gillespie 1942:170–174).

Forty years later, we read these accounts and look in vain for a Rosetta stone. But even at the time, Kardiner was impressed by the confusion of categories and symptom states associated with the traumatic neuroses: "[T]here is a vast store of data available on these neuroses, but it is hard to find a province of psychiatry in which there is less discipline than in this one. There is practically no continuity to be found anywhere, and the literature can only be characterized as anarchic" (Kardiner 1959:245). The war neuroses were not unique in this respect and, toward the end of the war, the War Department created a committee whose task was to bring order to the psychiatric nomenclature (Menninger 1948:258–265). By classifying the various neuroses on the basis of "dynamics of psychopathology," rather than by clearly defined symptoms (War Department 1946:180), the new nomenclature simply contributed to the "anarchy." The prevailing outlook in American academic psychiatry and the War Department committee was psychodynamic, and this explains the lack of interest in devising lists of criterial symptoms. It was assumed that while underlying psychodynamic processes would be constant (and therefore useful for discriminating disorders), their reactive symptom formations would be polymorphic, shaped by the patient's history and the particularities of his situation.

In the War Department's classificatory system, cases of traumatic neuroses were grouped under the general term "psychoneurotic disorders":

> The chief characteristic of these disorders is anxiety, which may be either "free-floating" . . . [or] unconsciously and automatically controlled by the utilization of various defense mechanisms (repression, conversion, displacement, etc.). . . .

> [The anxiety] is a danger signal felt and perceived by the conscious portion of the personality (ego). Its origin may be a threat from within the personality—expressed by the supercharged repressed emotions. . . . The various ways in which the patient may attempt to handle this anxiety result in the various types of reaction. (War Department 1946:181, 182)

The system contained no specific subcategory for "traumatic neuroses," and cases originating in traumatic events were classified according to their reactions: anxiety reaction, dissociative reaction (including amnesia), phobic reaction, or conversion reaction (pain, paralyses, anesthesia, etc.). While most of the cases described in Kardiner's *Traumatic Neuroses of*

War would correspond to "mixed reactions," the nomenclature did not allow this category. Each case was to be classified according to the predominant kind of reaction, thus dispersing cases of war neuroses over multiple categories (War Department 1946:182–183).

A Standardized Nosology

What is missing during this period is something that we now take for granted, namely, a *standardized psychiatric nosology*, a system of classification based on lists of criterial features and Aristotelian principles of inclusion and exclusion. The first standardized nosology in American psychiatry is the third edition of the *Diagnostic and Statistical Manual of Mental Disorders* (*DSM-III*), published in 1980. What follows is an account of the history of this nosology and the place that was created in it for the traumatic memory.

In 1974, the Council on Research and Development of the American Psychiatric Association appointed a Task Force on Nomenclature to begin work on a new edition of the APA's *Diagnostic and Statistical Manual of Mental Disorders*. In 1979, the council and the APA board of trustees approved the final draft of this edition (*DSM-III*), and it was published the following year.

DSM-III is an inventory of about two hundred named mental disorders. According to the manual's introduction, each of the entries is "a clinically significant behavioral or psychological syndrome or pattern that . . . is typically associated with either a painful symptom (distress) or impairment in one or more important areas of functioning (disability)" (Amer. Psychia. Assoc. 1980:6). The disorders are grouped into categories, based on shared features: mood disorders, anxiety disorders, sexual disorders, chemical substance use disorders, schizophrenia, and so on. Each malady is defined by a list of features that are individually necessary and collectively sufficient for the diagnosis. An accompanying text identifies associated (noncriterial) features of the disorder, its natural history, the impairments and complications that it might produce, its prevalence and predisposing factors, and other mental disorders with which it might be confused.

Prior to the publication of *DSM-III*, the various theoretical and clinical orientations of American psychiatry shared no common nosological language. The introduction to *DSM-III* claims that the manual provides a diagnostic metalanguage: a way of talking about mental disorders that is not particular to any theoretical orientation because it is based on features— overt behaviors, biochemical markers, cognitive deficits, and so on—that should be visible to any competent observer (Amer. Psychia. Assoc. 1980:4–5).

DSM-III is the product of fourteen advisory committees, consisting of experts in the major categories of mental disorders. Each committee prepared a succession of drafts, which were read and commented on by a liaison committee created by the Executive Assembly of the American Psychiatric Association. (The assembly is composed of representatives from the APA's regional branches.) Drafts were also circulated to other groups—the American Association of Chairmen of Departments of Psychiatry, the American Academy of Child Psychiatry, the American Psychological Association (representing clinical psychology), the Academy of Psychiatry and the Law, and three associations representing the views of psychodynamic psychiatry. Comments were solicited and passed on to the manual's editor, Robert Spitzer, and to members of the relevant committees. Drafts were likewise debated at annual meetings and conferences organized by the major psychiatric organizations. The proposed diagnoses were submitted to a series of field trials in order to identify problem areas in the classification and to try out solutions to these problems. "In all, 12,667 patients were evaluated by approximately 550 clinicians . . . in 212 different facilities, using successive drafts of DSM-III" (Amer. Psychia. Assoc. 1980:2–5).

It would seem unlikely that so many people, representing so many different points of view, would simultaneously find it in their interests to adopt a new psychiatric language. Nor is this what happened. Although a wide range of views was solicited and incorporated into the various drafts, the ideology of psychiatric pluralism extended mainly to the process of shaping the manual's *content*, that is, to the choice of the diagnostic criteria that would identify particular disorders. What was entirely nonnegotiable, according to Spitzer, was the feature intended to distinguish *DSM-III* from previous languages, namely, its *architecture*, the way in which its classifications would be constructed. The structure of *DSM-III* had been decided in advance by a small circle of people (including Spitzer) who identified themselves, in a self-conscious way, with the nosological perspective of the famous German psychiatrist Emil Kraepelin (1856–1926) (Blashfield 1984:31–37).

Kraepelin's approach to psychiatric classification was based on three ideas:

1. Mental disorders are best understood by analogy with physical diseases. Kraepelin's view was that medicine's progress against infectious disease took off only after received ideas about generic causes and processes were rejected, and researchers redirected their attention to discovering the specific causes of specific syndromes. Medicine's historical first step consisted of classifying the different kinds of diseases. Psychiatry must begin here also, if it is going to progress beyond its present undeveloped condition.

2. The classification of mental disorders demands careful observation of visible phenomena. It is only by systematically recording, collecting, and comparing case histories that it is possible to identify the clusters of symptoms that go together from case to case, follow a discernible course over time, and lead to a predictable outcome. Inferences based on etiological theories that lack solid empirical evidence or that invoke the operation of invisible mechanisms have to be rejected.

3. Empirical research will eventually show that the serious mental disorders have organic and biochemical origins. While relatively little is known about these causes at the present time, this presents no obstacle to classifying mental disorders. On the contrary, classification is a necessary first step to uncovering these etiologies (Kraepelin 1974 [1920]; Spitzer and Williams 1980:1043).

Earlier in this century, Kraepelin's "descriptive approach" to mental disorders was popular in American psychiatry. By the 1950s, however, his influence had become marginal. Most American psychiatrists remembered him either for his pioneering research on schizophrenia or, as Franz Alexander put it, as a "rigid and sterile codifier of disease categories" (Alexander and Selesnick 1966:211–214). Alexander, like Freud before him, found no common ground between Kraepelin's descriptive, "antipsychological" approach and his own psychodynamic views. Their respective positions were not merely different; they were antithetical. The most obvious difference is in the way they look at symptoms. The language of psychodynamic psychiatry had evolved to describe and diagnose unconscious conflicts and defenses. In the clinical encounter, attention is focused on

> the particular way the patient molds and distorts the interview situation in order to make it conform to his or her deeply ingrained (usually unconscious) fantasies, attitudes, and expectations about interpersonal relationships. The nature of these transference phenomena will be noted in order to predict future behavior in the treatment setting and to shed light on the patient's early developmental experiences and the conflicts that underlie the current disturbance. (Amer. Psychia. Assoc. 1980:11)

Symptoms are not intrinsically interesting in this context, and the nosological vocabulary of psychodynamic psychiatry is correspondingly simple. In Kraepelinian psychiatry, on the other hand, symptoms are interesting for two reasons: they are (or may be) tokens of underlying pathological structures and pathophysiological processes, and they are components of a system of meanings based on rules of inclusion and exclusion. In this system, the meaning of symptoms lies on the surface, within the semiotics of disease; that is, a symptom is meaningful because it is juxtaposed with other symptoms in stable formations (syndromes). In psychodynamic discourse, on the other hand, symptoms are polymorphic expressions of processes that are played out beneath the surface (Rycroft 1968:35). Taken to their limits, the two languages, psychodynamic and descriptive, are mutually unintelligible.

During this same period, Kraepelin's ideas continued to have a dominant influence on academic psychiatry in Europe and were incorporated into the official psychiatric nosology produced by the World Health Organization, the *International Classification of Diseases* (Klerman 1986). Nor did his ideas disappear entirely from American psychiatry: they continued to be represented in a popular psychiatric textbook of the period (Meyer-Gross et al. 1954) and, more significantly, found a home in the Department of Psychiatry at Washington University, St. Louis. It is to this department that Spitzer and many of his neo-Kraepelinian supporters trace their intellectual roots (Robins and Helzer 1986:410–414).

The rediscovery of Kraepelin by American psychiatry in the 1970s was preceded by a series of technological developments affecting clinical practice. Up to the 1950s, there was a tendency for psychiatrists to relinquish responsibility for a disorder once research connected it to an organic or biochemical etiology: "Vitamins were discovered, whereupon vitamin deficiency psychiatric disorders no longer were treated by psychiatrists. The spirochete was found, then penicillin, and neurosyphilis, once a major psychiatric disorder, became one more infection treated by nonpsychiatrists" (Goodwin and Guze 1984:xi). Similarly, once the efficacy of barbiturates and phenytoin (Dilantin) for the treatment of epilepsy was confirmed, psychiatry abandoned this malady to the exclusive attention of neurology (Klerman 1984:539). It was by purging itself of these biologically based disorders that psychiatry remained essentially psychological. Consequently, clinicians and researchers had little incentive for raising questions about the systematics of classification. But in the early 1950s, the situation changed, when a variety of psychoactive drugs of demonstrated efficacy, including chlorpromazine (Thorazine), became available. In contrast to what had occurred previously, the new treatments were incorporated into psychiatric practice, despite some resistance. By the next decade, psychiatrists were routinely employing four classes of drugs, each ostensibly specific to a category of mental illness: psychosis, depression, anxiety, and manic-depressive disorder.

The effectiveness of these drugs had implications that were recognized by large segments of the psychiatric community (Klerman 1984). Here was compelling evidence that disorders of affect (depression) were distinct from disorders of thinking (schizophrenia), and that the so-called "neurotic" conditions, which psychodynamic and psychosocial perspectives had attributed to generic processes, actually consist of discrete and discontinuous types of disorders, viz., phobias, obsessive-compulsive conditions, anxiety and panic states. It was not only the already convinced neo-Kraepelinian who could see that this response pattern paralleled classifications proposed by Kraepelin on the basis of his descriptive approach to symptomatology (Klerman 1984:539–540). Although the historic impact of the new drugs on psychiatric thinking is undeniable, in retrospect it seems that claims concerning their specificity had been overstated:

> While they were a significant factor [in changing psychiatric attitudes], they were not a major factor, because most psychiatric medications were helpful across supposedly discrete syndromes. However, the *potential* for more effective medications that might target discrete syndromes held great promise. . . .
>
> Medications, in other words, helped to create a need for a more experimentally based psychiatry; explicit diagnostic inclusion and exclusion criteria were essential in this endeavor. Equally important, however, was the growing awareness of the obvious lack of efficacy of psychodynamic psychotherapy for the more severe psychiatric disorders. (Wilson 1993:403–404; also Shepherd 1994)

American psychiatry lacked a nosology consistent with the needs of an experimentally based science, however. The first edition of the *Diagnostic and Statistical Manual of Mental Disorders* (*DSM-I*) had appeared in 1952 with the imprimatur of the American Psychiatric Association, but it did not fit the bill. Its system of classification was shaped by its famous editor, Adolph Meyer, who saw diagnostic groups as representing the *quantitatively* different reactions of the human personality to a unitary set of causes: psychological, social, and biological. In contrast to Kraepelin's descriptive approach, Meyer's "reactive" or "environmental" conception of the field of psychiatric disorders pictured a single gradient stretching from "mental health" at one end to severe "mental illness" at the other. Because of its orientation toward psychological process, Meyer's system was entirely consistent with psychodynamic premises (Amer. Psychia. Assoc. 1952:1; Weissman and Klerman 1978:707). *DSM-I* had a second limitation as a foundation for a research-based science, for its system of nomenclature was not universally accepted. Large numbers of American psychiatrists and institutions simply refused to adopt its language. Nor were authors of articles in major psychiatric journals required to stick to its terms and distinctions (Spitzer and Williams 1980:1045; Grob 1991).

DSM-II (1968) departed from the first edition in several respects, partly because it was intended to be compatible with the nosological system employed in the concurrent edition of the *International Classification of Diseases* (*ICD-8*). (For a history of the *ICD* system, see Kramer 1985.) Nonetheless, from the neo-Kraepelinian perspective, serious problems remained, for while Meyer's idea of "reaction" had been dropped from the nomenclature, the term "neurosis"—defined as an intrapsychic conflict that results in symptoms that unconsciously serve to control anxiety—was retained (Amer. Psychia. Assoc. 1968:39). In addition, the manual's diagnostic categories were "flawed by their failure to provide formal criteria for determining the boundaries of their diagnoses." Consequently, diagnosticians were forced to rely on "global descriptions of disorders that frequently entailed etiological assumptions" (Bayer and Spitzer 1985:189).

In 1974, the board of trustees of the APA decided that a further revision of the diagnostic manual was needed and appointed Spitzer, then a psychiatrist at Columbia University, to be head of its Task Force. As Spitzer recounts these events, he informed the board that it would be difficult to make any real progress on the revision so long as the *DSM-III* Task Force included people who had helped draft *DSM-II*. He wanted a clean slate, and the board accepted his conditions (Millon 1986:29). With this authority in hand, Spitzer then chose "a group of psychiatrists and consultant psychologists committed to diagnostic research and not to clinical practice," men and women with "intellectual roots in St. Louis instead of Vienna, and . . . intellectual inspiration derived from Kraepelin, not Freud" (Bayer and Spitzer 1985:188).

Exactly what the clean slate would mean for psychiatric nosology can be discerned in the introduction to *DSM-III*. "From the beginning, the Task Force functioned as a steering committee to oversee the ongoing work. All of its members shared a commitment to the attainment in DSM-III of . . . [ten] goals." The seventh of these goals is "avoiding the introduction of new terminology and concepts that break with tradition, *except when clearly needed*." This is followed by three goals that explain why, in many cases, a conceptual break with the previous editions was indeed needed, for

- reaching consensus on the meaning of necessary diagnostic terms . . . and avoiding the use of terms that have outlived their usefulness;
- consistency with data from research studies bearing on the validity of diagnostic categories;
- suitability for describing subjects in research studies. . . . (Amer. Psychia. Assoc. 1980:2–3)

In Spitzer's own words, the new edition was to be based on two principles: theories of pathogenesis would be confirmed by "principles of testability and verification," and each disorder would be identified by criteria accessible to empirical observation and measurement. Because the new classification system would be based on atheoretical and operational criteria, people of different theoretical perspectives would be able to write and talk about the same set of disorders, and researchers would be able to communicate directly with clinicians (Talbot and Spitzer 1980:27–28).

Spitzer's priorities signaled a shift in "the essential focus of psychiatry . . . from the clinically-based biopsychosocial model to a research-based medical model. . . . [R]esearch investigators replaced clinicians as the most influential voices in the profession" (Wilson 1993:400). The shift also meant purging *DSM-III*'s language of references to the unconscious:

> Freud used the term [psychoneurosis] both *descriptively* (to indicate a painful symptom in an individual with intact reality testing) and to indicate the *etiological process* (unconscious conflict arousing anxiety and leading to the maladaptive use of defensive mechanisms that result in symptom formation). . . .
>
> Although many psychodynamically-oriented clinicians believe that the neurotic process always plays a central role in the development of neurotic disorders, there are other theories about how these disorders develop. . . .
>
> Thus, the term *neurotic disorder* is used in DSM-III without any implication of a special [and invisible] etiological process. (Amer. Psychia. Assoc. 1980:9–10)

The culture of psychiatric science says that no meanings are immune from scrutiny, that all interpretations and definitions are tentative and subject to revision. But things are a bit different in everyday life, for researchers and clinicians are not ordinarily inclined to examine their basic terms. Most of them are pragmatists and puzzle solvers. Their eyes are fixed on getting results: producing knowledge, making correct diagnoses, achieving desirable outcomes. Splitting epistemological hairs is unrewarding and unnecessary. Everyday words ought to be tools for building knowledge, not weapons for subverting it. In the ordinary course of events, outside the rare scientific or institutional revolution, bedrock language changes only slowly and incrementally, without calling attention to itself. But *DSM-III* differs from its predecessors in precisely this respect: its appearance demarcates a revolutionary turn in American psychiatry.

Since Spitzer was advancing a project that was, from its start, hostile to the ideas and interests of psychodynamic psychiatry (especially psychoanalysis), it is only to be expected that it would be criticized from this quarter. These critics leveled three main arguments against the new edition. They rejected the notion that any diagnostic language could be atheoretical (Faust and Miner 1986:965). *DSM-III* was derided for its "cookbook approach"—that is, for making mental disorders equivalent to the aggregate of their symptomatic parts, whereas the (authentic) diagnostic process presupposes "an integrative gestalt based on clinical experience"; and for giving equal weight to each criterial feature within a classification, "even though there is no evidence that all have equal importance" (Frances and Cooper 1981:200). To take one example of *DSM-III*'s lack of nuance, according to critics it fails to distinguish between essentially different kinds of anxiety, viz., anxiety as a medium for expressing despair and appealing for help versus anxiety originating in threats to a patient's ego integrity (Vaillant 1984:544–545). Third, there were critics who rejected *DSM-III* on the grounds that it represented nothing less than a coup d'état, carried out by a circle of conspirators: "The task force that forged

DSM-III was not representative of the interests, the values, or the theoretical diversity of the profession. It was composed of an invisible college that is only one college in the university of American psychiatry" (Michels 1984:549).

Spitzer and his colleagues were unmoved by these objections. According to them, *DSM-III*'s lists of criterial features are not theory-impregnated, because they are grounded on statistical procedures (following the field trials) that are indifferent to vagaries of content and context. Saying that the manual takes a "cookbook approach" to mental disorders is empty rhetoric. The mission of the manual is to establish a scientific nosology, and this means devising criteria that will be explicit enough to obviate any need for interpretations based on tacit knowledge or theoretically informed "global appraisals." Finally, if these critics—mainly psychoanalysts—are unhappy with the work of the (Kraepelinian) "invisible college," they have no one to blame but themselves. According to Spitzer, he had conscientiously solicited comments from the major psychoanalytic organizations during the drafting process, but they had been unwilling to "confront forcefully the challenge of *DSM-III*" and failed to cooperate. And, he added darkly, their attempt to obstruct the manual's adoption was only partly doctrinal: it was also rooted in economic self-interest. "Psychoanalytic practitioners . . . feared that a change in psychiatric nomenclature might result in a challenge by third-party reimbursement sources seeking to limit payments to patients receiving long-term therapy." It was "no coincidence" that the Baltimore and District of Columbia Society for Psychoanalysis proved to be so conspicuous in its opposition to the final approval of *DSM-III*, since large numbers of federal employees live in this area and are entitled to "generous coverage for psychotherapeutic treatment [under the Federal Employees Health Benefits Program]." A standardized nosology would have put the psychoanalysts head-to-head with their more cost-efficient rivals: the eclectic practitioners, cognitive therapists, and so on (Bayer and Spitzer 1985:191–192).

If psychoanalysts had economic motives for opposing the adoption of *DSM-III*, other segments of the psychiatric community seem to have had compelling economic reasons for supporting it:

APA Medical Director Melvin Sabshin has recalled . . . that [by the 1970s] psychiatry was perceived by the federal government and private insurance companies as a "bottomless pit"—a voracious consumer of resources and insurance dollars—because its methods of assessment and treatment were too fluid and unstandardized. . . .

Under these unfavorable professional conditions, the psychosocial model, as the dominant organizing model of psychiatric knowledge and the source of many of these problems, would have to be significantly altered, if not jettisoned altogether. (Wilson 1993:403)

The dispute over *DSM-III* is ancient history now. "[P]ockets of resistance to the idea of categorical diagnoses . . . still remain among both social scientists and psychodynamic psychiatrists," but the "historic shift from an anti- to pro-diagnostic stance" is complete (Robins and Helzer 1986:409; also Williams, Spitzer, and Skodol 1985; Kirk and Kuchens 1992: chap. 7). Outside of analytic circles, everyone routinely speaks and writes in the language of *DSM-III*. And even though it is not the mother tongue of analytically oriented therapists, most of them have adopted it as a lingua franca for communicating with other psychiatrists.

Calculating Reliability and Validity

Once *DSM-III* had been adopted by the APA, researchers and clinicians found good reasons to adopt its conception of mental disorders. Unlike previous editions, this one is an authoritative text and sanctioned by key institutions, including the National Institute of Mental Health. By the early 1980s, American medical schools and residency programs routinely expected students and physicians to pass examinations based on *DSM-III* criteria. Both referees and editors expected manuscripts submitted to scholarly journals to be written in its language, and it was simply assumed that psychiatric research proposals would conform to its conventions. Researchers and clinicians who resisted these conventions could count on being excluded from these arenas and their resources.

The success of *DSM-III* is only partly explained by these social developments, however. The classifications and symptom lists also acquired a certain facticity, a quality based on standards, borrowed from epidemiology, of "reliability" and "validity." A diagnostic technology is described as reliable if it induces a diagnostician to (correctly) label a disorder when he observes it in a given patient on different occasions (*test-retest reliability*), and it induces multiple diagnosticians to give the same diagnosis to the same patient on a given occasion (*inter-rater reliability*). Test-retest reliability is especially important for diagnosing disorders such as PTSD, where symptoms oscillate or fluctuate over time.

It would be a mistake to see "reliability" as simply a technical issue, although this is the way in which it is usually represented in psychiatric discourse (Kirk and Kuchens 1992:28–30, 48–56). Reliability is, by convention, a precondition for aggregating like cases for research purposes. The aggregation of large numbers of cases is, in turn, a precondition for employing statistical technologies and for generalizing findings beyond single research sites. It is these same technologies, together with their guarantees of reliability, that permit researchers to liberate cases from contexts and to decompose distinctively messy lives into uniform and universal constitutive elements. In this way, a technique's reliability is linked to its objectivity.

Researchers have three ways to enhance the reliability of a classification or technology: diagnostic criteria can be made explicit enough and complete enough to identify all true cases and to distinguish them from cases of other disorders that are similar; technologies can be structured so as to ensure that diagnostic criteria are uniformly employed; and steps can be taken to recruit interviewers whose skills, knowledge, and experience equip them for utilizing these technologies. It is not necessary to maximize all of the elements—criteria, technology, personnel—in order to get high reliability. Researchers can attain it one of two ways. The *high-skill option* is to recruit personnel who have extensive clinical experience and have been socialized into a homogeneous professional culture. (Culture is important, since personnel of different theoretical persuasions are more likely to reach conflicting conclusions.) With this option, it is possible to achieve high reliability even though the diagnostic criteria require some interpretation. The limitations of the high-skill option are its high costs and the problem of finding suitable personnel. The *low-skill option* is to employ briefly trained lay personnel. This is feasible only if the diagnostic criteria are explicit, and interview technology is highly structured and easy to master. The most serious obstacle to employing the low-skill option is the availability of sets of satisfactory criteria and technologies.

The interest of American psychiatry in the reliability of diagnoses can be traced back to World War II, when it was discovered that a large fraction of men called up for military service were rejected on psychiatric grounds. This unexpected finding stimulated an interest in learning the prevalence of mental problems in the general population. In 1949, the U.S. Congress established the National Institute of Mental Health, which made this epidemiological question one of its earliest priorities. Over the succeeding decades, NIMH funded several major epidemiological projects. The first projects, which included the Midtown Manhattan and Stirling County Studies, focused on the prevalence of psychiatric impairment rather than specific disorders. There were several reasons for this decision. The psychiatric mainstream was psychodynamic and environmental at this time and had no compelling interest in collecting data on specific diagnoses. In addition, the organizers of these postwar studies preferred to avoid existing psychiatric nosologies. The most widely used nosological system during this period had been commissioned by the U.S. armed forces and was tailored to their particular requirements. Its classifications were believed to be unreliable and a source of potential dispute among the various factions represented in American psychiatry. Finally, by focusing on impairment rather than diagnosis, researchers were able to avoid the expensive high-skill option, which, at this time, would have meant employing psychiatrists (Weissman and Klerman 1978:706).

About the same time, researchers at Washington University had begun to develop a protocol that would make it economically feasible to undertake large-scale epidemiological studies of specific disorders. Their first project was to produce explicit, rule-guided diagnostic criteria for fourteen psychiatric disorders (Feighner et al. 1972). The protocol was subsequently expanded to cover twenty-five major diagnostic categories and renamed the Research Diagnostic Criteria (RDC). The RDC was developed as part of a collaborative study of the psychobiology of depression funded by NIMH, and Spitzer was an active participant in this study. It was in the course of Spitzer's association with this group, which overlapped his tenure as chief of the *DSM-III* Task Force, that he discovered his prototype for the new diagnostic manual (Spitzer et al. 1978; Robins and Helzer 1986:412–414; Wilson 1993:404). Spitzer was also working with colleagues at Columbia University to tailor a statistic, called "kappa," that was ultimately used for measuring and calibrating the reliability of the RDC criteria (Spitzer et al. 1967; for a historical account and critique of the kappa statistic, see Kirk and Kuchens 1992:37–45, 56–62.)

Access to reliable diagnostic technologies is necessary for establishing the "validity" of a classification. The term "validity" recurs throughout the psychiatric research literature, where it indicates that a given classification possesses intrinsic unity: it is neither a random phenomenon nor an artifact of the techniques through which it is detected, treated, experienced, and studied (Robins and Guze 1970; Spitzer and Williams 1980).

When psychiatric researchers and clinicians refer to a classification's validity, they usually have three standards in mind:

1. A disorder is said to have *face validity* if its criterial features are consistent with the clinical impressions and experiences of experts. Since clinical impressions are strongly influenced by clinical training and socialization, what passes for an expert consensus in one psychiatric circle may be rejected by another. According to psychoanalysts, for example, many of *DSM-III*'s disorders possess face validity merely because the task force ignored the "theoretical diversity" of American psychiatry; that is, face validity is an artifact of excluding psychodynamic clinicians from the task force (Michels 1984:549). Other critics have expressed doubts about the face validity of those classifications, including the sleep disorders, that had entered the *DSM-III* system following simple majority votes of their subcommittees and over the opposition of (equally expert) minority members (Amer. Psychia. Assoc. 1987:xx).

2. A disorder is said to have *predictive validity* if untreated cases are known to follow a consistent course over time: symptoms unfold or fluctuate in predictable ways; the disorder produces predictable degrees of impairment, and so on (Goodwin and Guze 1984:ix; Robins and Helzer 1986:427).

3. Finally, a classification acquires *independent validity* when research findings are believed to establish an underlying cause or process. Evidence includes family data, usually based on the psychiatric histories of the individual's first-degree biological relatives; biochemical data, such as unusual blood chemistry that identifies underlying pathophysiology; treatment data, as when a given diagnosis responds in a distinctive way to a particular intervention; and environmental data, as when onset follows exposure to an identifiable class of stressors.

The three kinds of validity compose a hierarchy. Face validity is on the lowest rung, because it is the most easily compromised by contingency, for example, a shared cultural bias that leads experts to pathologize behaviors that are normative within certain ethnic communities. Predictive validity is a step up, since it is generally a product of statistical operations and, in principle at least, is less vulnerable to contingency. Independent validity allows the strongest claims, particularly when it is based on biological evidence.

Reliability and validity are closely connected. When the reliability of diagnostic criteria and technologies is low, the validity of the disorders that they identify is moot. That is, without reliable resources, researchers cannot establish, to the satisfaction of their public (mainstream American psychiatry), the homogeneity (shared identity) of the aggregated cases (diagnosed patients) on which they are basing their statistical evidence.

By itself, strong reliability cannot be a sufficient condition for establishing validity, since it is possible to design a set of criterial features that are highly reliable but invalid according to the canons of current psychiatric thinking. For example, a "disorder" diagnosed by occasional episodes of abrupt awakening accompanied by autonomic arousal would be both highly reliable and without validity (cf. Amer. Psychia. Assoc. 1980:84–86).

Reliability and validity are further connected by the fact that attempts to increase reliability by defining criteria more precisely may simultaneously reduce the validity of diagnosed cases (Robins and Helzer 1986:423). For example, the Diagnostic Interview Schedule (DIS) is currently the most widely used instrument in epidemiological research. The DIS was created to operationalize the Research Diagnostic Criteria for use in a national epidemiological study, funded by the National Institute of Mental Health. It was designed for high reliability, so that a lay interviewer and an expert should reach the same diagnosis for the same patient. Interviewers using the DIS employ lists of standardized questions, keyed to the diagnostic criteria for the various disorders. The list for each disorder includes no significant omission (every criterial feature is tapped), and respondents are limited to choosing from simple, fixed responses, (yes/no). This is a "structured interview technique," borrowed from the psychometric tradition of personality assessment. The scoring of the responses is mechanical: it follows "clear diagnostic algorithms" and entails no interpretation (Robins and Helzer 1986:423–424). The DIS has been designed so that it can be used in conjunction with computers, further reducing interviewer bias, according to its authors. "The computer, unlike the clinician, looks only at those items to which the program directs it in making a diagnostic decision" (Robins et al. 1981:384–385).

Although the Diagnostic Interview Schedule attains high reliability, questions have been raised about its "specificity," that is, its ability to weed out false positives (superficially true cases), and its "sensitivity," that is, its ability to identify false negatives (ambiguous but nonetheless true cases). Problematic specificity and sensitivity undermine the validity of the samples (cases) that researchers use to represent a given disorder and the validity of research based on these samples—research that might be used, in turn, to confirm the validity of the classification in question, for example, by identifying a biological marker. During the 1980s, a compromise solution was proposed, consisting of a semistructured interview that would permit interviewers to clarify their questions, challenge negative responses, and confront apparent discrepancies in the respondents' statements. The Structured Clinical Interview for *DSM* classifications (SCID) is the most widely used of these semistructured technologies, but it has predictable limitations: it requires clinically experienced interviewers (psychiatrists, clinical psychologists, psychiatric social workers) and can be very expensive. Further it has significantly lower reliability than the DIS for certain disorders (Spitzer et al. 1992:627–628).

Now, remember that the neo-Kraepelinian epistemology gives primacy to scientific truth over clinical reality, to noncontingent and generalizable forms of knowledge over local knowledge. Psychiatry is about mental disorders, and its scientific truths are predicated on the facticity of these disorders; to be more precise, on the classifications and criterial features that are used to represent these disorders. Facticity, in turn, depends on access to reliable technologies, like the DIS, to operationalize the classifications. As we have seen, reliability is not sufficient by itself: validity must also be established. The problem here is that attempts to improve reliability and validity tend to pull in different directions: measures taken to increase reliability (and improve the cost/effectiveness of large-scale epidemiological research) limit validity, and vice versa. In theory, there is a simple solution to this problem, and it is to establish the independent validity of the disorder in conjunction with some highly reliable signifier, consisting of a test and a telltale outcome. In the case of PTSD, the search for this solution has led to two developments: the continuing pursuit of a credible biological marker and the creation of packages of psychometric devices. In chapters 4, 5, and 8, I will analyze these solutions and their implications. For the time being, I will simply conclude with an observation that anticipates the argument that is found in these chapters. It is this: the diagnostic technologies just reviewed (*DSM-III*, DIS, SCID) are an integral part of the historical formation of some of the disorders (including PTSD) that they now identify and represent. Ian Hacking makes an analogous point in the context of laboratory science:

> Theories are not checked by comparison with a passive world with which we hope they correspond. We do not formulate conjectures and then just look to see if they are true. We invent devices that produce data and isolate or create phenomena, and a network of different levels of theory is true to these phenomena. . . . Thus there evolves a curious tailor-made fit between our ideas, our apparatus, and our observations. (Hacking 1992:57–58)

How PTSD Entered *DSM-III*

"Post-traumatic stress disorder" entered the psychiatric nomenclature with the publication of *DSM-III*. The manual's first edition (*DSM-I*) included a superficially similar diagnosis, "gross stress reaction," defined as a psychoneurotic disorder originating in an experience of intolerable stress. Unlike PTSD, however, gross stress reaction was identified as a transient response, and its symptomatology was vague, consistent with *DSM-I*'s system of classification. The following edition, *DSM-II*, dropped this diagnosis. The closest that *DSM-II* comes to PTSD is "transient situational disturbances": "If the patient has good adaptive capacity his symptoms usually recede as the stress diminishes. If, however, the symptoms persist after the stress is removed, the diagnosis of another mental disorder is indicated" (Amer. Psychia. Assoc. 1968:48).

In the next chapter, I examine the diagnostic features of PTSD in detail. For the moment, it will be enough to note three points on which PTSD, as described in *DSM-III* (Amer. Psychia. Assoc. 1980:236–238), parts ways from these earlier stress disorders:

1. *DSM-III* specifies that the etiological events for PTSD should be "outside the range of usual human experience" and should evoke "significant symptoms of distress in most people."

2. *DSM-III* specifies a set of observable post-traumatic symptoms, consisting of persistent and distressful reexperiences of the traumatic event, such as dreams, flashbacks, and intrusive images; symptomatic numbing, such as emotional anesthesia or loss of interest in activities previously found pleasurable; a tendency to avoid situations that might trigger recollections of the traumatic experience; and increased physiological arousal, evidenced in sleep disorders, difficulty concentrating, irritability, and so on.

3. *DSM-III* distinguishes subtypes of PTSD, based on whether the onset of symptoms occurred more (or less) than six months after the traumatic event and whether the duration of the symptoms was more (or less) than six months.

In the following account of how PTSD entered *DSM-III*, I have relied on Wilbur Scott's recent work (Scott 1990), which is based on interviews with the principal actors.

The origins of the PTSD diagnosis are inextricably connected with the lives of American veterans of the Vietnam War, with their experiences as combatants and, later, as patients of the Veterans Administration (VA) Medical System. Throughout the early 1970s, the American news media reported what seemed to be an epidemic of suicides, antisocial acts, and bizarre behaviors committed by Vietnam War veterans (Dean 1992). Psychiatric authorities were said to have discovered unexpectedly high rates of mental health problems and self-destructive behaviors among these men, including alcoholism and drug addiction. The "crazy Vietnam vet"— angry, violent, and emotionally unstable—had become an American archetype.

> Mental health professionals across the country assessed disturbed Vietnam veterans using diagnostic nomenclature that contained no specific entries for war-related trauma. . . . VA physicians typically did not collect military histories as part of the diagnostic work-up. Many thought Vietnam veterans who were agitated by their war experiences, or who talked repeatedly about them, suffered from a neurosis or psychosis whose origin and dynamics lay outside the realm of combat. (Scott 1990:298)

The experiences of Sarah Haley, a psychiatric social worker at the Boston VA Medical Center, are emblematic. In the early 1970s, Haley interviewed a patient who appeared to be extremely anxious and agitated. The veteran reported that he had been a member of an infantry platoon that had murdered hundreds of civilians in the village of MyLai. Although he had been present, he claimed that he himself had not fired on the villagers. Following the massacre, he had been warned by fellow soldiers that he would be murdered if he revealed the event to anyone. The man had kept silent since then but had recently had a mental breakdown. He was overcome by feelings of terror, had problems sleeping, and now believed that members of his platoon were plotting to assassinate him. Haley, who was unfamiliar with the MyLai massacre at this time, accepted the veteran's story at face value. A short time after the interview, she presented her case notes and conclusions at a VA staff meeting:

> The staff assembled to discuss all the information and reach a diagnosis and treatment plan. When we met, the intake log already had a diagnosis [for this patient] filled in: paranoid schizophrenic. I voiced concern. The staff told me that

the patient was obviously delusional, obviously in full-blown psychosis. I argued that there were no other signs of this if one took his story seriously. I was laughed out of the room. (S. Haley, quoted in Scott 1990:298)

Robert Lifton, the author of a widely praised book on the victims of Hiroshima, *Death in Life* (1968), had read a newspaper account of the MyLai massacre and, with Haley's cooperation, interviewed this veteran. Lifton's subsequent book, *Home from the War* (1973), is based on his con-versations with Haley's patient and other Vietnam War veterans. In testi-mony before a Senate subcommittee, Lifton described the psychopatho-logical effects of serving in this "dreadful, filthy, unnecessary war." This theme is reiterated in *Home from the War*, together with Lifton's indict-ment of the moral and clinical failures of military psychiatry: "He singled out two articles in the *American Journal of Psychiatry* . . . for special criti-cism. Both articles boasted of psychiatry's effectiveness in containing war neurosis and returning troubled soldiers quickly to the battlefield. Espe-cially contemptible, Lifton argued, was the stance of military psychiatry as an advocate of the military's interests rather than those of the soldier-patient" (Scott 1989:302).

Lifton was a member of a loose network of veterans and clinicians inter-ested in publicizing the disconcerting effects of the Vietnam War on those who fought it and the unsettling homecomings they often received from their families, their former friends, their potential employers, and the VA. In 1972, the *New York Times* published an article on its op-ed page by a psychiatrist who belonged to this network. The author, Chaim Shatan, de-scribed a psychiatric disorder that he called "post-Vietnam syndrome." Ac-cording to Shatan, the disorder was produced by a "delayed massive trauma," and it accounted for many of the psychiatric problems then afflict-ing Vietnam War veterans. The men's typical symptoms—guilt, rage, psy-chic numbing, alienation, feelings of being scapegoated—constituted a dis-tinctive, but officially unrecognized and untreated, disorder. Shatan, Lifton, and others presented these same arguments at a series of panels that they organized at the annual meetings of psychiatric organizations.

In 1975, Spitzer began preparations for *DSM-III*. Lifton and Shatan ap-proached Spitzer, asking him to form a subcommittee on the "post-Viet-nam syndrome." Spitzer initially rejected their proposal, telling them that knowledgeable researchers had concluded that no new diagnostic clas-sification was needed to account for the psychiatric disorders affecting Vietnam War veterans. The mental health of these veterans was a politi-cally sensitive issue, though, and Spitzer agreed to reconsider his decision if persuasive evidence could be found to support the post-traumatic diagno-sis. Shatan organized a working group to collect data on the syndrome, which he renamed "post-combat disorder." The working group gradually expanded to include researchers on other categories of traumatic experiences, and a leading authority on stress response syndromes, Mardi Horo-witz, associated himself with the project (Horowitz 1976).

At this point, Spitzer appointed a Committee on Reactive Disorders to review the data that was being collected by the working group. Once the committee had completed its review, it would report its findings to the *DSM-III* Task Force. The committee consisted of six people. Three mem-bers (including Spitzer) were selected from the task force, and the remain-ing members were recruited from the working group: Lifton, Shatan, and Jack Smith, a Vietnam War veteran who headed an advocacy group, the National Veterans Resource Project, organized under the aegis of the Na-tional Council of Churches.

> The appointment of Smith to the Committee on Reactive Disorders was highly unusual. Of the roughly 125 experts serving on the various advisory committees [to the *DSM-III* Task Force] . . . , only six of the appointees were not M.D.'s. Of the six, only two were not Ph.D.'s, and of the two, only Smith did not have a graduate degree. In fact, as Smith recalls: "I didn't even have a bachelor's de-gree." (Scott 1990:306)

The working group collected evidence from four sources: published reports describing the clinical sequelae of natural and man-made catastrophes, case histories collected directly from Vietnam War veterans, clinical accounts provided by sympathetic practitioners, and endorsements from experts on pathogenic stress.

The working group proposed that the committee endorse the diagnosis, now renamed "catastrophic stress disorder." ("Post-combat stress reaction" would be included as a subcategory.) Influential neo-Kraepelinians con-tinued to oppose such a classification on the grounds that its symptoms were glued together by a dubious etiology. The disorder would reintroduce precisely the sort of nonoperationalized mechanism that *DSM-III* was in-tended to eliminate. Minus this etiological process, the disorder's symp-tomatology coincided entirely with the symptoms of already established diagnoses—depression, generalized anxiety disorder, panic disorder, and paranoid schizophrenia—and would thus be superfluous.

According to Scott's informants, the efforts of the working group con-centrated on the committee's leader, Nancy Andraesen, a psychiatrist who had extensive experience treating victims of severe burns. The working group succeeded in persuading Andraesen that she had been observing similar post-traumatic psychological reactions among her burn patients (Scott 1989:306). An additional effort was made to demonstrate that the proposed disorder enjoyed considerable face validity at the VA, even though it had been banished from the official nosology:

> I went through the [Boston VA] records of all the Vietnam veterans we had seen in a year. . . . What I looked at was the official, the DSM-II, diagnosis. . . . But

then in parentheses [for some] was a working diagnosis . . . usually "traumatic war neurosis." And so what I said was, "Look it, Nancy [Andraesen], we had to give these guys diagnoses [consistent with *DSM-II*] but if you look at what [some] clinicians are actually doing . . . they're basing their treatment on the fact that they recognize in these fellows similar traumatic war neurosis as they saw in the Second World War and Korean War veterans." (Andraesen, quoted in Scott 1990:307)

In January 1978, the working group, consisting of Lifton, Shatan, and Smith, presented its final report to the Committee on Reactive Disorders. A month later, the committee sent the final draft of its own position to the task force, recommending that the new classification be included in *DSM-III*, in the section on anxiety disorders.

War-Related PTSD

Andraesen, writing afterward, described PTSD as "a disorder that has long been recognized in clinical psychiatry but for which official recognition has been minimal, late in arriving, and long overdue" (1980:1517).

> [O]ne might have expected a . . . description of it in DSM-II when it appeared in 1968. On the contrary, however, the [*DSM-I*] category of gross stress reaction was inexplicably dropped. . . . No category was provided in its place. . . . The fate of the category seems to have been tied to the history of warfare. DSM-II was compiled during the relatively tranquil interlude between World War II and the Vietnam conflict. Perhaps in the absence of military conflict and in the presence of a rather foolish optimism that did not contemplate its recurrence, the category no longer seemed necessary. The Vietnam War . . . provided convincing evidence for such a need. The description of post-traumatic stress disorder in DSM-III was written to fulfil this need. (Andraesen 1980:1518)

Today, two decades after the fall of Saigon, scientific research on PTSD continues to be based largely on veterans of the Vietnam War. And the Veterans Administration remains the most significant source of research funds and cases (diagnosed veterans) for studying the disorder.

Neither *DSM-III* nor the revised edition, *DSM-III-R*, published in 1987, identifies war-related PTSD as a subtype of the PTSD classification:

> The single classic post-traumatic syndrome—involving recurrent nightmares, anxiety, numbing of responsiveness, insomnia, impaired concentration, irritability, hypersensitivity, and depressive symptoms—has been described in response to an enormous variety of stressful situations—prisoner-of-war camps, death camps, combat, auto accidents, industrial accidents, such mass catastrophes as Buffalo Creek [a deadly flood] and Hiroshima, rape, and accidents in the home.

The DSM-III definition highlights the fact that *these various stressors all tend to produce a single syndrome that appears to be the final common pathway in response to severe stress.* (Andraesen 1980:1518, my emphasis)

The reference to a "common pathway" notwithstanding, it is not difficult to show that cases of PTSD occurring among Vietnam War veterans are distinctive in several respects.

The standard tour of duty in Vietnam was twelve to thirteen months. A majority of diagnosed veterans served a single tour of duty, although some served two or even three. The frequency with which these soldiers were exposed to PTSD-level events (as identified by the *DSM-III* stressor criterion) varied greatly. It tended to be high in elite combat units and in medical units treating severely wounded soldiers but generally intermittent or infrequent in support units. Veterans who served in high combat zones report being exposed to the entire range of traumatic stressors mentioned in the diagnostic manuals: threats to one's own life; the threatened or actual loss of a body part or essential function through disfiguring wounds, injuries to the brain and spine, loss of limbs, and other major physical injuries; and intimate contact with the death, mutilation, and physical suffering of other people, including close friends. Many veterans report that they were (also) self-traumatized—by extreme violence that they themselves inflicted in morally forbidden ways, such as torturing prisoners, mutilating the dead, and killing civilians and fellow Americans (Haley 1974; Laufer et al. 1985; Fontana et al. 1992).

Most Vietnam War veterans returned home between 1964 and 1975. This means that, on the average, a decade elapsed between the time of discharge and the time that the PTSD diagnosis entered the official nosology, and an even longer period before the VA offered specialized diagnosis and treatment for PTSD. According to the advocates of the diagnosis, the VA's failure to recognize and treat PTSD during this period led to a variety of pathogenic consequences. Adduced as evidence for this is the high rates of concurrent mental disorders among veterans now diagnosed with PTSD (research indicates that a majority of these veterans also suffer from combinations of depression, general anxiety disorder, panic disorder, and chemical substance use disorders) and high rates of impairment, especially the social and economic sequelae of self-destructive and self-defeating behaviors (Laufer et al. 1981; Sierles et al. 1983; Yager et al. 1984; Kulka et al. 1990a and 1990b; Davidson et al. 1990; Jordan et al. 1991; Mellman et al. 1992).

These characteristics—the distinctive historical and moral meanings associated with the veterans' traumatic events, the chronicity of their disorders, their high levels of impairment—help to explain the VA's lack of enthusiasm for the new diagnosis in the years leading up to *DSM-III*. (On VA resistance to PTSD following the publication of *DSM-III*, see Atkinson et al. 1982:1119–1121). "The Veterans Administration does not, and subject to appropriations by law cannot, provide equal services for each of the 28 million living veterans" (Fuller 1985:6). Top priority goes to providing health care and benefits for disabilities incurred in the course of active duty, that is, for "service-connected" disabilities. And there lay the rub, for there was no doubt that PTSD would qualify as a "service-connected" disorder once it entered the official nosology. In contrast to the psychiatric disorders with which its symptoms overlap—depression, anxiety disorder, panic disorder, and so on—PTSD's criterial features (as advocated by the working group on reactive disorders) specified an etiological agency. While the onset of these other disorders could not be connected to any clear exogenous origin (*pace* the neo-Kraepelinian spirit of the new nosology), the PTSD diagnosis required a connection and allowed no time limit between it and the onset of its effects (Klerman 1989:29). In the case of the Vietnam War veterans, the prima facie connection would be to out-of-the-ordinary events in Vietnam.

It was clear to everyone that the proposed diagnosis would have important fiscal and manpower implications for the VA. It would make the VA the primary care provider for all present and future cases of PTSD affecting veterans, including the many false positives who would aspire to this diagnosis. The federal government and its designated agent, the VA, could anticipate substantial compensation claims from large numbers of veterans for chronic impairments plausibly attributed to PTSD. Further, VA regulations would permit veterans diagnosed with PTSD to seek retroactive compensation, going back to the time of their initial psychiatric (mis)diagnoses. These changes would require additional funds from Congress and/or a redistribution of the VA's resources among its clienteles:

> The political realities of the ebb and flow of veterans' benefits in Washington center on what is aptly called the "Iron Triangle" formed by the Congress, the Veterans Administration, and the Veterans Service Organizations [representing the traditional, pre-Vietnam War veterans groups]. Little can be accomplished without the acquiescence of each. Depending on politics and budgets, each corner of the triangle pushes and tugs at its neighbors until a consensus is reached. . . .

> [I]n the minds of the traditionalists, . . . proponents [of the PTSD diagnosis] were part of the most visible and activist symbol of the antiwar movement in the United States: the Vietnam Veterans Against the War, who were "probably all crazy before they got into the service in the first place." (Fuller 1985:6)

According to Fuller's informants, the established interest groups, representing the veterans of earlier wars, perceived the advocates of the PTSD diagnosis as "self-serving psychologists and psychiatrists" who, having opposed the Vietnam War, were now out to milk the VA (Fuller 1985:6).

The same issues that provoked comments like these in the VA and the House and Senate Veterans Affairs Committees had the effect of mobilizing support for PTSD in other circles, including the American Psychiatric Association. The case for adopting the diagnosis was twofold. Clinical evidence supporting the diagnosis could be traced back to the nineteenth century and Oppenheim's description of the traumatic neurosis (Andraesen 1980:1517). It was unreasonable for critics to ask for evidence based on experimental or epidemiological research when, until now, research funds were unavailable. In addition, the advocates were able to make a compelling *moral* argument for PTSD, albeit one that fell on deaf ears in the VA and the traditional veterans' organizations. The failure to make a place for PTSD would be equivalent to blaming the victim for his misfortunes—misfortunes inflicted on him by both his government and its enemies. It would mean denying medical care and compensation to men who, in contrast to their more privileged coevals, had been obliged or induced to sacrifice their youths in a dirty and meaningless war. Acknowledging PTSD would be a small step toward repaying a debt.

The handwriting was on the wall once the *DSM-III* Task Force accepted the report of the Committee on Reactive Disorders, recommending PTSD for inclusion. At about the same time, an important change of leadership was taking place within the VA, "with the confirmation of Max Cleland as V.A. Administrator during the Carter Administration. As the youngest V.A. administrator in history, he joined the ranks of his younger counterparts and progressives in the Congress to make improvements in a wide range of programs, with Vietnam veterans as his primary objective" (Fuller 1985:8). In 1979, the Senate Veterans Affairs Committee filed the report that authorized the VA to recognize PTSD among Vietnam veterans, and the following year the VA accepted PTSD, delayed type, as a potentially compensable disorder (Atkinson et al. 1982:1118; Fuller 1985:3; Scott 1990:307).

Diagnostic Criteria

In 1987, the American Psychiatric Association published a revision of *DSM-III*. The definitions of PTSD in the two editions appear in tables 1 and 2 at the end of this chapter. They are similar but not identical, and there are some points worth commenting on.

In the 1980 version, PTSD is identified with four categories of features:

   A. Traumatic event
   B. Reexperiences of the event
   C. Numbing phenomena
   D. Miscellaneous symptoms

In the 1987 revision, the disorder is identified with a slightly different set of categories:

A. Traumatic event

B. Reexperiences of the event

C. Attempts to avoid situations that might trigger reexperiences or traumatogenic distress, *or* a generalized numbing of responses

D. Forms of arousal

The main effect of this change is to further unify the PTSD diagnosis. In *DSM-III*, categories B and C are tied to the diagnosis only through their association with category A. In the 1987 version, B and C are also connected to each other within a single process. That is, painful reexperiences (B) are the implicit cause of the avoidance phenomena (C), and the adoption of efficient avoidance strategies now accounts for the absence or modulation of reexperiences during periods when patients appear to be asymptomatic for PTSD, notably during the "latency" period that sometimes occurs between the time of the traumatic event and the onset of symptoms (Horowitz 1986:85–102).

Similarly, in *DSM-III*, category D consists of a collection of heterogeneous symptoms that are tied to the diagnosis through their unidentified associations with category A. In the 1987 version, the symptoms in category D are connected to one another by an underlying physiological feature, autonomic arousal, that also explains their connection with category A. The change has been achieved by moving two symptoms out of category D. One of these—"avoidance of activities that arouse recollection"—is moved to the avoidance category (C). The other symptom—"guilt about surviving when others have not, or about behavior required for survival"—is moved into the descriptive text and demoted from a diagnostic criterion of PTSD to an "associated feature."

Both editions claim to be atheoretical. In customary usage, the term "theory" includes efforts to unify a congeries of features (a syndrome) by means of an underlying but invisible cause. If this definition is accepted, then these two accounts of PTSD cannot be considered atheoretical. Indeed, *DSM-III-R* tightens up the theory that is already tacit in *DSM-III*. If the two accounts are "atheoretical," it is only in the special sense that they are able to accommodate a variety of theories: psychoanalytic, neuroendocrinological, behavioralist, and so on.

Conclusion

The *DSM* theory of PTSD is simple. In both editions, it is simply taken for granted that time and causality move *from* the traumatic event *to* the other criterial features and that the event inscribes itself on the symptoms. Because the traumatic event is the cause of the syndromal feelings and behaviors, it is logical to say that it precedes them. If this were not true, if it were acceptable for syndromal features to occur *before* the traumatic event, then the term "reexperience" would lose its accepted meaning. PTSD would be a distinctive classification but also an incoherent one, since effects (symptoms) would precede their own cause (traumatic event). One can imagine another possibility: the patient's symptoms precede his (or his clinician's) discovery of a particular traumatic event/memory and are, in some appreciable sense, the cause of this discovery. The discovered memory is now the explanation (post hoc) for the onset of his symptoms. This possibility is coherent: it does not violate our commonsense notions of time and causality. On the other hand, it means that PTSD is no longer a distinctive nosological category: the syndrome would now be indistinguishable from combinations of already established psychiatric disorders, such as depression, generalized anxiety disorder, and panic disorder (Breslau and Davis 1987).

As a formal system of knowledge, the official nosology allows neither of these possibilities: its classifications must be simultaneously coherent and distinctive. The canons of psychiatric science (research) likewise permit neither option: an incoherent classification will be dismissed out of hand; cases falling into a nondistinctive classification will be reassigned to distinctive diagnostic categories.

In practice (diagnosis, research), the PTSD classification *is* coherent and distinctive. If one simply accepts this fact and looks no farther, then my remarks concerning time and causality are superfluous, merely flogging the obvious. In the next chapter, I look farther and reach a conclusion that is by no means obvious. I shall argue that the sense of time that is now firmly attached to PTSD does not emerge spontaneously from the facts. Rather, it is an *achievement*, a product of psychiatric culture and technology.

TABLE 1
Diagnostic Criteria for Post-Traumatic Stress Disorder in *DSM-III* (1980)

A. The individual experienced "a recognizable stressor that would evoke significant symptoms of distress in almost anyone."

B. The traumatic event is reexperienced in at least one of the following ways: (1) recurrent, intrusive, and distressful recollections of the event; (2) recurrent distressful dreams of the event; (3) sudden acting or feeling as if the traumatic event were recurring, because of an association with an environmental or ideational stimulus.

C. There is a numbing of responsiveness to the external world, or reduced involvement in it. This is evidenced by at least one of the following: (1) markedly diminished interest in one or more significant activities; (2) feelings of detachment or estrangement from others; (3) constricted affect.

D. At least two of the following symptoms were *not* present before the trauma: (1) hyperalterness or exaggerated startle response; (2) sleep disturbance; (3) guilt about surviving when others have not, or guilt about behavior required for survival; (4) memory impairment or trouble concentrating; (5) avoidance of activities that arouse the recollection of the traumatic event; (6) intensification of symptoms after being exposed to events that symbolize or resemble the traumatic event.

TABLE 2
Diagnostic Criteria for Post-Traumatic Stress Disorder in *DSM-III-R* (1987)

A. The individual has experienced a *traumatic event* that (1) is "outside the range of usual human experience" and (2) would be "markedly distressing to almost anyone."

B. The traumatic event is persistently *reexperienced* in at least one of the following ways: (1) recurrent and intrusive distressing recollections of the event; (2) recurrent distressing dreams of the event; (3) sudden acting or feeling as if the traumatic event were recurring; (4) intense psychological distress when exposed to events that symbolize or resemble an aspect of the traumatic event.

C. The individual persistently *avoids* stimuli associated with the trauma or experiences a *numbing* of general responsiveness. To meet this criterion, a person has to evidence at least three of the following: (1) efforts to avoid thoughts or feelings associated with the trauma; (2) efforts to avoid activities or situations that arouse recollections of the trauma; (3) an inability to recall an important aspect of the trauma; (4) a markedly diminished interest in significant activities; (5) feelings of detachment or estrangement from others; (6) a restricted range of affect; (7) a sense of a foreshortened future.

D. The individual experiences persistent symptoms of increased *autonomic arousal* not present before the trauma. The person must exhibit at least two of the following: (1) difficulty falling or staying asleep; (2) irritability or outbursts of anger; (3) difficulty concentrating; (4) hypervigilance; (5) exaggerated startle response; (6) physiological reactivity when the individual is exposed to events that symbolize or resemble an aspect of the traumatic event.

Four

The Architecture of Traumatic Time

POST-TRAUMATIC STRESS DISORDER is part of a *monothetic* system of classification. Within this system, each classification (disorder) is identified
with a list of criterial features that are individually necessary and collectively sufficient for including or excluding a case from the classification. A
case that overlaps the boundaries between classifications is handled in two
ways. Either the patient is given concurrent diagnoses—the case is diagnosed as belonging to both categories X *and* Y—or the patient is said to
have a "mixed disorder," which includes features of two different disorders, such as schizo-affective disorder.

Polythetic Classifications

DSM-III included one entry, schizotypal personality disorder, that did not
conform to the monothetic rule. To qualify as a schizotypal personality, a
patient needed to exhibit any four features from a list of eight. This is a
polythetic classification, meaning that it is possible for two correctly diagnosed cases to have no features in common. In *DSM-III-R*, the number of
polythetic classifications expanded, and it now includes the personality
disorders, behavior disorders, and chemical substance use disorders. While
DSM-III did not call attention to its departure from the monothetic rule,
DSM-III-R does (Amer. Psychia. Assoc. 1987:xxiv; Livesley 1985:355).

When, in its introduction, *DSM-III-R* discusses polythetic classifications, it is referring to an *explicitly* polythetic system (Widiger and Frances
1988:615). In order for a case to qualify as a member of a polythetic diagnostic category, it needs only to cross an indicated threshold value. Membership in the classification is based on overlapping features (family resemblances), as in the following example, where cases get the same diagnosis
("X") if they possess three out of six attributes (A to F):

Case	Attributes	Diagnosis
1	A-B-C	"X"
2	B-C-D	"X"
3	C-D-E	"X"
4	D-E-F	"X"

In such a system, the unity of a classification is *imposed from outside*, through the routinized use of diagnostic criteria listed in the official manual. Cases 1 and 4 get the same name, but there is no implication that they share anything else. This view is consistent with the Kraepelinian origins of *DSM-III* and *DSM-III-R* and with an editorial injunction against imputing any feature to any disorder (whether monothetic or polythetic) that is not observable or firmly established through scientific research.

So long as one sticks to the regulations that are supplied by the diagnostic manual, there are only these two sets of rules: explicitly monothetic rules and explicitly polythetic rules. In a moment, I am going to argue that diagnosticians have a hard time sticking to the rules, and that it is easy to find occasions when a putatively monothetic classification is expanded to include cases related through family resemblances only. If we want a name for these classifications (which include PTSD), we might call them either "quasi-polythetic" or "quasi-monothetic," since they follow neither monothetic nor polythetic rules exactly. Like the polythetic classifications described in *DSM-III-R*, they include members that share no single overt feature, but unlike these classifications, their unity is not imposed by convention, from the outside. Rather, cases are connected by something that is intrinsic to the classification but that does not appear on the official symptom list: a submerged feature, a tacitly understood element that allows diagnosis to include "partial" cases of a disorder (cases lacking one or more criterial features) within the rule-defined classification. (On "feature theory" and "contrast theory" in polythetic classification, see Rosch 1977:40–41; Atran 1985:302–303; and Armstrong et al. 1983.

In the case of PTSD, the submerged feature is the traumatic memory. The traumatic memory is not merely hidden below the surface of the *DSM* text; it is also polymorphous, showing itself in diverse presences (creating an equivalence between outbursts of anger and sleeplessness) and signifying absences (products of symptomatic avoidances). It is the traumatic memory's protean character that gives cases the appearance, on the surface, of sharing only a family resemblance. This is what Ribot, Charcot, Janet, and Freud wanted to call attention to when they compared the traumatic memory to a mental parasite. Lawrence Kolb evokes a similar image when he suggests that PTSD "is to psychiatry as syphilis was to medicine," that is, that both maladies mimic other disorders through their heterogenous symptomatology (Kolb 1989:811).

The Signifying Power of Symptoms

Beginning with *DSM-III*, the *DSM* system regards all symptoms as equally important for diagnostic purposes. Within a given classification, all criterial features have the same weight. If the text indicates alternative features, they each get the same weight. When the text is put into practice, however, and we move from the system's formal requirements to the cognitive processes affecting actual diagnostic judgment, diagnosis turns out to be more complicated and less rule-driven. Some symptoms are regarded as *typical* of their disorders and, because of this, they have a signifying power that other symptoms lack. The presence of a typical symptom colors the meaning that is given to other symptoms, and it may, under the right circumstances, even compensate for the absence of other criterial symptoms (Horowitz et al. 1981; Livesley 1985; Cantor and Genera 1986; Widiger and Frances 1985; Mezzich 1989; Blashfield et al. 1989; Frances et al. 1990).

PTSD's defining feature is its etiological event. Evidence of a traumatic experience is adduced from the patient's *active memory* of the event; from his *embodied memory* of it (consisting of traces of the event, mirrored in symptoms); and from *collateral information* that places the patient in circumstances severe enough to qualify as traumatogenic. A superficial reading of the *DSM* texts gives the impression that the nature of the traumatic event is clearly mirrored in post-traumatic symptoms. For example, *DSM-III-R* alludes twice to mirroring events in symptomatic reexperiences: a reference is made to patients who dream about their traumatic events, and a reference is made to patients who feel as if they were reliving the events. In practice, the ideational content of these symptoms is often open to multiple interpretations and consistent with alternative diagnoses.[1]

However it is obtained, evidence of a credible etiological experience transforms nonspecific symptoms into tokens of PTSD. Ruminations that would otherwise indicate a mood disorder are now changed into "reexperiences"; behaviors that resemble common phobias are turned into PTSD "avoidance behavior"; and episodes of irritability are redefined as "symptoms of autonomic arousal." It is in this special sense, of investing other symptoms with a degree of significance that they might not otherwise possess, that *the etiological event is typical of PTSD*. (For a parallel instance in the relationship between the schizophrenia classification and hallucinations, see Frances et al. 1990).

Even among typical features, there can be important differences in signifying power. When a typical feature occurs frequently in members (correctly diagnosed cases) of one classification but infrequently in members of others, it is useful for screening out false positives (cases that may appear to warrant the diagnosis but really do not) and false negatives (true cases mistakenly excluded from the diagnosis). In other words, the feature has *specificity* and/or *sensitivity* (Livesley 1985:354). The traumatic event—that is, the range of events routinely accepted as traumatogenic—has weak specificity, since many people who experience these events do not develop the symptoms of full-blown PTSD. On the other hand, a high degree of sensitivity is built into the *DSM-III-R* classification. Looked at this way, a false negative would be a case in which a patient has appropriate post-trauma symptoms but does not get the PTSD diagnosis because the disorder's typical feature, the traumatic event, is missing. Given a knowledgeable diagnostician, this outcome is highly unlikely since the post-trauma symptoms will themselves provide evidence of the typical feature, via the patient's embodied memory. There is no way to identify, or even to conceptualize, a false negative without this evidence (telltale symptoms) or without an independent source of evidence of PTSD (which does not yet exist).

Natural Kinds

The previous chapter described how ideas about "validity" contribute to the coherence of the official nosology. You will recall that psychiatric workers give "validity" several different meanings, and that one of these meanings makes a distinction between classifications whose criterial features are connected intrinsically and classifications for which this connection cannot be assumed. Inside psychiatry, a monothetic classification that has intrinsic validity is assumed to carve nature at the joint. Put into other words, there is a correspondence between the classification's boundaries (equivalent to the criterial features given in the diagnostic manual) and the boundaries of the natural phenomenon that it is supposed to represent.

In the case of post-traumatic stress disorder, the assumption would be mistaken: this monothetic classification is the *product or achievement* of psychiatric discourse, rather than its discovery. (A similar argument can be made concerning other psychiatric classifications, e.g., Hacking 1986:222, 228–230, 234.) On the surface—what we read in the *DSM* texts—the classifications are parts of an explicit system of meanings. Each classification is identified with a list of features, and the formal characteristics of the list will determine whether the classification is monothetic or polythetic. For such a system to work, diagnosticians must be able to match the lists of features with an external (nontextual) reality; and for this purpose they require auxiliary sets of features, since the criterial features are, on their own, generally ambiguous. Beyond knowing that in PTSD a "traumatic event is persistently reexperienced" (a criterial feature), the diagnostician needs to know the sorts of phenomena (auxiliary features) that qualify as "reexperiences": recurrent distressing dreams of the event, dissociative episodes called "flashbacks," and so forth. The auxiliary features are only a degree less ambiguous than the criterial features, but, in contrast to the latter, they are not provided with sets of defining features (see Frankel 1994 for a critical review of the literature describing the "flashback" phenomenon). Even if the text provided them with defining features, it would merely postpone the moment for tacit knowledge, since the defining features themselves require interpretation. Of course, this regress is in the nature of all definitions.

Criterial features are stable. They remain constant from one clinical or research site to another, because the text has provided them with auxiliary features. In addition, criterial features operate on a binary principle: either inclusion or exclusion. Even so, there may be differences among members. In psychiatric epidemiology, these differences are usually described in terms of "caseness." Caseness has no standard meaning: it can mean that a given case has features that are *typical* of its classification, or that the case is *exemplary* of the classification (possessing a high proportion of alternative and associated features), or that it is both typical and exemplary (Copeland 1981; Kendell 1988; Vaillant and Schnurr 1988). Caseness is also used to indicate the point at which particular symptoms have crossed a threshold of salience, such as a clinically significant level of distress or impairment.

In contrast to criterial features, meanings that are attributed to auxiliary features are tacit rather than explicit and fluid rather than stable. Moreover, these meanings are products of *analogical reasoning* rather than following rules (Bloor 1976:49; Barnes and Held 1982:70; Lakoff 1990:53). Analogy is a way of making inferences about a less familiar thing, a *target*, on the basis of what is already known about some more familiar thing, its *source* (figure 3). Sources of analogy in psychiatry include precedents (the history of a previous patient is used to diagnose a current patient, for example), schemas, prototypes, paradigms, metaphors, and so on (Hesse 1966, Needham 1980:chap. 2).

Two kinds of analogy are used in psychiatry. In *external analogy*, the source and its target are perceived as belonging to different domains. (Metaphor is the most obvious example.) The analogy's target parallels its source, but the person who is making the connection does not suppose that the target is an instance of the source. For example, there are psychiatric writers who would argue that the term "trauma," as used in "psychological trauma," represents an external analogy. It parallels a mental domain (the target) with a somatic domain (the source), as in Andraesen's (1985) comparison between the PTSD traumatic stressor and "the role of force in producing a broken leg."

In *internal analogy*, source and target are connected within a single domain. This occurs when a clinician diagnoses a current case (target) through his knowledge of published case histories, etiological and physiological schemas, and personal recollections of exemplary cases (sources); or when a researcher cites someone's published research findings (targets) to confirm or amplify his own findings (source); or when someone uses an

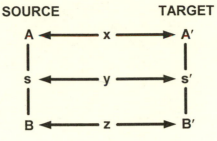

A, A', B, B'	=	features
s, s'	=	circumstances connecting features
x, y, z	=	circumstances connecting source and target

Figure 3. The structure of analogy

account of a successfully completed experiment (source) to plan and conduct his own current experiment (target).

In each of these events, the source and its target consist of an indeterminate number of features, and every analogy allows an indeterminate number of connections, even though at any particular time only some of these features and connections are salient and penetrate into awareness. (A connection can be positive, where target and source are similar; negative, where target and source are unlike; or neutral, where similarity or difference remains to be established.) Because of this indeterminacy, categories produced by analogy are poor containers of meanings, in contrast to Aristotelian classifications produced by rule-driven diagnostic criteria. In the absence of rules, analogy does a better job at proliferating meanings than conserving them, and semantic leakiness creates a situation in which meaning (and membership) can change from time to time, and from speaker to speaker, depending on the circumstances. This is sometimes called "the problem of meaning variance" (Hesse 1988:322; Lakoff 1990:39, 42, 53–57). The "problem" in this case is that knowledge based on analogy is likely to be *local knowledge*, shaped by the particular set of features and connections that are favored by a particular research team or clinical unit.

Local knowledge, untransformed by regulations and standards, is a potential source of polythetic categorization. That is, when diagnoses produced at different sites are aggregated, the result may include members (cases) that share only family resemblances. Because analogical reasoning is unavoidable in making diagnoses, the problem of meaning variance is not something that can be eliminated from a classification. Rather, it is something that is controlled by limiting, through the development of standardized technologies or "instruments," the subset of analogical features and connections that are employed for making diagnoses and by the emergence of a terminus a quo for these features, a point beyond which it is no longer necessary or rewarding to scrutinize meanings or definitions. (I address this subject in chapter 5).

The Meanings of the Traumatic Event

The workability of the PTSD classification as a source of stable and universal knowledge depends on effective control over the meaning of the traumatic event criterion, its most typical feature. The disorder's other criteria are distinctive only because of their relationship to the traumatic event. *DSM-III-R* defines the traumatic event as follows:

> The person has experienced an event that is *outside the range of usual human experience* and that would be *markedly distressing to almost anyone*, e.g., serious threat to one's life or physical integrity; serious threat or harm to one's children, spouse, or other close relatives and friends; sudden destruction of one's home or community; or seeing another person who has recently been, or is being, seriously injured or killed as the result of an accident or physical violence. (Amer. Psychia. Assoc. 1987:250; my emphasis)

The traumatic event is defined by the two underlined features. I begin with the second feature, the idea that the etiological events are markedly distressful to almost anyone. The meaning would seem to be unambiguous and immediately accessible to even untutored readers of *DSM-III-R*. But the literal meaning of "almost everyone" is routinely ignored by diagnosticians, for the reality is that there are otherwise normal individuals, perhaps entire subcultures, who are not markedly distressed by experiences that are routinely accepted as prima facie traumatogenic. Freud and Rivers made a similar point when they contrasted the psychological vicissitudes of civilian soldiers during World War I with the sangfroid of professional soldiers and mercenaries. In practice, diagnosticians are generally willing to waive the "almost anyone" condition, if they can establish that the event in question was markedly distressing to the current patient, and that many other people would be distressed by this kind of event (Kulka et al. 1990a:38).

The "markedly distressing" condition would seem to be more discriminating than the "nearly everyone" condition. But is it? The *DSM* text does not specify the range of emotional content that identifies "distress" in the context of PTSD, but it mentions fear and guilt, and it gives examples (rape, torture) that may lead one to think of grief, shame, anger, sense of loss. What is more significant is that the text does not prescribe *when* the distress is supposed to occur. At first glance, the answer seems obvious:

distress coincides with the traumatic event. This was the understanding of Erichsen and Page when they connected fear to nervous shock so as to make fear an element intrinsic to the traumatic moment. Freud was also thinking along these lines when he wrote of the role played by fright in creating a traumatic breach in the brain's stimulus barrier. Kardiner takes up the same idea in his book on the war neuroses. These writers also refer to the emotional distress that follows the traumatic event: Page mentions depression, and Freud, Rivers, and Kardiner refer to high levels of anxiety. In each of these accounts, the two moments of distress—one integral to the event, the other following it—are carefully separated. Beginning with *DSM-III*, the situation changes, and the distinction between the two vectors of emotion—emotion as cause and emotion as effect—is blurred.

Put into other words, the obvious interpretation of "distress," that is, an emotion integral to the traumatic event, does not match the range of events that are in practice accepted as traumatic. Take the example of epidemiological and experimental research conducted on Vietnam War veterans. In these studies, seven classes of events are accepted as traumatogenic: (1) the patient was a direct or indirect victim of unusual violence; (2) he perpetrated unusual violence unintentionally; (3) he perpetrated unusual violence intentionally but in a culturally acceptable context (e.g., to survive); (4) he perpetrated unusual violence intentionally as part of his military duties, but his acts were personally or culturally reprehensible (e.g., torturing prisoners to obtain information); (5) he perpetrated unusual violence intentionally because it was pleasurable (e.g., rape, killing prisoners, mutilating bodies); (6) he actively witnessed similar events (e.g., because he found them interesting or satisfying); or (7) he passively witnessed similar events (e.g., he happened to be present on these occasions).

This list of traumatogenic events reflects two departures from earlier diagnostic practices. The first is an expansion of the affect logic of the traumatic memory, with respect to *guilt*. During the two world wars, fear is the most commonly cited traumatogenic emotion (see chapter 2). Shame is also mentioned in earlier accounts but mainly in connection with traumatic hysteria (Janet, Breuer, and Freud). Rivers describes a case of pathogenic revulsion involving an officer whose mouth came into intimate contact with a putrefying corpse. But there are few references to guilt, and they are similar to William McDougall's account of P.M., an invalided British infantryman:

> [During] the Battle of the Marne in 1914, he came to close quarters with a German soldier who fired his rifle at him, but missed him. P.M. promptly struck him in the abdomen with his bayonet and killed him. He felt rather proud of this achievement and laughed over it with his two chums. Soon afterwards these two chums were killed, and P.M. began to see them come to his bedside at night and

would hear them talk. . . . Since being in hospital he has slept very badly because
every night he sees the ghost of the German soldier whom he killed. . . . During
the night this figure appears suddenly in the ward, points his rifle at P.M., says—
"Now I've got you. Now you can't get away," and fires point blank at him.
(McDougall 1920b:150)

In this case, P.M.'s guilt is attached to an event in which he inflicted violence in a culturally acceptable context: the soldier's struggle to survive
and his duty to engage and destroy the enemy. In terms of the seven possibilities listed above, P.M.'s situation corresponds to category 3. Likewise,
when Freud writes in *Beyond the Pleasure Principle* that nonprofessional
soldiers can be adversely affected by having to obey "outrageous demands"
during wartime, he is thinking of situations that would fall into categories
3 and 4. It is only following the Vietnam War, the publication of *DSM-III*,
and the adoption of PTSD by the VA as a service-connected disability that
guilt events based on pleasure rather than necessity, categories 5 and 7,
become attached to the traumatic memory.

This list of seven possibilities permits a second departure from previous
diagnostic conventions. Some diagnosed Vietnam War veterans claim (remember) that their etiological events were *not* especially distressing when
they took place—particularly events in categories 5 and 7, but also sometimes those in categories 2, 3, 4, and 6. Such cases fall into two types. In the
first, the patient claims that he was not disturbed by the original experience
but that he is now distressed by memories of the event, for example, when
he is asked to recall it during diagnostic interviews or when images of the
event now intrude into awareness. In the second type, the patient reports
that he has never experienced any overt distress in connection with the
indicated event—neither then nor now—but his diagnostician finds evidence of distress inscribed in the patient's symptomatic behavior, for example, in "psychic numbing," "emotional anesthesia," "self-dosing" with
alcohol or drugs, or other kinds of avoidance behavior. The interpretive
process operates the same way in both of these cases. The content of the
patient's current distress, either his expressed emotion (grief, guilt, etc.) or
his embodied distress, is projected back, over time, to the traumatic moment. In this way, the projected distress infuses and connects the morally
and experientially heterogeneous events (1 through 7) with a new and homogeneous meaning.

I started this section by pointing out that *DSM-III-R* identifies the traumatic event, a criterial feature for PTSD, with two auxiliary features. I have
examined one auxiliary feature, the idea that the etiological events for
PTSD are markedly distressing to nearly everyone, and I have pointed out
that, in many diagnosed cases, the key term, "distressing," is associated
with events that were affectively positive (pleasurable) or neutral at the

time they were experienced. The distressing events acquire negative affect, such as grief and guilt, only years later. In some of these cases, the attributions (negative affect) are not made directly by patients but through the interpretive practices of diagnosticians. I turn now to the second auxiliary feature, the idea that the traumatic event is "outside the range of usual human experience."

Here again, there is a difference between the literal meaning of the text and actual diagnostic practice. There is no problem coming up with a list of events that are *universally* outside the range of usual human experience and that are *universally* experienced as overtly distressful when they occur. The list would be relatively short, though, and it would exclude many events that are now routinely diagnosed as the points of origin for PTSD. Once beyond the short list, the meaning of "unusual" human experiences becomes *contingent*, changing from culture to culture and, sometimes from subgroup to subgroup within a culture. When diagnosticians and researchers refer to traumatogenic events as being "outside the range of usual human experience," they are, for the most part, employing a culture-specific and therefore contingent notion of what is "usual" in human experience, not a set of values, experiences, and cognitive structures that is shared by (nearly) all human beings in all human societies. Put into practice, "outside the range of usual human experience" is used by diagnosticians to bracket together a variety of experiences and acts, including exposure to imminent death or dismemberment; exposure to grotesque and disturbing situations, such as handling dismembered bodies and seeing and touching the internal organs of dying people; perpetration of morally forbidden acts, such as rape and torture; witnessing or otherwise learning about the sudden death, serious injury, or mutilation of close friends (Laufer et al. 1984 and 1985; Ursano and McCarroll 1990; Green et al. 1989; Hendin and Haas 1991; Yehuda et al. 1992).

Before deciding whether or not an indicated event is outside the range of usual experience in the contingent sense, one would have to know something about the meaning of the event in context: the frequency with which it occurs or occurred in the person's community; the community's moral order, that is, its ideas about what is good, tolerable, forbidden, and salient; the degree to which the person understands and identifies himself emotionally with this moral order; his level of self-awareness, empathetic awareness, and moral autonomy; and so on. These questions are particularly important when we look at the circumstances surrounding events in Vietnam. During wartime, many military units (especially combat units) develop into moral communities, tightly organized around their particular codes of conduct and systems of sanctions and rewards. In Vietnam, the local moral order was often very different from the official moral code—the rules established by the Department of Defense and the four military services—

and certainly different from the soldiers' preservice moralities. During the Vietnam War, the local morality and the official morality coincided in some military units: acts like rape, killing civilians, and torturing prisoners were forbidden and punished, and they were consequently infrequent. In other units, however, such acts were a part of everyday life and were tacitly accepted by officers and men, even though they were also officially condemned. In these units, there were a minority of men who preserved a sense of outrage and shame when they witnessed atrocities. For them, such events remained outside the range of usual human experience, because "usual human experience" continued to be defined by them in terms established in their prewar lives and moral communities. To the other men in these same units, the experiences were no longer shocking or extraordinary. For these soldiers, many of whom were subsequently diagnosed with PTSD, the events had been brought temporarily within the range of usual human experience.

Collective Memories

A brief summary: The advent of *DSM-III* established a distinctive *system* of meanings. For the system to function, diagnosticians must be able to match the lists of criterial features with an external (nontextual) reality. To do this, they require second-order features, attributes that will enable them to identify particular criterial features. Second-order features are provided by the manual and include definitions such as "an event outside the range of usual human experience" to mark the "traumatic event" (a criterial feature). At this level, of criterial features and their attributes, it is possible to speak about monothetic and polythetic classifications in an unambiguous way, since the system of meanings is self-contained. When the system is mapped onto clinical realities, however, an additional order of features is required, in order to identify the second-order features—in order to know, for example, when an actual event is outside the range of usual human experience.

These third-order features are *not* included in the manuals. Rather, they are products of analogical reasoning, through which the second-order feature (target) is defined through similarities to and differences from some more familiar phenomenon (source). The history of the traumatic memory is a chain of analogies: between posthypnotic suggestion and pathogenic secrets, between surgical shock and nervous shock, between victims of unusual violence and perpetrators of unusual violence, and so on. But analogy does a better job at proliferating meanings than containing them, and it it introduces the problem of meaning variance.

Meaning variance was neither a problem nor a problematic for Freud, Rivers, or Kardiner, because their work proceeded by means of the anecdotal method: research hypotheses were tested on small numbers of clinical patients, and theories were advanced through exemplary cases. The anecdotal method thrives where local knowledge predominates. *DSM-III* changed things in this respect by creating conditions for the emergence of a *collective* traumatic memory, beyond biography and anecdote. "Collective memory" can mean at least two things. As Maurice Halbwachs proposed, it can be the images, ideas, and feelings that are shared by people who belong to some group: "[I]t is individuals as group members who remember. While these remembrances are mutually supportive of each other and common to all, individual members still vary in the intensity with which they experience them" (Halbwachs 1980:48; also Connerton 1989). The collective memory can reside not only in individual minds but in practices, standards, apparatuses, and social relations, and in the calculations and documents that these things produce. In Halbwachs' collective memory, ideas and images emerge contingently, in the course of history, and diffuse and congeal according to their own logic.

The other kind of collective memory is also a product of history, but it is intentional and the work of a particular discipline: psychiatric epidemiology. Here memory making is a controlled process, consisting of three well-defined stages. Stage one is the *aggregation* of diagnosable people, from whom information is elicited or otherwise obtained via interviews, questionnaires, bodily substances, and so on. Stage two is the *disaggregation* of knowledge, breaking meanings free from their original biological and biographical contexts, transforming them into standardized bits and pieces of data. Stage three consists of *reaggregation*, the organization and assimilation of these data into meaningful arrays of numbers and into new and collective contexts.

At this point, I want to illustrate some of the practices through which psychiatric epidemiologists glue these collective memories together. There have been five major epidemiological studies of PTSD: the Vietnam Experience Study (VES), conducted by the Centers for Disease Control (1987); the National Vietnam Veterans Readjustment Study (NVVRS), paid for by the Veterans Administration and conducted by the Research Triangle Institute (Kulka et al. 1990a and 1990b); the Epidemiologic Catchment Area study (ECA study), a national psychiatric survey conducted by researchers based at Washington University (St. Louis) (Helzer et al. 1987); a regional study (Detroit) conducted by researchers based at the Henry Ford Hospital and the University of Michigan (Breslau et al. 1991); and a regional study (North Carolina) conducted by researchers at Duke University (Davidson et al. 1991).

These are large-scale surveys, involving thousands of informants and large sums of money; the NVVRS alone cost over $9 million (Kulka et al. 1990a:xxiii). They cover overlapping sections of the PTSD terrain: the

VES sample consists of Vietnam War veterans and war-era veterans (men and women who did not serve in Vietnam during this period); the NVVRS sample includes war veterans, war-era veterans, and civilians; the ECA study and the Davidson samples are drawn from the general population; and the Breslau sample consists of a single cohort of the general population, men and women between the ages of twenty-one and thirty years.

Where the studies have collected epidemiological data on similar populations, it is possible to compare their findings, and, as we can see from table 3, these prove to be remarkably divergent.

The first set of numbers in the table contrasts the VES and NVVRS on current prevalence of PTSD among Vietnam War veterans: NVVRS figures are *six times* greater than VES figures. (The VES figures are based on symptoms experienced in the previous month; NVVRS, the previous six months. Even so, there is a "stark contrast" between the two findings and it is acknowledged by the NVVRS researchers [Kulka et al. 1990b:E-13].)

The second set of numbers in the table contrasts the VES and NVVRS rates on the lifetime prevalence of PTSD among the veterans sampled. NVVRS rates are *double* those reported by VES.

When the figures in sets one and two are compared, a third disparity emerges. The VES reports that, of veterans ever diagnosed with PTSD, 15 percent currently suffer from this disorder, while the NVVRS percentage is more than *three times* greater.

The third set of numbers compares statistics on the prevalence of PTSD in the general population. Four divergent findings emerge:

1. The NVVRS reports that civilian men are more likely to get PTSD than are women (4:1). The ECA study and the Breslau data show precisely the *opposite* (ECA study, 1:7; Breslau, 1:2). (The NVVRS figures are for current cases, while the ECA study and the Breslau figures are for lifetime prevalence. However, there is no obvious reason to suppose that the *ratios* between men and women would be strongly affected by this difference.)

2. The NVVRS rates for current prevalence of PTSD among civilian males is *seven times* greater than the ECA study rates for lifetime prevalence for the same population.[2] (Lifetime rates indicate whether an individual has had the disorder at any time in his or her life. Therefore, lifetime rates should exceed current rates.)

3. Breslau's figure for lifetime prevalence among women is *nine times* that in the ECA study; her figure for men is *thirty-five times* that reported by ECA.

4. Davidson's figure for lifetime prevalence among women is more than *five times* that in the ECA study.

These differences have not escaped the attention of the various researchers, who have offered a variety of explanations (Helzer and Robins 1988; Kulka et al. 1990a:xxvii, 137–138; Breslau et al. 1991:221). In the next

TABLE 3
Prevalence of PTSD as Reported in Five Epidemiological Surveys

1. Current Prevalence of PTSD among Vietnam War Veterans

VES	(symptoms within the previous 1 month)	2.2 %
NVVRS	(symptoms within the previous 6 months)	15.2 %

2. Lifetime Prevalence* of PTSD among Vietnam War Veterans

VES	14.7 %
NVVRS	30.9 %

3. Prevalence of PTSD in the General Population (United States)

		males	females	ratio m : f
NVVRS	(current %)	1.2	0.3	4 : 1
ECAS	(lifetime %)	0.17	1.3	1 : 7
Breslau	(lifetime %)	6.0†	11.3†	1 : 2
Davidson	(lifetime %)	0.9	1.5	3 : 5

* "Lifetime prevalence" = one or more diagnosable episodes at any time, past or present.
† People between the ages of 21 and 30 years.

paragraphs, I set out three possible explanations. The first is my own. The second and third come from different teams of researchers and center on differences between the NVVRS rates and the rates published by the ECA study and the VES.

Here is my account. The divergent rates are the result of meaning variance associated with the traumatic event. Researchers have succeeded in controlling meaning variance *within* each study by employing standardized instruments and by training their workers to use these instruments in uniform ways, but the research teams employed different sets of standardized instruments and trained workers to different regimes. Moreover, they possessed neither a shared mechanism nor a compelling incentive to control the meaning variance *between* their respective studies.

The next account is part of an exchange between members of the NVVRS team (Keane and Penk 1988) and the ECA study team (Helzer and Robins 1988) in the pages of the *New England Journal of Medicine*. Keane and Penk trace the differences between the ECA study data and the NVVRS data to the ECA study's diagnostic protocol. Both teams employed the Diagnostic Interview Schedule (DIS), a widely used protocol based on the symptom lists in *DSM-III* and *DSM-III-R*. The ECA study used the standard *DSM-III* version and the NVVRS used a modified version. The NVVRS version simplified the questions and limited each question to a single symptom. According to Keane and Penk, the NVVRS instrument is better suited for capturing false negatives, because its questions are easier to understand. Conversely, the ECA study instrument can be expected to underestimate the rates.

Helzer and Robins respond that the ECA study instrument asks respondents if they have experienced each of nine symptoms associated with PTSD. "Thus they are asked about the occurrence nine separate times—a procedure as likely to overestimate as underestimate the reporting of traumatic episodes" (1988:1692). Further, if Keane and Penk are correct and the ECA study technology underestimates rates for men (ECA study lifetime rates for men are one-seventh of NVVRS current rates), why is the effect absent in the case of women? (ECA study rates for women are over four times the NVVRS rates.)

At this point Keane and Penk shift the grounds of the debate from differences in technology to differences in expert knowledge:

> [I]n the process of constructing reliable and valid measures for post-traumatic stress disorder for the National Vietnam Veterans Readjustment Study, *a panel of scientific experts* found that the section of the Diagnostic Interview Schedule on post-traumatic stress disorder inadequately addressed the first criterion of the diagnosis—exposure to a traumatic event. That section required extensive modification before the scientists would agree to its use. . . .
>
> *Accordingly*, we have concerns that the Diagnostic Interview Schedule used by Helzer et al. may have underestimated the prevalence of post-traumatic stress disorder. (Keane and Penk 1988:1691; my emphasis)

The panel of scientific experts consists of clinical psychologists and psychiatrists contracted by the Research Triangle Institute for the NVVRS project. In reality, their scientific credentials are no different from those of Helzer (a psychiatrist) and Robins (an epidemiologist), both of whom you may remember from the previous chapter, which describes their collaboration with Robert Spitzer in the origins of *DSM-III*. In their response, Helzer and Robins simply ignore the reference to the scientific expertise of the NVVRS team.

Keane and Penk shift again, now moving to take the moral high ground. Commenting on the ECA study's comparatively low rates for men, they write that

> [P]ost-traumatic stress disorder is not a de novo diagnosis whose entrance into DSM-III was politically motivated, but rather it is a serious psychiatric condition affecting many men and women, who have survived life threatening events, including combat, rape, political torture, and other disasters that occur in the course of world events. Its effects should not be minimized. (Keane and Penk 1988:1991)

The reference to "politically motivated" behavior is an allusion to the doubts that Helzer and Robins might continue to harbor concerning the legitimacy of the PTSD diagnosis. There is a subtext here, and it refers back to the events leading to the publication of *DSM-III* and the resistance of influential neo-Kraepelinians to including PTSD in the new nosology, and to the struggle by veterans and veterans' advocates against this opposition. The quotation's reference to the terrible effects of rape, torture, and disasters is an effort to conflate doubts about the NVVRS rates with doubts about the magnitude of the mental suffering that these events inflict on people. (For a similar debate, see the exchange between Lindy et al. 1987:271–272 and Breslau and Davis 1987:262–263.)

In their response, Helzer and Robins imply that the NVVRS researchers may have some biases of their own to explain. After all, the ECA study researchers set out to collect epidemiological data from a general population in order to learn the prevalence of a comprehensive range of psychiatric disorders. PTSD was only one of these disorders: "[The] fact that the [ECA study] inquiry was not addressed to a sample known initially to be at special risk and that the disorder under study was treated no differently from the others inquired about largely protected it from biased reporting" (1987:1633). The reference to "biased reporting" is a rhetorical thrust in the direction of the NVVRS.

The final account is buried in a technical appendix to the NVVRS report (Kulka et al. 1990b). According to this account, the NVVRS researchers were at first puzzled by the discrepancy between NVVRS and VES rates for war-related PTSD. The NVVRS rates are seven times the VES rates, even though both studies were "conducted according to high scientific standards." The most obvious sources of the divergence would be significant differences between NVVRS and VES diagnostic instruments, significant but undetected differences between the NVVRS and VES populations, or a combination of these two factors (Kulka et al. 1990b:E-17).

The search for a solution began with the research technologies. The VES employed a single instrument, a version of the Diagnostic Interview Schedule (DIS). The NVVRS employed multiple instruments, which included the version of the DIS that I mentioned a moment ago. So the first step in comparing the NVVRS and VES technologies would be to match the DIS findings collected by each team. When this was done, the difference between the rates narrowed down to 177 percent.

The NVVRS and VES versions of the DIS were similar but not identical. In addition, the teams scored key questions differently. The NVVRS workers had been instructed that an intrusive symptom could be identified as a PTSD intrusive symptom only if its content (e.g., a dream motif) evidenced a connection with the respondent's reported traumatic event. However, the other categories of symptoms did not have to be overtly connected in order to qualify as PTSD symptoms. In contrast to this, the VES rule was that *all* categories of symptoms had to be overtly connected to qualify as PTSD symptoms. The different rules mean that some people who qualified for the

NVVRS diagnosis would have failed to qualify for the VES diagnosis. To compare the DIS results produced by the two teams, the NVVRS researchers now asked their computers to reconfigure their DIS findings according to the VES scoring standards. When this was done, the NVVRS prevalence rates were equivalent to the VES rates.

Having succeeded in bringing the discrepancy down to zero, the NVVRS writers concluded that divergence was a product of differences in technologies rather than in the populations (Kulka et al. 1990b:E-17, E-20). Having established this point, the NVVRS writers move on to provide grounds for preferring their technology over their rival's. They take a position similar to the one argued by Keane and Penk in their exchange with the ECA study researchers, Helzer and Robins. Their argument is that the NVVRS technology is superior, because (1) it incorporates *multiple* PTSD indicators, (2) the multiple indicators can be assumed to measure *different* features or dimensions of the disorder, and (3) the findings that are produced by these different indicators tend to *converge* (Kulka et al. 1990b:D-2, D-12, D-14, D-17, D-27). In this context, convergence is equivalent to what might be called a "triangulation effect."

The presumption here is that a triangulation effect produced *within* a phenomenal domain (psychometrics) can be compared to the intersection of findings *between* phenomenal domains (psychometrics and biochemistry). This is a common analogy in psychiatric epidemiology, and it is used to ground claims regarding a classification's independent validity, based on the correlation of independent sources of information. The convergence of findings among techniques can be measured by a "consistency coefficient." The ability of a package of techniques to make facts about PTSD is tied to this coefficient. To make facts, it should be high, showing agreement. At the same time, there are advantages if the coefficient is less than perfect (100 percent), that is, produces a some relatively small gap or divergence. The total absence of a gap can be interpreted as evidence that the multiple techniques are simply asking the same questions in different ways, and that they are therefore interchangeable. If this were the case, there would be no reason to employ multiple methods, the logic of triangulation would be undermined, and the grounds for claiming the superiority of the NVVRS multimethod over the VES monomethod would evaporate. (A perfect coefficient is a problem only if the techniques operate within the same phenomenal domain.)

In one sense, the development of a package of measures is an effort to produce a small but powerful gap, which can be incorporated into the production process as one of its constituent elements. The possibility of making valid data could come afterward. A well-made gap is a double signifier. As we have seen, it signifies that the multiple technologies are measuring different dimensions of the disorder. It also signifies a "latent trait," a submerged something that is not yet within the epidemiologist's grasp. As it is interpreted by researchers, the latent trait can stand for either of two things. It can be an unidentified feature or dimension that waits to be captured by a future instrument. This is how statistical experts generally talk about latent traits. However, PTSD researchers tend to interpret it as evidence of a natural kind, an authentic monothetic category. Remember, the symptoms of PTSD are eminently visible—rages, phobias, anhedonia, and so on—but the pathogenic process that connects these symptoms together is not. In the latent variable, researchers and their audiences glimpse, *as if by accident*, a mathematical picture of this process (Duncan-Jones et al. 1986:391; Gould 1981: 310, 314–315).

Aside from a few skirmishes between rival teams, meaning variance has had no significant practical effect on psychiatric discourse on PTSD. The "problem" of meaning variance has congealed into a problematic: "At this point, we really do not know which technique is 'best,' but it should be acknowledged that differences in instrumentation may result in considerable prevalence variability" (Davidson et al. 1991:720).

The future of the collective traumatic memory would seem to belong to psychometrics rather than epistemology.

Time and Causality

The description of PTSD in *DSM-III-R* presumes that time moves *from* the etiological event *to* the post-traumatic symptoms. But clinicians know that, in some cases, the relation of time to symptoms and events can be ambiguous:

> One of the striking features of post-traumatic stress disorder (PTSD) is the degree to which a past event comes to dominate the patient's associations. As all roads lead to Rome, all the patient's thoughts lead to the trauma. A war veteran known to us can't look at his wife's nude body without recalling with revulsion the naked bodies he saw in a burial pit in Vietnam, can't stand the sight of children's dolls because their eyes remind him of the staring eyes of the war dead. . . .

> The ultimate gravitational attraction in the physical universe is represented by the black hole, a place in space-time that has such high gravity that even light cannot pass by without being drawn into it. . . . PTSD patients struggle to avoid thoughts, activities, or situations associated with the trauma . . . , *not only because they are so painful but also because they are so absorbing.* (Pitman and Orr 1990:469–470; my emphasis)

Pitman and Orr paint a picture in which time flows in two directions: from a significant event out to its symptoms (the *DSM* conception of PTSD) and from a person's current psychological state back to the event, where it acquires a genealogy and a discrete set of meanings. From this idea of time going in two directions, it is only one more step to another idea, namely that, in certain cases, time may be flowing *mainly* from current psychological states back to the events. Actually, it is an old idea, going back to Ribot and his conception of the self as a made object that continues to remake itself by appropriating the past. In recent times, this idea has been limited almost entirely to cases of "factitious" PTSD, involving people who are believed to have invented their traumatic events (Jackson 1990, Perconte and Gorenczy 1990, Sparr and Pankrantz 1983). The typical case is a veteran of the Vietnam War period who saw no combat service, has a history of psychological problems, and has participated in Vietnam veterans' "rap groups" organized by VA outreach centers. Sitting in these groups, he has socialized with combat veterans, learned about PTSD, and eventually acquired a traumatic event, modeled on someone else's authentic story. With his event, he renarrates his disappointments and failures and gains entry into a treatment program, where he is able to live out his fantasy. The event becomes part of his autobiography, and he is now unable to separate himself from it. In the last few years, questions about factitious PTSD have also been raised in connection with the so-called "false memory syndrome": cases in which therapists help "victims" to rediscover repressed childhood memories that often incriminate close relatives, most commonly in acts of traumatic sexual abuse (see Herman 1992 and Terr 1994 on recovered memory; cf. Holmes 1990, Loftus and Ketcham 1994, Ofshe and Watters 1994, Wright 1994, Yapko 1994 for critiques).

Earlier writers, from Erichsen to Rivers, also mention cases of spurious traumatic events, usually involving attempts by railway passengers to obtain compensation (railway spine) or by soldiers to obtain invalid status (shell shock). Equally often, however, these writers are referring to cases where the reported events are real, but the patient's physical and emotional after effects are judged to be disproportionate. In such cases, the pathogenic mechanism is generally described as being either autosuggestion or heterosuggestion (the unintended influence of the therapist) and not nervous shock. Babinski's theory of pthiatism, in which he equates traumatic hysteria with the range of symptoms that might be lifted by countersuggestion, was the most widely accepted of these accounts during the World War I period. It was also commonly assumed that the people who develop these disorders have psychological or neurological diatheses. You may recall from chapter 2 that this was exactly how Rivers explained the onset of traumatic neuroses among the lower ranks. Officers, he contended, might suffer similar symptoms, but the path to sickness was different, a result of physical and nervous exhaustion rather than congenital or acquired weaknesses.

In recent times, researchers have shown relatively little interest in this category of people. The most obvious reason is that research on the traumatic memory continues to center on Vietnam War veterans with multiple psychiatric diagnoses and significant psychosocial impairments. Now, decades after the war, this population is entirely unsuitable for investigating the role played by premorbid diatheses, suggestibility, retrospection, and contingency in the production of current symptoms (Kulka 1990a).

Among the few recent studies that have investigated this subject, the best-known research is based on populations and events in Australia (A. C. McFarlane) and Scotland (D. A. Alexander and A. Wells). McFarlane studied the psychological effects of a destructive brushfire on fire fighters. In the course of the fire, the men were exposed to life-threatening situations that met the PTSD stressor criterion. Following the fire, the men were given clinical interviews, and clinical evaluations continued over the next two and a half years. Because numerous fire fighters worked at each site, reports of personal experiences could be independently corroborated. McFarlane also had access to the men's psychiatric histories. In each of these respects, the fire fighters in McFarlane's study were markedly different from the Vietnam War veterans who participate in PTSD research.

McFarlane advances four main conclusions based on his own research:

1. Severity of exposure was *not* the major determinant of post-traumatic morbidity (McFarlane 1989:224). Some men "decompensated with a minimal or low exposure to the disaster." Further, the fire fighters' descriptions of their events and their assessments of the threats to which they were exposed often "had little to do with their actual experiences" (McFarlane 1986:10).

2. "Distress" can affect anyone following life-threatening experiences, but it should not be confused with "morbidity," a state in which distress takes the form of an emotional preoccupation with the experience. Many fire fighters who reported being "extremely distressed" after the brushfire did not become "psychiatrically impaired" (McFarlane 1988:138).

3. Only a minority of men, identified by McFarlane as "anxiety-prone" individuals, developed the "cognitive and emotional preoccupation with the trauma." Premorbid factors, including a history of psychiatric problems, "accounted for a greater percentage of the variance of disorder than [did] the impact of the disaster." In some cases, morbidity may be an "indirect marker of neuroticism," a preexisting personality disposition (McFarlane 1988:138; 1989:227; see Eysenck and Eysenck 1975 on "neuroticism").

4. Finally, PTSD researchers have a hard time divorcing themselves from the psychological immediacy of the experiences that people describe to them. This may explain the many "graphic and detailed statements . . . about the impact of the event" found throughout the PTSD literature (McFarlane 1986:12).

Let us take a quick look at the Scottish research (Alexander and Wells

1991). In 1988, an oil rig exploded in the North Sea. Several months later, the rig's dormitory module was retrieved from the sea bed. A team of police officers searched this "dangerous and unpleasant environment" and recovered seventy-three bodies, many in badly deteriorated condition. The bodies were taken to a mortuary, where they were stripped, washed, and photographed by a second group of officers, who also assisted pathologists in dissecting the bodies. The situation parallels the Australian study: people are exposed to PTSD-level events, their reported experiences are corroborated, they are assessed soon after the event and then again at intervals, and so on. Alexander had the additional advantage of having assessed the mental status of this same police force shortly before the disaster, as part of an occupational health study. Various standardized measures of personality and mental health had been employed, and so it was possible to make precise pre-event and post-event comparisons. The researchers' conclusions are similar to McFarlane's.

The essence of McFarlane's argument is that people exposed to PTSD-level events can be sorted into three categories:

Category 1 consists of people whose responses are distressful rather than syndromal. Given the situations that they faced, the experience of distress is normal and part of the human condition. Unfortunately, McFarlane continues, we lack a standard psychiatric vocabulary for talking about these distinctions: "DSM-III-R does not provide a definition for the term distress and no major psychiatric textbook discusses this concept at all despite its importance to psychiatric diagnosis. It seems to be a phenomenon which everyone knows about but one that presents many definitional problems" (McFarlane 1993:422).

Category 2 consists of people whose responses *are* pathological but *are not* triggered directly by exposure to PTSD-level events. The effect of exposure was to trigger major depressive disorder and anxiety disorders, and it was these disorders that were responsible for the men's PTSD-like symptoms, notably their symptomatic reexperiences: "Once anxiety symptoms and/or depression have become established, a *feedback effect* begins to occur where the intensity and frequency of the memories of the disaster are increased" (McFarlane 1993:424; my emphasis). If category 2 people were evaluated according to the prevailing interpretation of PTSD, they would be given the diagnosis of "PTSD with concurrent major depression and anxiety disorders." However, the same set of symptoms can be explained by McFarlane's category 2 diagnosis: major depression and anxiety disorders *plus* the feedback mechanism. (The possibility of a feedback effect is also examined by Breslau and her colleagues [1994a and 1994b]. This analysis, based on the epidemiological data mentioned earlier [Breslau et al. 1991], indicates that people with a prior history of major depression are 3.3 times more likely to develop diagnosable symptoms of PTSD following exposure to a PTSD-level event than are people without this history. The PTSD "hazard ratio" for people with a prior history of an anxiety disorder is 2.2.)

Category 3 consists of people whose responses coincide with the *DSM-III-R* account of PTSD: the event triggers the symptoms. This response contrasts with category 2, where exposure triggers psychological or psychobiological processes that induce the person to invest attention and emotion in memories of the event (McFarlane 1993:427–428).

For the moment, let us assume that McFarlane is correct and that we can expect to find these three categories of people following PTSD-level events. This would have important implications for current PTSD research. It would mean that the collective memory that I described in connection with PTSD epidemiological research on veterans (the NVVRS and the VES) rests on a significant "category error," since the researchers would have found it impossible to discriminate between categories 2 and 3.

To understand the full significance of this point, one needs to keep in mind that McFarlane's research is focused on rapid-onset PTSD: the interval between the identified event and the onset of symptoms is very brief. Most Vietnam War veterans have "delayed onset PTSD," meaning that the interval between the event and the symptoms may be measured in years. Typically, the individuals' concurrent disorders—depression, anxiety disorders, and chemical substance use disorder—emerge during this interval. In many cases, the interval is also full of social and interpersonal problems, and individuals experience demoralization. This combination of circumstances points to a possibility to which McFarlane only alludes, namely that category 2 includes two subcategories. Category 2-A consists of people whose depression and anxiety disorders are triggered by exposure to the identified events (and the feedback mechanism accounts for their emotional attachment to memories of the event). This subcategory would include McFarlane's fire fighters. Category 2-B consists of cases of delayed-onset PTSD. These are people whose depression and anxiety disorders are *not* triggered by the identified events. Rather, they are triggered by (or exacerbated by) events and circumstances of a later period—in the case of veterans, by situations that evolved after they have left the combat zone. Once established, the disorders would operate through the same feedback mechanism. Category 2-B would include cases, mentioned earlier in this chapter, where patients remember experiencing no feelings of distress at the time of the presumably traumatogenic event.

Only a small minority of people working in the PTSD field seem interested in examining the implications of McFarlane's work. The majority of PTSD researchers are untroubled by his findings, simply because the data can be easily reinterpreted in ways that make them consistent with prevailing assumptions. The majority would argue that categories 2 and 3 are not separate but are positions on a continuous variable, corresponding to PTSD's underlying mechanism.

The easiest ways to illustrate how McFarlane's findings can be reinterpreted is with the use of Mardi Horowitz's "stress response syndrome" schema (Horowitz 1976; 1986). Horowitz describes it as an information processing model, but the schema's origins go back to Janet. It consists of a series of "response phases" that are set in motion by a distressful life event. Phase one is the outcry phase: the person "quickly processes the implications of the event, has an alarm reaction that interrupts ordinary activities, and expresses warning signals." Phase two is a denial phase: the person refuses to face the memory of the event. Phase three is an intrusion phase: the person is affected by unbidden thoughts and images of the event. (Phases two and three correspond to the avoidance and intrusion features of the *DSM-III-R* classification.) Phase four consists of working-through:

> [I]t involves some kind of decision about what it all means for the overall self-organization. The process is thus an evolutionary, survival-of-the-species function, preparing the person to make new commitments after a loss or injury, and to accept the self and the world for the true view of what the new and present situation now consists of. Before this adaptive end of the stress review process, there is not adequate differentiation of fantasy from reality. (Horowitz 1993:53)

Phase five is the completion phase; it marks the "relative end" of active processing of the traumatogenic event. In this schema, pathology is continuous with normality. PTSD consists of getting stuck before reaching completion, so that one recycles between the denial phase and the intrusion phase.

We can see how Horowitz's schema rediagnoses McFarlane's fire fighters. Category 1 consists of people who have moved expeditiously through all five phases to completion. The people in categories 2 and 3, on the other hand, are stuck in the process. The differential response to stressful events is explained by a "diathesis stress model": "[It] assumes that extreme events may elicit crisis in almost everyone, but depending on the intensity of the stressor and the individual's threshold for tolerating it, the stressor will be either contained homeostatically [category 1] or will lead to an episode of the disorder [categories 2 and 3]" (Jordan et al. 1991:214). As for the contrast between people in categories 2 and 3, this would be explained by the preexisting neurotic character of those in category 2: "[Their] character structure will make ambivalence, and hence conflict, more likely during the stress review process of the working-through phase. Abnormal schemas of self and other, and tendencies to image-distorting or information-distorting control processes in order to avoid emotional pain" (Horowitz 1993:53). As for the contrast between people in categories 2-A and 2-B, this can be explained by the argument, mentioned earlier, that "delayed-onset PTSD" refers only to appearances. That is, the interval is characterized by both (1) the absence of visible symptoms and (2) the presence of PTSD's invisible feature, the traumatic memory. Indeed, the interval is not asymptomatic at all, but rather a period of successful avoidance strategies, phase two in Horowitz's schema (see also Horowitz et al. 1990:64–69).

Although I have identified these schemas with labels and authors, they are the tacit knowledge of every competent PTSD researcher and diagnostician. Operating within a self-vindicating system of diagnostic technologies and routines, the net effect of such schemas is to get traumatic time to run in the right direction: *from* the etiological event *to* the post-traumatic symptoms.

The Traumatic Memory, 1994

Every psychiatric discourse starts with a defining object. The discourse and its object evolve together, each affecting the other. In time, each discourse also acquires a genealogy, a consensual account that tells the history of the discourse and its object. The discourse on PTSD has a genealogy built on two themes. The first is the continuity of PTSD: PTSD exists, as an entity, prior to and independently of the ways in which psychiatric experts diagnose it, study it, and treat it. The second theme is chronology, the chain of events that leads up to current knowledge of PTSD. The earliest link in the chain is the spontaneous recognition of the traumatic memory, dating back to *The Epic of Gilgamesh*. This is followed, millennia later, by the earliest medical recognition of the post-traumatic syndrome, by Erichsen and Page. Then comes the codification of the disorder's syndromal features by Kardiner, followed in turn by the political struggle that makes a place for PTSD in the official nosology. And now we are in a period devoted to the patient accumulation of empirical knowledge of this disorder and to the refinement of diagnostic and therapeutic technologies.

In the first four chapters of this book, I constructed a history of the traumatic memory that departs from its conventional genealogy at several points. The genealogy represents the traumatic memory as a found object, a thing indifferent to history. Research into this memory and the associated pathogenic secret is portrayed as a process of discovery. I have argued for something else: the traumatic memory is a man-made object. It originates in the scientific and clinical discourses of the nineteenth century; before that time, there is unhappiness, despair, and disturbing recollections, but no traumatic memory, in the sense that we know it today.

The conventional genealogy is correct in describing the current era as a period of accumulating facts and findings about traumatic memory. But because it treats the traumatic memory as if it were an immutable object, it ignores another recent development, namely, the enlargement of the memory's active principle, the pathogenic secret.

The pathogenic secret starts off in the nineteenth century as something located in the mind of individuals. Typically, it extends no further than the patient and the therapist. But in the 1970s, the pathogenic secret begins to enlarge, moving outward. At first, it is simply a matter of mental contagion: the secret replicates itself in the therapist's mind and in the minds of the patient's spouse and offspring. The more interesting development consists of a process in which the secret expands through mirroring rather than contagion.

The psychological pathology of the individual, the microcosmos, has a mirror image in the moral pathology of the collectivity, the social macrocosmos. The collective secret is a willful ignorance of traumatic acts and a denial of post-traumatic suffering. Patients are victims twice over: victims of the original perpetrators and victims of an indifferent society. The therapeutic act of bringing the secret into full awareness is now inextricably linked to a political act. Vietnam War veterans are the first traumatic victims to demand collective recognition, and they are followed by victims of other suppressed traumas, such as childhood incest and domestic rape.

> Fifty years ago, Virginia Woolf wrote that "the public and private worlds are inseparably connected . . . the tyrannies and servilities of one are the tyrannies and servilities of the other." It is now apparent also that the traumas of the one are the traumas of the other. The hysteria of women and the combat neurosis of men are one. . . .

> With the creative energy that accompanies the return of repressed ideas, the field [of trauma studies] has expanded dramatically. . . . Now each month brings forth the publication of new books, new research findings, new discussions in the public media. . . .

> But history teaches us that this knowledge could disappear. Without the context of a political movement, it has never been possible to advance the study of psychological trauma. (Herman 1992:32)

The growing interest in various categories of victims notwithstanding, writing and research on PTSD continues to be weighted toward the experiences of Vietnam War veterans. This situation reflects, in no small part, the concentration of resources in the Veterans Administration Medical System: its internal sources of funding, its experienced PTSD researchers, and its large and accessible patient population. The next four chapters examine this population and this medical system.

Part III

POST-TRAUMATIC STRESS DISORDER
IN PRACTICE

Five

The Technology of Diagnosis

THIS CHAPTER describes a series of diagnostic sessions conducted during 1986 and 1987 at a Veterans Administration psychiatric facility, the National Center for the Treatment of Post-Traumatic Stress Disorder (a pseudonym), specializing in the diagnosis and treatment of war-related PTSD. The center's origins and operations are described in some detail in chapter 6.

The veterans who come to be diagnosed at the center are drawn from two main sources. Approximately two-thirds of the men are referred from other VA psychiatric units, usually alcohol and drug abuse inpatient units, acute psychiatry inpatient units, and mental hygiene outpatient units. About a third of the veterans are self-referred. These men usually arrive with the encouragement and advice of other veterans—often men who have been previously diagnosed and treated at the center—and are sometimes familiar with PTSD diagnosis through self-help literature, such as *The Vietnam Vet Survival Guide* (Kubey et al. 1986: 100–116), *Post-Traumatic Stress Disorder—V.A. Disability Claims and Military Review* (Lepore 1986), and *The Veteran's Self-Help Guide on Stress Disorder* (a pamphlet published by the Veterans Education Project, a veterans' advocacy group).

Diagnosis at the center is based on several sources of information. (1) Most of these veterans have previous psychiatric diagnoses, made at other VA units. Their VA psychiatric records are reviewed at the center, and their therapists and case workers are contacted if necessary. The previous diagnoses tend to be read as assessments of the severity and chronicity of the patient's psychiatric problems rather than as classifications that need to be taken at face value. (2) Veterans are given a structured interview that asks for information about preservice history (including family life, social life, school performance), military history (including exposure to combat stressors), postservice history (including marital history, employment history, current income, disability payments or claims), and medical and psychiatric history (including past and current drug and alcohol use). (3) Diagnostic information is collected by a standardized diagnostic protocol, keyed to the *DSM-III* and (after 1987) *DSM-III-R* criteria for PTSD. (4) Observations are made concerning each man's current status, based on his appearance, behavior (including posture, facial expression, body movements, speech, ability to interact with the interviewer), appropriateness of emotions and predominant mood, and intellectual functioning (including orientation to his surroundings, memory, insight, stream of thought, indications of delusions or hallucinations). (5) Diagnosticians want to see each man's military records. These usually consist of the DD 214 form that men are given at the time they are discharged from active duty. If the form is complete, and often it is not, it identifies a man's military specialties, the ranks he has held, the units and places in which he has served, his decorations, and any disciplinary actions taken against him. (6) When it is necessary and possible, an attempt is made to collect additional information concerning the veteran's symptomatology and current curcumstances from spouses, parents, or siblings. (7) Each man is expected to answer sets of questions about his symptoms and psychological states. The most important of these for diagnostic purposes is the MMPI (Minnesota Multiphasic Personality Index).

The MMPI used at the center (and most psychiatric units) consists of 399 questions, each asking for a yes or no answer. The questions are keyed to ten clinical scales. Each scale is supposed to detect clinically significant features, as outlined in table 4. In 1984, Keane and his associates proposed an MMPI subscale for detecting PTSD (see table 5). Their subscale consists of forty-nine MMPI questions; a score of thirty out of forty-nine was found to provide the "optimal cutting point" for separating men in the PTSD and control groups. (At the center, subscale questions are grouped into clusters, each keyed to a PTSD diagnostic feature described in *DSM-III*.) In the same article, Keane and his colleagues also identify a characteristic PTSD profile, consisting of peaks and valleys stretching across the ten clinical scales of the MMPI. The PTSD group was found to score significantly higher than the control group on every scale except masculinity-femininity (no difference) and to peak on the depression and schizophrenia scales (Keane 1984; the development of a second PTSD subscale is reported in Keane et al. 1988).

The center also requires veterans to complete the Impact of Event Scale (see table 5). This consists of sixteen questions and is intended to measure frequency of reexperiencing and avoidance symptoms (Horowitz et al. 1979). Avoidance and reexperiencing symptoms are dynamically related in the *DSM-III-R* text. Avoidance behavior is an attempt to defend oneself against reexperiences of the trauma. Therefore, the absence of avoidance behavior concurrent with the presence of symptomatic reexperiences (or vice versa) during the designated period prior to testing can itself be symptomatic (Brett et al. 1988:1233).

The questions making up the PTSD subscale and the Impact of Event Scale are nonspecific and go over the ground covered by the diagnostic interview based on *DSM-III*. The MMPI includes a scale, the F (frequency) scale, that is supposed to help diagnosticians assess the truthfulness or ac-

TABLE 4
MMPI Scales Used for Diagnosing PTSD

I. THE CLINICAL SCALES:

1. *hypochondriasis:* detects attempts to gain sympathy by exaggerating (vague) physical complaints

2. *depression:* detects low morale, feelings of hopelessness, sorrow, brooding, mental dullness, psychomotor retardation

3. *hysteria:* detects a disposition to use (somatic) symptoms as a means of solving conflicts or avoiding responsibilities when under stress

4. *psychopathic deviate:* detects antisocial acting out of impulses; sub-scales include social alienation and problems with authorities

5. *maculinity-feminity:* detects problems of sexual identification, indicated, e.g., by the rejection of gender appropriate occupational roles

6. *paranoia:* detects feelings of persecution, grandiosity, and pervasive suspiciousness

7. *psychoasthenia:* detects excessive doubt, unreasonable fears, compulsions and obsessions

8. *schizophrenia:* detects unusual thought processes, lack of deep interests, apathy, peculiarities of perception; sub-scales include feelings of social alienation, emotional alienation, bizarre sensory experiences, defective inhibition, lack of ego mastery

9. *hypomania:* detects high activity or excitement level, expansiveness; sub-scales include amorality, ego-inflation, psychomotor acceleration

10. *social introversion:* detects problems in social interaction, limited social skills

II. THE F (frequency) SCALE: detects fictitious symptoms, cry for help, etc.

III. THE MACANDREW ALCOHOLISM SCALE

Source: Lachar 1974

curacy of statements. In practice, the F scale can be used to draw contradictory conclusions: that someone is deliberately faking mental symptoms for secondary gain, such as compensation, *or* that he is exaggerating real problems and symptoms in order to get attention, that is, a "call for help." High F scale scores also identify "unconventional" thinkers, and people experiencing extreme stress or suffering from confusion and personality disorganization accompanying acute psychotic reactions. (See Lachar 1974 and Graham 1987 on the MMPI. For a review of efforts to diagnose PTSD through the MMPI, see Penk et al. 1988.)

Diagnostic sessions at the center are held twice a week. A typical session is attended by seven staff members, including the center's clinical director (a psychiatrist), its psychometrician, the members of the outpatient staff who interview and present these cases, and the heads of the inpatient unit and the nursing staff. Three of these men have doctorates in clinical psy-

TABLE 5
Sample Questions from Tests Used for Diagnosing PTSD

I. THE PTSD SUBSCALE (Keane et al. 1984): 49 questions

REEXPERIENCING:
    15. Once in a while I think of things too bad to talk about.
    241. I dream frequently about things best kept to myself.

NUMBING AND ALIENATION:
    8. No one seems to understand me.
    366. Even when I am with people I feel lonely much of the time.
    389. Whenever possible I avoid being in a crowd.

HYPERAROUSAL:
    43. My sleep is fitful and disturbed.
    144. Often I feel as if there were a tight band around my head.

MEMORY AND CONCENTRATION PROBLEMS:
    32. I find it hard to keep my mind on a task or job.

GUILT AND DEPRESSION:
    61. I have not led the right kind of life.
    76. Most of the time I feel blue.
    106. Much of the time I feel as if I have done something wrong or evil.

FEAR OF LOSS OF CONTROL:
    39. At times I feel like smashing things.
    139. Sometimes I feel as if I must injure either myself or someone else.
    182. I have sometimes felt that difficulties were piling up so high that I could not overcome them.

II. IMPACT OF EVENT SCALE (Horowitz et al. 1979): 16 questions
Which of these comments about the designated life event were true for you during the past seven days?
    1. I thought about it when I didn't mean to.
    6. I had waves of strong feelings about it.
    7. I had dreams about it.
    8. I stayed away from reminders about it.
    12. Other things kept making me think about it.
    14. I tried not to think about it.

---

chology; the others have master's degrees in either psychology or psychiatric social work. Four are Vietnam War veterans.

The diagnostic sessions follow a routine format. First, the presenter reviews the information that he (or, occasionally, she) collected during the interview. He follows the sequence of questions in the protocols and usually finishes by proposing a diagnosis. Next, the psychometrician presents findings based on the structured clinical interview, the MMPI, and the Impact of Event Scale. The MMPI data are displayed by means of an overhead projector, and the patient's profile is compared with the typical PTSD profile (Keane et al. 1984) and (sometimes) the profiles of veterans with whom the staff have had clinical experience. Finally, the staff members discuss this information, and the session ends when they can collectively answer the following questions; Does this man have PTSD? If he does not, then what is his diagnosis? If he does have PTSD, does it originate in a service-connected traumatic event? Does he have a concurrent psychiatric diagnosis? If he has service-connected PTSD, should he be treated as an inpatient or an outpatient, at the center or somewhere else? If he is going to be treated at the center, does he have to meet prior conditions, such as completing an alcohol or drug detoxification program? In a formal sense, the staff members participate as equals. In practice, the clinical director has a determining influence on the final decision.

To sum things up, the diagnostic process at the center is comprehensive and thorough. It is conducted by a group of experienced experts, utilizing techniques and procedures that are comparable to the most rigorous diagnostic methodologies reported in the PTSD literature (e.g., Kulka 1990a: chap. 3).

The following four case studies are accounts drawn from the diagnostic sessions that I attended at the center. I have chosen sessions in which men are diagnosed as suffering from service-connected PTSD. Each man is determined to have the correct set of symptoms, the right MMPI profile, a satisfactory score on the PTSD subscale, and at least one appropriate traumatic event. Although I have attempted to be faithful to the form and content of the actual proceedings, these are not verbatim accounts. They are taken from my notes, written hurriedly while the sessions were taking place. Also, I have altered the accounts in minor ways to disguise the identity of the patients. The names of the persons and VA units mentioned in these accounts are pseudonyms.

The Case of the Easy Diagnosis

MALLOY [who conducted the diagnostic interview]: I'm presenting Brian Murray. He's thirty-four, white, [and] he shows effects of long alcohol abuse. He's got 30 percent disability for PTSD. He came [to this medical center] asking for alcohol treatment, but they put him in 32A [the acute psychiatry ward]. In 32A, he was uncooperative and they decided to discharge him. After he was discharged, he went back to living in the woods [a state park not far from the hospital], in a tent or some kind of a shack with another vet who's an alcohol abuser. He lived there for a while longer but deteriorated badly and had to be readmitted, this time to 26A [the alcohol and drug detoxification ward]. This is where he is now.

Presenting symptoms: He's alcohol dependent. His chief complaint is nightmares. He also talks about not fulfilling his responsibilities to his wife, to his kids, and to society. He says he wants to work, and that he's worried that the U.S. is on the brink of getting into another war. He says he's worried about where he stands with God. He also says he's unhappy about not participating in life, that he spends his time boozing out in the woods. He said he had a nightmare the night before he came from 26A for the interview with us.

His dreams are about his military police job in Vietnam. He was in the marines. He trained as an infantryman, but when he got to Vietnam, they made him a military policeman. His first job was to patrol the perimeter of a military base. On this base, there was a team of U.S. [marines] and Vietnamese interrogating military prisoners and civilians. When Brian was on patrol, he could hear screaming from the rooms where the interrogations were taking place, and he knew that very bad stuff was going on. Next thing that happens to him is that he's taken off the patrol duty and told that his new job is to assist with these interrogations. This is where he has his traumatic experience. His officer tells him that his job is to monitor the interrogations, to make sure that they don't violate the Geneva Conventions. He didn't want to have any part of this, but they didn't give him a choice. Once he's in the room where they're doing the interrogations, it turns out that they don't want him to interfere. He's very upset by what he sees—people being tortured—and he can't do anything because he's outranked by everyone. At the end of the first interrogation, one of the interrogators, a marine sergeant, pulls him over and says to him, "You've only been in-country for four months and you don't know anything. You're on-team here. This is real stuff, not bullshit. You keep your mouth shut." He claims that he tried to make them comply with the Conventions a couple of times, but they ignored him, and he was eventually ostracized by the other marines. He says that, on one occasion, they were torturing a Vietnamese nurse, and he actually pulled his rifle on them and said, "I'll kill the next guy who does something else to her."

The things he saw during these interrogations have been a constant source of conflict to this man. He was brought up in a strict Catholic family; he was one of seven children. He has an excellent premorbid history. A good childhood. He went to church, he belonged to the CYO [Catholic Youth Organization]; he helped organize dances. He was a very social guy. He played on a football team and in Little League. He was also patriotic and says he went into the service because he thought "being a marine was the most honorable thing a guy could do." He idealized the Corps. There were several veterans in his family: his father, his brothers, his uncles. And he felt no conflict over the Vietnam War—no "Peace Now" stuff. He enlisted in the marines, and he expected to do whatever they asked him to do. Things began to change in Vietnam.

Brian says he didn't drink while he was in the service. His nightmares are extremely severe. He says he tries to stay awake because he's afraid of going to sleep.

He had a classic homecoming. His girl was waiting at the airport; one of his brothers came from out of town to welcome him home; the family made a party and invited relatives and neighbors and friends, et cetera. But he says he was in a state of shock when he got back to the U.S., that he couldn't stand being around these people while they were partying, and that he left his own party. For a while, things went along okay. He resumed his preservice job, working in a local supermarket. But he says that he couldn't take anything seriously and that he wanted to laugh at the people around him. After a while, he started sneaking away from work at odd times. The girl at the bar would know he was coming and she'd have two drinks and two beers waiting for him. At first, he was drinking only at lunch. Then the drinking accelerated: he'd drink more often and more alcohol each time. As a result of the drinking, he got into several [automobile] accidents and got three DWIs ["driving while intoxicated"]. He says his drinking botched his chances for moving up to a better job, and it screwed up his marriage.

He's honest and he didn't try to manipulate me during his interviews. He talks openly about his feeling of guilt. He says he feels he could have done things differently in Vietnam, that things didn't have to turn out the way they did. He thinks that drinking is one of his flaws. He says that, when he's drinking, he turns into another person and that he becomes violent. He's a little guy and slight, and he picks fights with big guys and he gets beat up a lot. According to him, he has a distorted consciousness of events when he's been drinking. He says he discovered himself on one occasion walking down the street, wearing cammies [camouflage fatigues]. When he became aware of where he was, he couldn't remember how he got there. Another time, he walked up to a recruiting station and broke the front window, because he had just seen something about Cambodia on TV, while he was in a bar. After a while, his wife started drinking excessively too. He had a severe flashback that he doesn't remember but that his wife told him about later. During this flashback, he held his wife and kid hostage, because he said he saw enemy soldiers in a tree line across the street from his house. I believe what he's telling me.

He says, "If I drink or I get high, I don't dream. The longer I'm sober, the worse things become."

Brian described to me the episode about the nurse being tortured. He gave details, and the event was horrible for sure. A while ago, he looked through the letters that he wrote to his mother when he was in Vietnam, and he says that he was able to piece together the chronology of events surrounding this episode by reading between the lines. He told me that he has documents [letters?] for everything that happened to him in Vietnam. He says he has the [tortured] nurse's name and that he plans to write to her. He also says he doesn't know what happened to her.

Brian reports a second traumatic event, involving an MP who was shot in a gun battle with a drug dealer. When Brian got to the scene, this MP was dying, and the Vietnamese guy was dead. Brian describes his body as being cut in half by the MP's bullets. This scene really knocked Brian for a loop, he says.

These are his current things: He feels bad about not going to his son's graduation, and about going downhill. When I first saw him, I thought he must be fifty, but he's only thirty-four. It's clear that he wants to cooperate [with us], even though he says that he mistrusts the VA. He's an articulate guy. He described for me a dream he had after drinking heavily. In this dream, he's escorting the president on a patrol. He says, "I felt ridiculous when I woke up." He says that when he had the money, he drank every day. He's been in and out of Alcoholics Anonymous since 1973.

After he left the service, he says he couldn't readjust. He decided that he had to reenlist in order to "get things straight." We see this with some of our guys: it's a typical attempt to master the [traumatic] event. So he reenlists, and he's assigned to teach guys riflery on the firing range. While he was on the range, a recruit messes up and kills another guy in his sector. He says that there was a congressional investigation to discover who was to blame. At this point, Brian felt as if life was repeating itself for him. But he was cleared by the investigation—absolved of responsibility.

Brian's having trouble with his short-term memory. He's dreaming a lot. He values social relations very much. He likes the idea of having neighborhood friends, people who recognize him and that he can socialize with on a day-to-day basis. He says he liked being part of a large family. He has a strong sense of ethics and a strong superego.

He had a third notable event in Vietnam. On this occasion, he applied mouth-to-mouth resuscitation to a Vietnamese, while the man's family watched him and prayed. He went to the dispensary later the same day to check up on the guy and was told that he died. The man was a victim of a Viet Cong satchel charge [explosives]. Afterwards, Brian wrote a letter to his mother saying, "I'll never forget this night." He says that when he rereads this letter now, he's surprised that he was such a strong person *then*. He says he can't believe how weak he's become. Brian told me, "I had a lot of contact with Vietnamese people as a result of my job, and I thought of them as being a very interesting culture. When I came back to the U.S., everything was silly bullshit. Everybody I came into contact with seemed so childish. I used to save lives in Vietnam and back in the U.S. I was stacking milk." He says that when he came back to the U.S., "the geography looked the same [as before], but the people were all clones." He says he went back to his job in the supermarket three days after returning from Vietnam.

Recommendation: Brian's a clear case of PTSD, and he should be admitted [to the center] as an inpatient. At this time, he's being treated for substance abuse [as an inpatient] on [ward] 26A.

GORDON [the unit's psychometrician]: His tests are strong too. He has a surprisingly low F scale. Strong ego strength. His 4-scale [psychopathic deviate] is a little high, centering on the clusters for social alienation and self-alienation. His 5-scale [masculinity-feminity] is high, [but it's] not clear why.

MALLOY: There's a big male inadequacy issue here. It's reflected in his pattern of picking fights with big guys.

DUROCHER [psychiatrist and clinical director]: And there is the identification issue with the tortured nurse: it's a memory filled with pathological tenderness.

GORDON [continuing with his review of the psychometric data]: His 9-scale [hypomania] is also up a bit. But altogether it [the MMPI results] looks good. His PTSD scale is thirty-four. His dissimulation scale is low for our guys. He peaks on depression [and within this scale, on] guilt. There's a modest peak on mistrust. He's low on somatics [hypochondria scale]. This is consistent with his high guilt, [and it shows] he's accepting responsibility.

MALLOY: He's one of the best candidates I've seen here.

DUROCHER: You say he was shattered by his experience?

SAMS [head of the inpatient section]: Where does he place what happened on the firing range?

MALLOY: He says he feels that he should have been able to prevent the guy's death.

[The session ends.]

The Case of the Proud Marine

WYLIE [who conducted the diagnostic interview]: I'm presenting Robert Treadaway. He's got a 10 percent disability [for a] Vietnam-connected head injury. I was impressed by the very great details that he gave when he filled out his forms. Also, he brought a lot of [war] memorabilia with him. At first I thought that he might be a "professional Vietnam vet." But I spent all day with him, and I'm convinced that he's authentic.

His presenting symptoms are headaches, dreams—not nightmares—[and] he's depressed. He says that he thinks about Vietnam all the time. There's also evidence of some acting-out [i.e., the expression of a repressed emotional conflict in action rather than words] and anger. He's currently taking two minor antidepressants. During the interview, he had difficulty staying on task.

His background: normal psychological development. No drug use before or after Vietnam. His father was an ex-marine, and he [Treadaway] joined the inactive [marine] reserves while he was in high school. He was an average student. When he heard about the Gulf of Tonkin Resolution, he decided to enlist.[1] He served two consecutive [thirteen-month] tours in Vietnam as a sergeant. His first tour began in 1966. He headed a community action team in a Montagnard [tribal highland] village. He says that his first tour was easy—no events worth reporting. He says he gradually adjusted to the new environment, that he learned the tricks of his trade—things like identifying booby traps, learning how to avoid places where ambushes could happen—and he taught these tricks to his men, and he was able to protect them. In the second tour, everything changed. He was a squad leader, under the command of a platoon leader and a company commander, and he was in the middle of a lot of combat against North Vietnamese regulars. It's in this second tour that he says he had problems.

He received two Bronze Stars [for heroism] and three Purple Hearts [for wounds]. He was recommended for the Navy Cross [a decoration for outstanding heroism].

During the interview, he spoke only about Vietnam. These are some significant quotes. He referred to the men in his platoon as "my men," and he hugged himself when he said this. He referred to himself as "their daddy." Off and on, he was tearful, but he couldn't explain why. Several times he told me, "I am not a warrior."

When I asked him to describe his most memorable event in Vietnam, he told me about several events. In the first one, he was in charge of a night ambush. He was leading a team of twelve men. After the ambush was over, he counted his men in the dark: when they passed him on the trail, he tapped each man on the shoulder. He counted twelve men, including himself. But he says he saw two more men coming along the trail. When he saw this, he says that he knew that his team had been infiltrated by enemy [soldiers], but he there was no way he could be sure which two of the men were enemy—whether they were in the middle or the end of the column. He says he could write a book about what went through his mind in the millisecond before he decided what to do. He shot the last two men dead, and they turned out to be the enemy.

He described a second incident in which he was accused by a marine lieutenant of mutilating a prisoner. The lieutenant and a sergeant were in charge of a documentary film crew that was filming combat involving Treadaway's company. According to Treadaway, after the main combat was over, he saw a Vietnamese soldier start to come out of a spider hole [a concealed firing position]. He says he saw this happening, but he didn't have a weapon close at hand. So he grabbed an entrenching tool and decapitated the Vietnamese with it [striking him from the back]. Treadaway was supposed to be brought up on charges, but his colonel did not press them.

On another occasion, Treadaway says he was accused by another marine of shooting a boy off of a water buffalo. He describes a fourth occasion, where he saw an old Vietnamese man and a boy coming along. He waved to them, indicating that they should continue to move on, that it was safe. As they moved along, a [helicopter] gunship came up behind them and shot both of them.

His demeanor was matter of fact during the interview. He cried a couple of times, but he couldn't identify a particular feeling. He also said that he participated in at least one "body race." He told me this in a very matter-of-fact way, like everybody knows what a body race is, the same way that everybody knows what football is. No explanation, no nothing. I stopped him and asked him to explain what a body race is. He said that after a battle with North Vietnamese troops, he and the guys in his unit collected the bodies of dead Vietnamese. They smashed their pelvises and spines with entrenching tools, tied up the bodies into balls with their belts and straps, and then rolled them down the hill. This is what he calls a "body race." When I asked him to talk about his feelings about this—about what he and the other guys did to the bodies—he said he doesn't have any particular feelings about it. I got the impression that this race had been some kind of amusement after the fighting. I asked him whether it could be called an atrocity, and he said that he doesn't call it an atrocity.

He describes another occasion when he was wearing an NVA [North Vietnam Army] hat that he had found in the pack of a dead enemy. His captain saw him wearing this hat and told him, "Take it off or I'll shove it up your ass." Treadaway took it off, but the captain was still angry and said to him, I think I'll shove it up your ass anyway."

He also described an occasion when he was bitten by a rat in Vietnam. He was sent to a field hospital and given rabies shots, but he returned immediately to be with his men. I have rarely met anyone who was so concerned to be with his men. This was also his reason for asking to be sent back to Vietnam when his first tour was over.

One of his Purple Hearts is the result of an injury he suffered when a rocket-propelled grenade exploded against a tree that he was lying next to. As a result of this explosion, he bled profusely from his eyes and ears, and he couldn't see or hear for a while. He says that his chronic headaches began soon afterward.

He also reports an incident in which he dropped a grenade down a spider hole. He says that he held it until it was ready to explode and then dropped it in. After it exploded, a Vietnamese began to emerge from the hole. Treadaway describes the man as dragging his upper body up from the hole and says that it was no longer attached to his lower body. He describes the man as "growling" at him. He says that he had "humongous" arms and tried to grab Treadaway's rifle, and that Treadaway shot him.

He brought notebooks full of details about his tours in Vietnam to the interview. The notes included the dates on which different men in his unit died and other details.

DUROCHER [psychiatrist and clinical director]: What do you think this means?

WYLIE: I don't know. I asked him what was the most traumatic event that happened after he returned from Vietnam. He described an incident in which he was thrown to the ground by a peace demonstrator who tore his uniform. Treadaway's comment was that the demonstrator's behavior was not consistent with his peace ideology.

He's been married for seventeen years. He has an excellent relationship with his wife and three children. Since Vietnam, he has had eight jobs. The longest job lasted seven years. He says that the continuous and severe pain that he suffers from his headaches interferes with his work, and it is because of this that he's had to quit all of these jobs. He has documentation from his former employers to support this claim about his chronic pain. He says that the headaches have grown worse over time. He is currently unemployed and on public support.

In 1975, he was working as a draftsman. He says that he'd been talking to someone at work about his time in Vietnam and this person asked him to bring in photographs that he said he took there. Treadaway brought in some pictures and let the man borrow them, but this guy let other people at work see the pictures without telling Treadaway. One of these photographs showed Treadaway holding the decapitated head of a Vietnamese, with Treadaway's fingers stuck into the eyes. From that time on, people at work avoided him—wouldn't talk to him or look at him. Treadaway says they didn't understand him.

I tried to reach Dr. Barrington, who referred Treadaway to us from the Cedar Valley VA. I also phoned Treadaway's wife, and she confirms everything he told me about his problems and his symptoms.

Treadaway had a confirmed flashback about six months ago. He says he was sitting at the bottom of a hill somewhere and a clump of earth rolled down. He thought [imagined] it was a grenade, and he jumped into a ditch and put his hands over his head. Also, he says that when he watches TV and sees Time-Life advertisements for books on the Vietnam War, he cries. [The advertisements include film clips showing U.S. combat troops.] But he can't say why. He says he feels different from other people, but he can't describe what his feelings are.

DUROCHER: On the other hand, he told you that he can write a whole book about the feelings he had when he was on his ambush. It seems that he can talk about his emotions only in the context of the moment they occurred—in the context of his war experiences. Do we know anything about his parents? Did one of them die recently?

WYLIE: His father is deceased.

. . . I was curious about his second Bronze Star. He received this in 1981 [for an event that took place in 1968], after the marines found some of his war records. I wondered whether getting this medal made him a small-town hero. His wife told me that when he got this [second] Bronze Star, Treadaway was unemployed and said to her that he was no hero if he couldn't get a job.

GORDON [the unit's psychometrician]: Can I give his test results now? His SCL-90 [symptom check list] is typical for PTSD. He is very high on intrusion. Arousal is also high, but not in a manic way. Somatization is high [and is] connected with his headaches. On the [MMPI] PTSD scale, he is thirty-nine. This is typical for our guys. There is little dissimulation but high alienation and a low sense of propriety—we see this reflected in his showing his Vietnam decapitation photographs at work. When we break his high alienation into subscales, we find no evidence of sociopathy.

WYLIE: Treadaway has had a recent neuropsychiatric evaluation of his headaches at Metropolitan [VA], but the evaluation is incomplete. We're still waiting to hear from his doctors at Cedar Valley [VA] about his physical condition. Treadaway says that the doctors either showed some interest concerning his spine, or they have some information about his spine—

DUROCHER: There's no connection between his spine and his headaches. There's no organic basis for his [pain] problems.

WYLIE: Treadaway was first interviewed and evaluated at the Northfield VA. This was before he started going to Cedar Valley. He says that the therapist there seemed to show a real interest in his problems and asked Treadaway if he could visit him at his home. Treadaway assumed the therapist thought that this would help him to identify and treat his problems, and he said yes. When the therapist came to Treadaway's house, he tried to sell him Amway products.

[Laughter]

MALLOY [who has also interviewed Treadaway]: Treadaway makes these frequent references to "my men," to his being "their daddy," to the fact that he "had to go back to Vietnam for a second tour of duty," et cetera. Compare these feelings with his sense of alienation since the war. I think that this is good evidence that he could benefit from being able to communicate his—

TIMKEN [another therapist in the outpatient unit]: He's one of the guys who went direct from combat back to his hometown in forty-eight hours.

MALLOY: He impresses me as being a mature guy. His decision to enlist, to go active, shows some planning, that there was decision making going on, [and that he was] not impulsive. His first step is joining the reserves, and then he's motivated to go active after he hears about the Gulf of Tonkin Resolution. Everything he says and everything we know about him makes it clear that he was a proud marine.

WYLIE: What is our recommendation, then? He needs inpatient treatment.

DUROCHER: What about medication? Is he receiving any?

WYLIE: Minor tranquilizers. I'll have to get in touch with the people at the Cedar Valley VA . . . with his psychologist.

MALLOY: You know, the context of his Vietnam experience is really different from many of the other guys who come to the center. He was proud of what he did. He has no great conflict regarding what he did.

SAMS [head of the inpatient unit]: Exactly what was his traumatic event? I'm not sure. Was it that he had to leave his men behind at the end of his tour of duty?

BAUER [a therapist in the outpatient unit]: He seems conflicted to me. He's a marine sergeant with lots of combat, he leaves Vietnam, and he's assaulted by a hippie in San Francisco, and he doesn't do anything about it!

WYLIE: He said at least twice—without any prompting from me—"I am not a warrior."

BAUER: He also committed atrocities . . .

DUROCHER: He seems to be an intelligent man who—

WYLIE: What about admitting him to [inpatient treatment at] the center?

DUROCHER: The only concern is his medication. He must know [from the information provided to him] that the structure of the [center's inpatient] program is to approach the [PTSD] problem in a different way [i.e., through drug-free milieu therapy.] He can't be using minor tranquilizers [while he is in the program].

MALLOY: Why is this a problem with him?

DUROCHER: How dependent is he on his medication [for managing his somatic symptoms]?

WYLIE: You are assuming that his headaches are psychosomatic.

DUROCHER: I am not assuming. He is suffering from an obsessive-compulsive character disorder. His story about the North Vietnamese hat [and the captain's threat] shows this. [Laughter follows this remark, but it is unclear whether Durocher intended to make a joke.] There is a failure on his part to deal with his aggression. That is why I asked about [the death of] his parents, to learn if he had suffered a recent loss. His decision to enlist on the basis of the Gulf of Tonkin Resolution [i.e., to strike back at the North Vietnamese] underscores his aggression. It is not pathological aggression, but it does lead to depression [in this case].

BAUER: Look, time is running out, and we have two other men to review today.

DUROCHER: Let him into [the inpatient program at] the center once he has given up the minor tranquilizers.

[This is followed by some discussion concerning where and how Tread-away should be taken off his medication, and the session ends.]

The Case of the Metaphorical Event

JENSEN [who conducted the diagnostic interview]: This is an update on Edward Hungerford. He's a thirty-three-year old white guy and has been admitted to 32A [the acute psychiatry ward] for attempted homicide and attempted suicide. He'd been at a bar and got into an argument with the bartender and went home to get a gun. Somewhere along the way, he tele-phoned his mother and told her that he was going to shoot up the bar and then kill himself. She called the police. The police arrived at the bar before Edward hurt anyone, and they took him to the VA.

Edward doesn't drive and his mother's got to chauffeur him around. He has a history of his mother sabotaging his appointments with us. Either she brings him up to the Metropolitan VA instead of down here or, when she does bring him here, it's on the wrong day and he can't see his therapist. And when he does meet with his therapists, she tries to draw them into their pathological relationship.

I think that he had some very serious problems before his combat-related PTSD. It seems that, when he was a child, he was sexually and physically abused by his stepfather. Edward has a lot of hostility for this man, but he denies having any anger against his mother. On the other hand, his mother gives me the impression that she feels guilty about the childhood abuse and tries to deal with it by projecting it onto his care-givers. Her usual com-plaints are that we're not doing enough for him, or that we're continuing to screw him up on drugs, or that we're not involving her in his treatment.

DUROCHER [psychiatrist and clinical director]: Are you implying that his [PTSD] reenactments—for instance, this episode at the bar—refer back to his childhood trauma with his stepfather?

JENSEN: That's correct. His reenactments began even before he got to Vietnam, and they led to a couple of Article 15s [disciplinary actions]. It was partly because of his acting-out behavior that he was sent to Vietnam when he was. And it wasn't too long after he got to Vietnam that his acting-out landed him in the stockade. But he did also have new traumas in Viet-nam. One was when he was locked in the stockade and the base received a rocket attack, and he had to sit it out. The second event resulted from his work as a medic.

DUROCHER: How do you sift the two sets of traumatic events? How do we know that the contents of the Vietnam events [as he experienced them] don't reflect the experiences with his stepfather?

JENSEN: Well I *do* see the Vietnam events in relation to the stepfather.

And I see it in Edward's attitude towards the VA. He says that he thinks that the treatment he is getting at the VA is punitive. He also says that his mother is forcing him to undergo this treatment and that she is trying to get him to sign a contract with her saying that he promises to comply. And at the same time, the two of them are collaborating with each other to sabotage his appointments.

I think he's treatable, but there are a couple of things that need to be recognized before we can get started. First of all, we need to understand that his Vietnam experiences are a metaphor for these other [childhood] traumas and whatever benefit he gets [here] from therapy isn't going to be for combat PTSD. Second, he's being released on a slew of medications [including antipsychotic drugs]. Building 3 [location of the acute psychiatry ward] wanted to release him as a paranoid schizophrenic, and I argued against this diagnosis. But it's clear that he does have terrific borderline personality [disorder] features. He has a new MMPI for this evaluation. His original diagnosis at the center was made by Kenny Morgan, when he was still here. It says Edward's got PTSD with borderline features.

GORDON [the unit's psychometrician]: The new MMPI is very consistent with borderline. His old MMPI was full of 8/4 stuff [peaking on the schizophrenia and psychopathic deviate scales] . . . low depression . . . high paranoia. The new MMPI is very similar. His PTSD scale is now forty-one—it's gone up a couple of points from the first MMPI.

DUROCHER: Does he have Vietnam intrusive imagery?

JENSEN: He says that he dreams about being a medic and having to tell his patients that they have lost their arms and legs. He told me that this was his actual job in Vietnam. But I was a medic [in Vietnam], and I never had to tell guys that they had lost their limbs. That was strictly the doctors' job.

DUROCHER: In other words, we have a contaminated nightmare. It is the degree to which his domestic trauma has influenced his PTSD imagery that should determine whether or not we take him as a patient.

JENSEN: Well, let's take a look at his flashback phenomena. He says that, on one occasion, he was out walking, two or three in the morning, without knowing it. On another occasion, he poured gasoline down the middle of the street and set it on fire. Neither of these episodes parallels any of his Vietnam incidents. But then there are his dreams, where he's being mortared or shelled; and he says that he was shelled while he was locked up in the stockade—so there is some connection here. He also has a second dream in which he sees people's faces and he wakes up sweating and yelling, but this sounds pretty vague.

DUROCHER: Does he have any nightmares about his childhood?

JENSEN: He won't give any information about his childhood.

DUROCHER: Should we make him an outpatient?

JENSEN: I wish I had your ego boundaries, Peter. If I did, then I'd send

Edward to some other VA. Let them handle him. But I feel that if we send him somewhere, they'll see that our diagnosis is PTSD, and that'll just cloud the issues, and his treatment will become even more enigmatic than it's going to be here.

DUROCHER: I may [eventually] reach the same conclusion. But even if he is withholding information about his childhood experience, it is obvious that he is giving preference to his Vietnam experience in what he tells us. It may be that he is using his childhood experience because it resembles his Vietnam experience. What he says and remembers about his relations with his stepfather could be an expression of his Vietnam experience. Look at the content of his imagery. At night, in his nightmares, we can see his spontaneous expression of his traumatic event. And what are the nightmares about? They're about Vietnam. This observation of the spontaneous material makes it possible to interpret the content of his daytime statements and behavior, over which he has more conscious control. And the [correct] conclusion is that his Vietnam experiences are being expressed through what he thinks and says about his childhood events.

JENSEN: Well, it's true that he does give preference to his Vietnam trauma. On the other hand, someone could say that, if he didn't express himself through the Vietnam trauma, then he would have to acknowledge his feelings of hate and anger toward his mother—and that's what he really wants to avoid. A lot of what Edward is saying about his Vietnam experiences sounds like a metaphor for his relations with his stepfather.

DUROCHER: I agree with you about the metaphor. The question is, What is being a metaphor of what? I think that we can admit him [into the outpatient program] as someone with Vietnam-related PTSD.

JENSEN: I don't see why he can't be both [a victim of childhood trauma and war-trauma].

DUROCHER: His PTSD is a *mixture* of both.

SAMS [head of the inpatient unit], [addressing Jensen]: Can you rule out psychosis?

JENSEN: There are no reports of a psychotic break with reality. He says he hears voices, but it's "quasi-psychotic," not the real thing. [Ward] 32A ruled out schizophrenia on these grounds and decided on borderline.

DUROCHER: We would have to reduce his medication [if we admit him] for outpatient. And if he had psychotic phenomena [in the future], we would have to refer him back to acute psychiatry. Are you [Jensen] ready for that?

JENSEN: Well, yes. He's got no frank psychotic symptoms, and he's been previously diagnosed for service-connected PTSD by us and by a [VA compensation] rating board. So we're obliged to treat him. And then, if he refuses to comply with our rules or if it turns out that his [borderline] behavior makes it impossible to treat him, we can transfer him.

DUROCHER: When you make contact with him, tell him that we expect him to keep his appointments.

JENSEN: Yes—and that we want urines too [to monitor his alcohol and drug use].

[The session ends.]

The Case of the Shady Event

MALLOY [who conducted the diagnostic interview]: My guy is Freddy Williams. He's thirty-five, black, and currently on the drug and alcohol treatment ward here. He's something of an enigma. That explains why so many of us have seen him here. In Vietnam he was a clerk, stationed near Saigon and out of any real danger. According to what he says, he has an excellent premorbid history. He was a good student, a bartender in a country club, an up-and-comer, ambitious. Then he went to Vietnam, and he gets hooked. One sad story after another. As soon as he gets to Vietnam, he's anxious, even though he's a clerk and he's living on a large military base away from the action. He can't get over the idea that there are people out there who want to kill him. His fear gets to the point where he's depressed and crying, and he wants to do anything to get out of Vietnam. He goes to his CO [commanding officer] and to several NCOs [sergeants], and no one wants to listen. Then he goes to a psychiatrist on the base. The psychiatrist tells him, "You'll have to take it like everyone else here." In our interview, Freddy tells me, "I was just afraid. First couple of months in Vietnam, I would hear choppers overhead and I'd think it was VC. I didn't know at that time that they didn't have choppers." After the psychiatrist turned him down, Freddy goes to his buddies. They turn him on to heroin, and his depression and anxiety lift. Then he gets deeply involved in the social life connected with drugs. He gets into the black market and says he squirreled away eighteen thousand. During this period, he has his most traumatic event, the death of his buddy, named Thomas. On this occasion, Freddy and Thomas go to a whorehouse and get some transactions going. They're also drinking. Next thing Freddy knows, there's a lot of shooting and the top of Thomas's head is blown off. Freddy doesn't know how this happened or who was involved—whether it was cowboys or Vietnamese hoods—but he runs out, leaving Thomas's body lying there, because he doesn't want to be connected with the criminal activities that were going on in this place.

This is 1972, and there's a lot of racial stuff going on and a lot of decadence. He finishes his tour in Vietnam, and just when he's about to be discharged, his drug problem is discovered and he's hospitalized for two weeks. From his release from the hospital in 1973 until today, he's into constant heroin use and other kinds of drugs. Also into antisocial behavior of every kind. Lots of fights. Either he's being beaten up or he's beating on someone else.

When he comes here, he presents himself like a zoned-out junkie. That's on the one hand. But when he dries himself out a bit, he appears to be an intelligent man, with insight into his behavior. He talks about his "passive aggressiveness," his need to buckle down and to get off drugs. He says that he doesn't know if Vietnam did this to him, but he knows he was a different kind of person before Vietnam. He says, "I took on another guy in Vietnam, and I'm still struggling to protect him—with drugs." He says he wants to get into the PTSD inpatient unit. He thinks that being with other vets will help to crystallize his Vietnam issues. When he says this, I guess he's talking about his guilt for being a clerk in the rear and about his connection with Thomas's death.

But all of this is giving him the benefit of the doubt. There are a lot of shady things going on here. He still hasn't brought us his DD 214. He says he got the Bronze Star [for heroism]. I said to him, "A Bronze Star for being a clerk?" His answer was vague—if you know what I mean.

BAUER [a therapist in the outpatient unit]: He told me he got it by doing favors for a colonel. That's not beyond belief, you know. He said he stashed away 18k. What did the colonel get?

MALLOY: Freddy's got some redeeming features: he can engage in a therapeutic context. But he's a tough cookie and he hasn't shown any improvement over years of treatment. He's got a lot of characterological problems; his life style is heavy drug abuse. I don't think that I can add much to Bob's original assessment.

BAUER: I did follow him up with Ted Michaelson. Ted was working in the drug and alcohol unit when Freddy was there. Ted says that Freddy uses jails and hospitals as "structures" to keep himself together, because he's a dependent person. When he doesn't have this structure, he's a goosey-loosey, off the walls. Also, Freddy spent some time in Chicago a couple of years ago in a methadone program. Ted thinks he left Chicago not so much because of problems connected with drug deals as because of a homosexual panic. He was living with a guy, and he had problems. I really doubt what he says about his premilitary life. Was he really a nineteen-year-old bartender? From his description, I think he was a town stud; he had lots of money and girls. Maybe I'm a cynic, but I think that he was into drugs and alcohol before he went into the military.

GORDON [the unit's psychometrician]: His premorbid account sounds like a story of a narcissistic loss: "I was on the right track, I was doing everything right, and then—Vietnam hits me." Unbelievable. But it's clear that he felt disastrously derailed.

MALLOY: You're right. And now he's grieving over his lost self, his lost ego ideal. He has nightmares . . . but considering his lifestyle—

TIMKEN [a therapist in the outpatient unit]: He sounds like a "street soldier."

GORDON: Can I tell you about his tests now? Okay, he's taken the

MMPI twice; the second time was this week. The first one was exaggerated, all over the place, completely uninterpretable. I asked him to take it again, and I told him that he doesn't have to exaggerate, that we need to know what's really bothering him, so that we can find out what's causing his problem, et cetera, et cetera. On the second test, his PTSD score went down from forty-three to forty. Also his F scale is down; his dissimulation is down. And now he's got a typical PTSD profile. The question now is whether there is characterological stuff [a personality disorder] or whether he's crazy [psychotic].

MALLOY: I think there's narcissistic loss. Before he goes into the service, he's a town stud, a big man. The fact is he *was* an up-and-coming guy, but [in Vietnam] it turns out he's a coward.

GORDON: His MacAndrew [MMPI Alcoholism Scale] is seventy-five. Very high. He's been in and out of alcohol programs.

BAUER: He does very well when he's in a program and there's structure. The problem is that he has no structure of his own, and so he reverts as soon as he leaves a program. He's a *very* dependent guy in terms of someone else structuring his life for him. The other question is, Does his MMPI tell us anything about sexual identity problems? Can he handle programs like ours okay?

GORDON: The testing doesn't show this.

BAUER: Maybe Ted was wrong about Chicago.

FOWLER [head of nursing]: Is he suicidal?

MALLOY: He has suicidal ideation in the past but no suicidal behavior.

GORDON: He had a dream of a man chasing him with a .38 pistol that would only spray hot water.

BAUER: Peter [Durocher, who is psychoanalytically oriented] would have a lot of fun with that. [Laughter] Freddy's compliant when he's in a program. Is there any compensation [disability pension] issue?

MALLOY: Currently, he has zero dollars. He has no service-connected disability, and he says he's not interested in compensation. But this guy's a con man from the word "go."

BAUER: I wouldn't take him from Hammond [the VA medical center in Freddy's town] until we get his compensation file from them. I'd like to see whether he got the Bronze Star.

MALLOY: He's living with his mother; he's never been married.

BAUER [looking at Freddy's file]: This guy was heavy into hallucinogens: two hundred trips in the early seventies."

MALLOY: What if he got the Bronze Star and he's got a clean compensation file? Do we take him into our inpatient program then?

FOWLER: He does well in programs, but he doesn't seem to benefit from them. He was in a program here for eleven weeks. Can we make a dent on this guy?

MALLOY: I'm a cynic. I'd rather not see guys like this altogether. Character disorder—drug abuser—needs constant limits. That's it, but I try to keep an open mind. What if we can help him examine this massive traumatic injury in Vietnam—when he runs out on his buddy who was killed and he discovers that he's a coward? We can get him started reconstructing this experience and get him started on resolving his conflict about his manhood in Vietnam [and do this] in terms he would respect. We can give him the benefit of the doubt. He does have a capacity for describing events that would work well in group [psychotherapy], and we have a structure of ideas he could use. But I don't see any reason for seeing him as an outpatient until he gets some cognitive notion about why he's coming here. I've seen him two times on an outpatient basis, and both times he's been fucked up.

BAUER: But these other programs he was in [also] have cognitive elements. If you're saying that he's got to deal with *underlying* cognitive stuff—for example, his guilt about Thomas—well, the fact is that he hasn't been able to cognitively transform anything.

GORDON: This will be his first time to be treated for PTSD, though.

BAUER: He's currently off heroin. He's getting a pharmaceutical substitute injected into his jugular vein.

SAMS [head of the inpatient unit]: I'm very dubious about him. How long is this guy likely to stick with the [inpatient] program? Also, his traumatic event didn't occur in combat. The guy gets involved in a felonious episode that he entices a friend into, and then the friend gets waxed. It's what Bill [Malloy] said: his intrusive events aren't about Vietnam, even if there are some Vietnam-like motifs. And this guy has lots of stuff in the present to intrude into his mind. There really are lots of people out there looking for him.

MALLOY: I don't think he's got PTSD.

TIMKEN: Does he meet diagnostic criteria for PTSD, Bill? [i.e., Does Freddy meet *DSM-III* criteria, based on the clinical interview?]

MALLOY: Sure! According to what he says about himself, he does.

BAUER: I don't know about the PTSD diagnosis. Marginally he meets the criteria—

MALLOY [To Gordon, the psychometrician]: Do blacks score differently on the MMPI for PTSD?[2]

GORDON: We don't know for the PTSD population. If you look at the 6-scales, 4-scales, and 8-scales which—

MALLOY: I think I'm telling 90 percent of the black guys I'm seeing here that they don't have PTSD. Not that I know what that means.

SAMS: It's a good question.

MALLOY: This isn't just a clinical thing, either. I think that every black I've told he doesn't have PTSD has written a letter to the president.

GORDON: I always go overboard with the black guys, so that no one can ever say there was racism.

BAUER: Well you can go too overboard too. I've got a black guy that I'm presenting here on Tuesday, and he's a hell of a lot better on PTSD than Freddy is. The guy's got legitimate stressors.

GORDON: Well, I want to call attention to the fact that there are selective things I attend to and that I avoid and—

BAUER: We all have those things. But in this case, we're seeing loud and clear—I'm sorry, I can't support a PTSD diagnosis. We could do [i.e., interview, test] a population in the state penitentiary and get the same guilt that Freddy's telling us.

MALLOY: I'm looking down the line. One day we're going to have people coming around here and they're going to ask us to account for the guys we've refused. In a lot of cases, we can't give a coherent explanation [to outsiders] about exactly how we came to the conclusions that we came to. How we chose which guys, why we rejected other guys.

BAUER: Let's say someone writes to his congressman, and there's an investigation. I don't think that our statistics are going to break down according to black versus white. We have a smaller proportion of blacks in the program for obvious and less-than-obvious reasons. The blacks I've seen [diagnosed] have split out the same way as the whites.

MALLOY: Well, what are we going to tell Freddy? We tie these guys up in a four-month-long assessment process, [and] a degree of expectation is set up and—

BAUER: I think we have this: you can tell him that in previous programs he was told to follow through the outpatient part, but he has never done this. I don't think that we should see him, because he didn't follow through with our recommendation last time, when we told him to get into the outpatient drug treatment program with Phil, at Hammond. He told us he had a "transportation problem," but his real problem was that he was in jail.

MALLOY: As Ralph [Sams] said, his stressors are qualitatively different from the stressors we see with our combat vets.

SAMS: I don't know if I can rule out PTSD. He may have it. But there's another point we have to clear up. His records show that he can't benefit from our program. I don't think he'll come clean [talk about his experiences in Vietnam] during group and individual [psychotherapy]. He'll talk all right, but we'll get nothing from him but sociopathic sincerity [i.e., characteristic of a sociopathic personality disorder].

GORDON: I feel real clearly that I can't rule out PTSD at this time. I compare his history with other guys we've accepted. His history isn't as bad as lots of guys we've accepted, and my feelings about whether he would benefit from our program are not nearly as negative as when we looked at Gary Roberts. Gary is in much worse shape than this guy.

Drugs . . . criminal activities . . . shady Vietnam traumas . . . failed programs. You name it, he's done it. You remember: he came here for the interview, his arm was broken in two places, his face was banged up, and he told us he had an "accident." Well, he's in the [inpatient] program a week now, and I'm surprised as hell. We'll see how much longer he lasts, but right now he's hanging in. I can't rule out PTSD [for Freddy], and I can't say he can't benefit from treatment from us. I feel very strongly about this.

TIMKEN: Ron [Gordon] makes a good point. The focus of Freddy's treatment over the years has been his chemical dependency and his characterological problems. If the guy's had a major trauma, it hasn't been the focus of anyone's attention. I find myself saying, His trauma occurred during a felonious activity, and I find myself asking, Does he deserve our attention? But I had the same questions with Marvin Johnson [another black veteran]. The important question is Freddy's premorbid adjustment. If this was good, then his chances for success in our program are good.

SAMS: The issue of the legitimacy of his traumatic event comes down to the question of how will he be able to interact on the ward. I can't see him going into group [psychotherapy] and saying to the combat vets up there, "My event happened when I was fucking around in a whorehouse." I'm looking at it from the [inpatient] ward's perspective.

TIMKEN: Yeah, it's true.

BAUER: You have to look at it from a program [i.e., center] perspective. I appreciate what you're saying, Chuck [Timken]. But I know the alcohol and drug program here well enough to know that he had some real chances. It's not just a detoxification program. I know Ted Michaelson for ten years and Bert Connery for over ten years [therapists in the alcohol and drug program], and I know that Freddy could have made some progress there. I just can't support him for inpatient here at this point. If you want him to continue as an outpatient—I'm not ruling this out. His PTSD is overlaid on this other stuff—*if* there is any PTSD. Also, he's having a real problem staying out of jail.

MALLOY: If we can't rule out PTSD, then I should see him as an outpatient.

BAUER: I have no problem with that—if he can stay sober.

MALLOY: I'm looking at the overall picture: political and—

BAUER: Sure.

MALLOY: If we say, "We can't rule it out"—then it sounds wishy-washy. We're supposed to be the experts. If we think he has PTSD, then we should give him our best shot.

GORDON: It sounds as if we've still got two questions. First of all, what are this guy's stressors? Are they legitimate enough? I mean, psychologically legitimate enough to warrant PTSD diagnosis? The second question is, If he does have PTSD, is it treatable, given his drug dependency, his sociopathic tendencies, and his postservice adjustment or whatever? Are these things going to make him untreatable?

MALLOY: I think it's clear that the prognosis is poor. It's not likely that he's going to follow through. But from a programmatic point of view, we should accept him for outpatient treatment and see how he progresses.

BAUER: Are you willing to see him [as his therapist] in outpatient?

MALLOY: Sure. It's better than me giving him some half-ass explanation about why—

FOWLER: Does he live with his parents, Bill?

MALLOY: With his mother.

FOWLER: So we could get more data—from his mother. He marginally meets the criteria for PTSD, we can't say he doesn't have it—

MALLOY: What we should do with marginal cases is make them outpatients until they select themselves out—

BAUER: Well, we're already doing that with Rademacher, aren't we? I'm seeing him on what amounts to an extended evaluation.

MALLOY: Administratively, we have to accept Freddy on an outpatient basis, even if the extended evaluation takes the form of therapy.

BAUER: The problem is that you can't say to someone, "This is an extended evaluation," when he starts spilling out his problems. You have to deal with them.

GORDON [Preparing to record the committee's final decision]: So is the consensus moving to the decision that he meets the criteria, that he does have PTSD?

BAUER: We said that he *marginally* meets the criteria. He wouldn't make a good patient, and so we'll treat him as an outpatient. I appreciate the diagnostic distinctions you [Gordon] are making, but for those of us who have to deal with it, as you well know—

GORDON: As I well know! I've gotten more than marginally burned on such cases—

SAMS: And upstairs [in the inpatient program] we have to treat whole people. We have to field whatever they throw at you.

BAUER: Our consensus is that Bill [Malloy] will treat him as an outpatient.

GORDON: For the record now, let me get this straight. He is marginally PTSD—

BAUER: With a lot of other problems—

GORDON: Drug dependence—in remission?

BAUER: Or is it "partial remission"?

MALLOY: Mixed personality disorder?

GORDON: With any special traits?

BAUER: Antisocial—

MALLOY: Special traits? Armed robbery . . . jailed several times. Also, when I interviewed him, he told me, "I have passive-aggressive attitudes."

BAUER: He sounds to me as if he's just plain aggressive. Guys like him go in and out of these programs, and they pick up the [psychiatric] terms. He doesn't know what they mean.

SAMS: [I] Recommend continued drug treatment.

MALLOY: Do we have to also monitor his pee-pee [conduct urine tests for drugs and alcohol]?

BAUER: That's correct. If he's going to be an outpatient, he can't be stoned.

[The session ends.]

Narrating PTSD

In the Case of the Easy Diagnosis, the presenter describes his patient, Brian, as being one of the best candidates he has evaluated at the center. The unit's psychometrician adds that Brian's MMPI profile is consistent with PTSD, except for an insignificant detail or two. There are a couple of perfunctory questions, after which everyone agrees that Brian has service-connected PTSD and should be admitted to the center as an inpatient. The session is over.

In these sessions, it is the presenter's opening account, based on his interview with the patient, that provides the context for everything that comes afterward. The *content* of the accounts changes from case to case, recounting different childhoods, different wartime experiences, different employment histories, et cetera. Their *structure* remains constant, however. Each account consists of three juxtaposed segments, corresponding to a man's premilitary life, military life, and postmilitary life. In accounts that end with a recommendation for a PTSD diagnosis, the three segments are connected by a transformative (traumatic) experience. The account is thereby transmogrified into a *narrative*, and its structure connects times, places, events, actions, and emotions into a signifying whole, with a beginning, middle, and end.

Listening to a presenter's opening account, it is easy to get the impression that narrative structure is intrinsic to the details provided by interviewees, and that the structure of a presenter's narrative is also the structure of his interviewee's life. In practice, the structure of these narratives exists prior to their content. Even before an interviewer has begun to collect his statements, the organization of his account is already in place, embedded in the composition and clustering of the questions making up his protocols. Even before his audience has heard all of the details, they know, in a general way, what is coming next and how it all fits together: the structure of the account is presupposed in their knowledge of *DSM-III*'s account of PTSD.

In the Case of the Easy Diagnosis, the structure and content of the presenter's narrative are congruent from beginning to end and the signifying whole is never in doubt.

1. The narrative's subject, Brian, preserves his moral and existential unity throughout Malloy's account. He is continuously and self-consciously responsive to a set of unchanging values that simultaneously define and confirm his sense of self. Also, the emotions that he experiences at the moment of the traumatic event (frustration, guilt, sorrow) are the same emotions that he reexperiences at the time of his interview. At no point in the narrative is there a question of "who" or "what" Brian is.

2. The segments of the account articulate on an inversion, in which the sociable innocent of the first part is transformed into the prematurely old recluse of the third. This device simultaneously unites the account into a signifying whole and verifies the magnitude of the energy released by the traumatic event.

3. The narrative is loaded with inscriptions. Some of these are nonspecific. There is the dissociative episode during which Brian is said to have seen enemy soldiers near his home, and there are references to Brian using alcohol as a sedative and soporific drug, to insulate himself from intrusive thoughts and images. But other inscriptions recall details from his traumatic experience: the point at which he reenlists in the marines in an attempt to remaster the traumatic event; the point at which he symbolically repeats this event on the rifle range; the graphic content of his dreams.

At first glance, the narrative that introduces the Case of the Proud Marine seems similar to Malloy's narrative in the Case of the Easy Diagnosis. A close look reveals a significant difference, however. In Malloy's account of Brian, structure and content are congruent, while there are puzzling incongruities in Wylie's account of Treadaway. Wylie's account makes three things clear: Treadaway participated in intense combat during his two tours in Vietnam; he reports that several of his Vietnam experiences were memorable; and Wylie suspects that one or more of these experiences was traumatic. But which experiences? Was it the patrol on which Treadaway shot two infiltrators? As Treadaway retells this experience, it is memorable mainly because he *avoided* a traumatic event, namely, killing his own men. Was it the (possible) atrocities that he mentions—the "body race," the decapitations, the boy who was shot off a buffalo, the two civilians who were gunned down by a helicopter? But Wylie reports that Treadaway was unaffected by these events when they occurred and that he was entirely matter-of-fact and unapologetic when he recounted them during his interview. Someone else at the session describes Treadaway as "proud of what he did. He has no great conflict regarding what he did."

After Wylie finishes his account and makes his recommendation, Sams asks him, "Exactly what was his traumatic event? I'm not sure. Was it that he had to leave his men behind at the end of his tour of duty?" Wylie's reply is forestalled when someone interjects another question. The discussion wanders off into a string of claims and counterclaims about Treadaway's psychological conflicts. Time runs out, and the session ends. Wylie never gets to answer Sams's question.

In the Case of the Easy Diagnosis, Brian's passage from a state of innocence to a state of guilt signals the continuity of the narrative's subject. But this is precisely where the Case of the Proud Marine seems the most problematic. In Wylie's account, Treadaway undergoes a gradual metamorphosis, taking him from student to reservist to active-duty marine to combat rifleman to squad leader. The metamorphosis is completed *before* Treadaway returns to Vietnam and experiences his nominated events. By the beginning of his second tour, he is in his mid-twenties, a sergeant, a skilled professional soldier whose identity is (on his own account) inseparable from the Marine Corps. At this point, he continues to respond to a set of values that define and confirm his sense of self, but they do not seem to be the traditional values with which the narrative starts, despite his claim that he is no "warrior." In short, Treadaway's experiences during the second tour were undoubtedly unforgettable and frightening, but it is problematic whether they were also so unusual or morally dissonant as to fall "outside the range of usual human experience," or at any rate, of Treadaway's usual experience.

Wylie's narrative includes three kinds of inscriptions: displays of emotion, a flashback, and symptomatic pain. Once again, there is a problematic correspondence between the narrative's structure and its content:

1. Wylie's narrative refers to emotions in three contexts. First, it refers to a lack of emotional arousal in connection with the nominated events. Second, it mentions points during the interview when Treadaway becomes tearful but cannot or will not give a context to his emotions. Finally, it describes points at which Treadaway's verbal content implies anger and frustration, even though his expression remains matter-of-fact. In this last context, Treadaway's commentary focuses not so much on nominated events as on various people who have misrepresented the part that he played in these events: the lieutenant who unjustly accused him of decapitating a prisoner, the marine who unjustly accused him of shooting a boy, and the co-workers who unjustly shunned him after they were shown a photograph of Treadaway with a decapitated head.

2. In his story of the flashback, Treadaway pictures himself sitting in a park. A clod of earth rolls down a hill, he imagines that it is a grenade and throws himself into a ditch. During the interview, Treadaway does not provide a particular context for his behavior, and the event seems to inscribe

nothing more specific than the enduring effects of traumatic energy. But months later, after he has been admitted to the center as an inpatient, Treadaway describes a parallel event from his second tour in Vietnam. According to Treadaway, his captain chose his squad for every dangerous job, and relations between the two men grew hostile. On one occasion, the company came under fire, and Treadaway's squad was ordered to assault the enemy position. After the attack, the company was taking a break on the slope of a hill. Treadaway was sitting near the crest and threw a clod of earth down in the direction of the captain, who was taken by surprise. He looked up, stared into Treadaway's face, and the two men shared the knowledge that the clod might have been a grenade and Treadaway could have thrown it.

3. Treadaway's presenting symptom is pain, consisting of headaches and back pain. Wylie's narrative says three things about the pain: it is severe and unremitting; it originates in a wartime explosion; and it makes Treadaway's long-term employment difficult or (now) impossible. In his narrative, Wylie reports that Treadaway had a neurological work-up at the VA, that it uncovered no evidence of organic damage, and that Treadaway says a neurologist expressed some interest in his spine. Durocher, the center's clinical director and its only psychiatrist, interrupts Wylie and says: "There's no organic basis for his [pain] problems." At this point, Durocher's claim is based on three ideas, and everyone attending the session knows what they are. First, Treadaway's symptoms, statements, and behavior have to be decoded. Their manifest meaning/content screens repressed memories and unbearable psychological conflicts that are the source of his disorders. Second, psychiatric training, psychoanalytic insight, and long clinical experience give Durocher privileged access to these hidden meanings. Third, eventually Treadaway's pain will be discovered to inscribe a traumatic act of commission, perhaps an atrocity. Wylie, on the other hand, reasons on the basis of his previous experiences that the absence of neurological evidence does not eliminate an organic etiology for the kind of pain Treadaway is experiencing. Toward the end of the session, he returns to this point and says that Durocher is assuming that Treadaway's pain is psychogenic. Durocher replies that his clinical judgments are not equivalent to assumptions. No one is inclined to argue this point. The session ends, and Treadaway is diagnosed with service-connected PTSD.

In summary, the two cases are only superficially similar. In one case, everyone recognizes that the content and structure of the presenter's narrative are congruent, and the unit proceeds to a quick consensus and an easy diagnosis. In the second case, content and structure are not transparently congruent. Even so, by the time the session is over, everyone is ready to accept that Wylie's account of Treadaway is consistent with the *DSM-III* account and the psychometrician's findings. In contrast with the first case, though, consensus is the product of the shared structure of these two accounts (Wylie's and *DSM-III*'s), augmented by Durocher's professional and institutional authority. The puzzling surface meanings supplied by Treadaway are deferred to his therapists.

The Case of the Metaphorical Event is special because it starts off with *parallel* accounts of sickness: a case of PTSD traced to traumatic events in Vietnam and a case of borderline personality disorder traced to traumatic childhood sexual abuse. The psychometrician echos the interviewer: Edward's MMPI is consistent with either PTSD or borderline personality disorder. This means that the unit has to consider three possibilities: Edward may have PTSD, or a borderline personality, or both. His final diagnosis boils down to choosing among his nominated events.

Edward's Vietnam events are said to be a good choice because they are a "focal" experience, that is, cognitively and emotionally salient. But no one, except Durocher, is sure what these events and emotions signify. At one point, someone describes Edward's Vietnam experiences as being a "metaphor" for his childhood experiences, in the sense that if Edward were unable to express his psychological conflicts through his Vietnam trauma, he would have to recognize the hatred and anger he actually feels toward his mother (who failed to protect him from his stepfather), and this would be unbearable. This is plausible because everyone at the session believes that nominated events occasionally screen or substitute for even more painful events that are being repressed. Edward may be speaking in metaphors: "The question is, What is being a metaphor of what?," interjects Durocher.

In Edward's case, inscriptions are no help. For one thing, either set of traumatic events—those in Vietnam or those in his childhood—can be read into his symptomatic behavior, for example, the barroom incident that landed him in the VA hospital. Second, if Durocher is correct and Edward's disorder originates in an undisclosed Vietnam event, hidden beneath the surface of both sets of nominated events, it is premature to read inscriptions.

The impasse is overcome when Malloy argues that since Edward has been diagnosed by a VA compensation rating board as having PTSD, the center is obliged to treat him, and since the center has to treat him for PTSD, its diagnosis ought to be consistent with its clinical mission. Malloy's logic is problematic, since the center occasionally makes diagnoses that are inconsistent with the rating boards'. But everyone seems grateful for the solution, and the Case of the Metaphorical Event is closed.

In the final session, the Case of the Shady Event, Freddy provides the unit with everything it needs for a PTSD diagnosis: a plausible account with a choice of traumatic experiences, a good MMPI profile, a strong PTSD subscale score, a set of responses indicating low dissimulation and low exaggeration. The problem is that no one seems to trust Freddy. At several points, his symptomatology is described as "characterological stuff." This is a shorthand way of saying that his primary diagnosis is not PTSD but, rather, one or more personality disorders overlaid with the life-style of a drug addict, petty criminal, and con man. During the session, someone asks whether Freddy meets the diagnostic criteria for PTSD. "Sure! According to what he says about himself, he does," is his inter-viewer's ironical response. (A person with antisocial personality disorder "has no regard for the truth, as indicated by repeated lying . . . or 'conning' others for personal profit or pleasure" [Amer. Psychia. Assoc. 1987:345].) Someone else comments that Freddy has anxiety dreams with violent, "Vietnam-like motifs." But how much do these old experiences count in comparison with his current violent life as a "street soldier"? Freddy has "lots in the present to intrude into his mind. There really are lots of people out there after him." And what about his "symptomatic guilt"? "We could do a population in the state penitentiary and get the same guilt that Freddy's telling us about."

The banter goes on partly because no one at the session has the resources needed for reducing the multiple interpretations and possibilities to a single account. Durocher is absent, and no one else can conjure the sort of author-itative inscriptions that ended the Case of the Proud Marine (where Duro-cher discovered the psychogenic origin of Treadaway's pain). Nor can any-one find a formula for organizing the kind of consensus that closed the Case of the Metaphorical Event. After three-quarters of an hour, the staff decide to give Freddy "the benefit of the doubt." There are enough reasons—test results, Malloy's narrative, their shared anxieties about accusations of rac-ism—to diagnose him with PTSD, with mixed personality disorder on the side.

These four cases are about unusual men talking about extreme situa-tions; their stories are filled with murder, mayhem, and mutilation. There is nothing out-of-the-ordinary about these cases in the context of Vietnam War–related PTSD, however. They represent the spectrum of cases pre-sented at the center. Nor are there any good reasons for supposing that the men diagnosed with PTSD at the center are significantly different from Vietnam War veterans diagnosed with PTSD at other VA hospitals.

According to their psychometric tests and standardized diagnostic inter-views, Brian, Treadaway, Edward, and Freddy have PTSD. They are fit for service in any of the research projects described in chapter 3. But when these men start talking about themselves, filling in details, things become unclear. A space opens up between the official account of PTSD (which connects the four men through shared criterial features) and their disparate narratives. The diagnosticians have ways of managing these difficulties, of narrowing the distance between narrative structures and narrative details.

But when the diagnosticians are finished, a residue of indeterminacy, the suspicion that this nominated event is "metaphorical" rather than etiological, or that putative inscriptions have been misread, remains.

How much simpler everything would be if researchers and diagnosticians had some way to bypass the things that men say about themselves and their pasts. I shall return to this point in chapter 8, where I analyze the biological science of post-traumatic stress disorder.

Six

Everyday Life in a Psychiatric Unit

This judge rolled everything up into one, starting
with the police-reports and the vagrancy, and
then presented it to Moosbrugger as his guilt.
But for Moosbrugger it all consisted of separate
incidents that had nothing to do with each other,
each of them with a different cause, which lay
outside Moosbrugger and somewhere in the
world as a whole. In the judge's eyes his acts
came towards him the way birds come flying
along. For the judge Moosbrugger was a special
case; for himself he was a world, and it is very
difficult to say something convincing about the
world. There were two kinds of tactics fighting
each other, two kinds of unity and logical consistency; but Moosbrugger had the less favourable
position, for even a cleverer man could not have
expressed his strange shadowy arguments. . . .

It was like a vapour that is always losing its
shape and taking on other forms. He might, of
course, have asked his judges whether their lives
were essentially different. But such things never
occurred to him.
  (*Robert Musil,* The Man without Qualities)

PTSD IS TREATED with both pharmacotherapy and psychotherapy. The
most commonly prescribed medicines are antidepressants and minor and
major tranquilizers, and they are used mainly against conditions that co-
occur with PTSD: depression, generalized anxiety, and alcohol and chemical substance abuse (Silver 1990:36; Solomon et al. 1992:634). Although
pharmacotherapy is less often used for symptoms specific to PTSD, several
drugs have been found to have a modulating effect on intrusive phenomena, such as nightmares and dream recollections, and symptoms of auto-

nomic nervous system arousal, such as irritability, aggressive outbursts, exaggerated startle response, and hypervigilance. Sedatives are also given to patients who have problems falling or remaining asleep, an arousal effect sometimes associated with PTSD (Silver 1990:36–37). Even when specific symptoms are alleviated by pharmacotherapy, patients characteristically continue to experience significant distress from other symptoms and their effects, most commonly guilt, social impairment, and somatic complaints. Thus, when drugs are employed, it is usually as an adjunct to psychotherapy (Silver et al. 1990:34; Solomon et al. 1992:634).

Psychotherapies for PTSD fall into three broad categories: behavior therapy, cognitive therapy, and psychodynamic therapies (including hypnotherapy).

The theoretical basis of *behavioral therapy* for PTSD is described in chapter 1. In a nutshell, behavioral therapists assume that the patient has been exposed to a frightening and/or painful stimulus together with some neutral stimuli. The experience conditioned him to the neutral stimuli so that when he is reexposed to them, he experiences anxiety and arousal, just as if he were reexposed to the frightening stimulus. The previously neutral stimuli have become aversive and are able to produce higher-order conditioning, so that additional stimuli, including words, thoughts, and images, now produce anxiety. The avoidance behavior characterizing PTSD constitutes the patient's efforts to insulate himself from these conditioned stimuli. Therapy aims to reduce anxiety and attendant avoidance behavior by a process of deconditioning, in which patients are exposed to stimuli that are anxiety-inducing but objectively harmless. Other benefits are also claimed, including reduction in intrusive thoughts, reexperiencing, nightmares, sleep disturbances, and depression (Keane et al. 1993:368–369).

Two procedures are employed. In systematic desensitization, the individual is subjected to gradually increasing exposure to some feared (conditioned) stimuli. When the therapy is effective, the stimuli lose their power to evoke a conditioned response. The other technique is implosive or flooding therapy. The aim is to get the individual to recollect and experience the traumatic memory as completely as possible. In order for him to gain access to this memory, he must be put in a high state of arousal. Recall is assisted by prompting him with an array of cues: visual (e.g., a videotape in which the camera-eye moves down a jungle river), auditory (e.g., recorded sounds of small-arms fire), olfactory (e.g., the smell of gunpowder), and tactile (e.g., the patient grips a rifle stock). "The patient is slowly and gradually presented with cues of the specific event, emphasizing and prolonging in imagination the elements of memory that are most thought-provoking" (Keane et. al. 1985:279). Over the course of this treatment, he is reexposed to disturbing "symptom-specific traumatic cues" until the anxiety effect is extinguished and the need for avoidance mechanisms is eliminated. Narcosynthesis is sometimes employed to achieve similar effects (Kolb 1985). PTSD patients also suffer from guilt, anger, shame, and dysphoria. Some therapists claim that implosive therapy can be used to treat these emotions also, since it involves "all possible conditioned stimuli, including hypothesized faulty beliefs and value systems (e.g., themes such as loss of control, eternal damnation, guilt, or humiliation and shame)" (Keane et al. 1993:367). However, the majority opinion is that these emotions are not extinguished in the same way that anxiety is. Thus, behavioral therapy (in common with pharmacotherapy) is used mainly as an adjunct to other kinds of therapy (Solomon 1992:636).

Cognitive therapies are intended to provide patients with skills for controlling disvalued emotional states that are associated with PTSD, especially anger and anxiety. There are many kinds of cognitive therapies—some aim at altering patients' habitual patterns of reactive thinking and self-arousal, some at improving communications skills, some at teaching muscle relaxation and breathing control, and so on—and they are often employed in combinations (Solomon et al. 1992:636). The most significant of these techniques in the treatment of PTSD is "cognitive restructuring," based on the ideas of Aaron Beck (1976) and Albert Ellis (1977). Beck's system presumes that emotional problems are often produced by distorted perceptions of reality based on erroneous assumptions. These perceptual distortions and consequent emotions are a source of self-defeating behavior and lead the individual to experience the world as a negative and punitive place. The individual can identify these maladaptive and unrealistic cognitive phenomena through introspection, and cognitive therapy provides him or her with techniques for testing and correcting these distortions.

In contrast to therapists of the psychodynamic school, Beck sees irrationality as being rooted in commonplace processes, such as making incorrect inferences on the basis of inadequate or incorrect information and failing to distinguish between imagination and reality (Beck 1976:12–21). Cognitive distortions include "all-or-nothing thinking" (seeing people and events in black-and-white terms), "overgeneralization" (discovering a never-ending pattern in one negative event), "mind reading" (concluding, without evidence, that someone is reacting negatively to you), "emotional reasoning" (assuming that your negative emotion mirrors the objective situation), and "should statements" (motivating yourself with "shoulds" and "musts," as if you had to be threatened with punishment before you could be expected to do anything) (Burns 1980:40–41).

Cognitive restructuring is targeted to various emotional states connected with PTSD. Of these, guilt, often associated with atrocities, is recognized to be particularly difficult to modify through this technique:

[T]he therapist focuses on the environmental contingencies and the situational specificity of behavior required in combat as opposed to the behaviors one would demonstrate under less stressful circumstances. Through this approach we do not attempt to excuse the acts committed, but rather offer a rational explanation for the veteran's behavior in combat. The intent is to prevent the patient from *over-generalizing* about his behavioral patterns. (Keane et al. 1985:287)

Psychodynamic therapy for PTSD concentrates on the contents of the traumatic memory—the details of the traumatic event, the subjective meanings associated with these details, and so on—and in this regard they are different from other psychotherapies. The therapeutic goal is to restructure the contents of the traumatic memory, to the point where the patient is able to integrate it into his ongoing view of the self and make it bearable in consciousness. Although psychodynamic therapies are divided among themselves by doctrinal differences, they are generally committed to bringing repressed or dissociated material into consciousness. "Rather than telling patients *not* to think about the trauma, one tells them to think about it, with the inference that once the process is over they can attend to other things" (Spiegel and Cardena 1990:42).

This chapter describes of how cases of PTSD were treated at the pseudonymous National Center for the Treatment of Post-Traumatic Stress Disorder, the Veterans Administration psychiatric unit introduced in the preceding chapter. In 1990, the National Center was reorganized by the Veterans Administration. Its various functions were distributed among several VA medical centers, and the unit in which I conducted my research was closed. When, in the following pages, I speak of the center in the present tense, it should be understood that this is the "ethnographic present"—it refers to how the center operated while I was there—rather than the conventional use of this tense.

The treatment program at the center is based on forms of cognitive therapy and psychodynamic therapy. Behavioral therapy is not practiced at the center, and pharmacotherapy is limited to the treatment of preexisting conditions. The choice of therapies has to be seen against the backdrop of general psychiatric attitudes toward efficacy. When asked to describe their clinical repertoires, a majority of American psychotherapists describe themselves as *eclectic* (Lambert et al. 1986:159). When asked to explain why they prefer an eclectic approach to a unitary clinical technology, psychotherapists take three positions:

1. Some psychotherapies do a superior job against certain disorders but are not better than rival psychotherapies against other disorders. For example, cognitive and behavioral therapies have demonstrated superior efficacy against depressive, phobic, and compulsive disorders; biofeedback

works especially well against migraine; relaxation therapy has a better than average effect on tension headaches; and so on.

2. No psychotherapy is intrinsically more efficacious than any other, and the choice of treatments should be determined on situational grounds rather than with respect to diagnosis. This position can be identified with the "demoralization thesis," described by Jerome Frank in *Persuasion and Healing* (1961, 1973, 1990). The thesis is that, when mental disorders are encountered clinically, they are associated with *two* categories of symptoms: core symptoms, which are distinctive to a specific disorder, and secondary symptoms, which are the same for all mental disorders. The secondary symptoms represent the demoralization effect and are rooted in the patient's perceived loss of "self-efficacy." He (she) feels unable to control his emotions, solve his problems, or master the everyday situations that other people handle with ease. His self-esteem erodes, and his constricted life space is dominated by four emotions: sadness, shame, anxiety, and anger. He is uncertain what his symptoms signify or how serious they are; he feels vulnerable and thinks that he may be going crazy.

3. The third position overlaps the demoralization thesis: there are core and secondary symptoms, and psychotherapy's effect is limited to the secondary. At the same time, psychotherapies *are* specific in the following ways: in order to be efficacious, psychotherapy has to produce a combination of affective, cognitive, and behavioral effects; each psychotherapy has a different potential in these three respects; and therefore a successful treatment program must incorporate an appropriate mix of psychotherapies. Affective arousal prepares patients for cognitive and behavioral changes, by unfreezing feelings and undermining psychological defenses. Cognitive techniques, such as insight-oriented therapies and rational-cognitive therapies of the sort described by Aaron Beck (1976), "use reason and meaning . . . over affect as their primary therapeutic tools and . . . attempt to achieve their effects through the acquisition and integration of new perceptions, thinking patterns, and/or self-awareness" (Karasu 1986:691). Similar techniques can be found around the world, and the distinctiveness of the "Western scientific therapies" (Karasu's term) lies not in their unique efficacy but rather in their interest in rooting out irrational beliefs, achieving authentic insight and rational understanding, and restoring autonomy. By themselves, insight and rationality are insufficient, and purely cognitive approaches have countertherapeutic effects when they encourage the patient to overintellectualize his problems, thereby strengthening his defenses and capacity to resist therapeutic change. In some patients, the pursuit of insight and self-awareness can be a source of anxiety, anger, guilt, and shame—strong emotions that are difficult to endure. Therefore, in order for cognitive approaches to produce desirable changes, therapy must include techniques for managing emotional arousal and transferring the new infor-

mation and modes of thinking from the safety of the structured clinical setting to the world outside.

The obvious way to compare the efficacy of competing positions is by outcome studies. The typical outcome study begins with a sample of people with the same or similar diagnoses. The sample is divided into two groups, and each group is treated with a different kind of psychotherapy. Combinations of diagnostic interviews, psychometric instruments, and psychological scales and are used to assess symptoms and levels of dysfunction and distress before treatment begins and at one or more points following completion of treatment. Outcomes between the groups, such as changes in symptoms and/or levels of dysfunction, are then compared. (Pretreatment data usually serve as the "control groups," but occasionally an actual control group is used.)

Outcomes are products of multiple determinants: the patient's personality, early life experiences, "psychological mindedness" (the capacity to reflect on one's motives, perceptions, thought processes, et cetera), psychiatric history, psychological beliefs and explanatory models of illness, concurrent life events, coexisting medical problems, prospects of secondary gain, et cetera. Outcomes are also influenced by factors associated with the clinician: his or her personality, technical skills, interactional style, social stereotyping, et cetera. Outcome studies usually limit themselves to a single set of determinants, namely, interventions. This is not necessarily a problem, say those who conduct meta-analytic studies of psychotherapy, if contingencies (other determinants) are randomly distributed among the experimental and control groups. To randomize these effects, large numbers of studies have to be compared, and this means comparing (with the aid of specialized statistical techniques) results produced by different psychometric instruments, each of which produces a distinctive set of outcome data. Analysis of these data confirms Frank's intuition: patients who receive any form of psychotherapy do better than patients who receive none; no matter what form of therapy patients receive, those who show initial improvement are likely to maintain it; and, although there are instances where some psychotherapies are initially more efficacious than others, these differences tend to diminish over time (Lambert et al. 1986; Klein and Rabkin 1986). These findings are therefore consistent with the second and third positions described above but not with the first (which claims that certain therapies are intrinsically more effective against certain disorders).

There is, however, a second difficulty in comparing and interpreting outcome studies. This time the problem is to find a way to translate across psychological doctrines rather than psychometric instruments, and it is epistemological rather than technological. Simply put, different doctrines can give different meanings to the same outcome. While behaviorists and cognitive therapists say that a technique is efficacious when it produces

enduring changes in disvalued behavior patterns, psychodynamic therapists, particularly clinicians oriented to psychoanalytic perspectives, locate
the meaning of altered behaviors elsewhere—in etiologies, symbolic content, and psychological processes. Simply reducing the intensity of symptoms can be countertherapeutic and may signal the formation of more effective psychological barriers to insight into etiological conflicts. Real
efficacy means releasing a potential for inner growth and maturation and
enhancing the ability to establish and sustain gratifying social relationships. In these circumstances, the behaviorist and psychodynamic valua-
tions would be not simply different but also incommensurable: they could
not be measured by a common set of standards.[1]

Both the patients and clinicians at the center talk a great deal about the
efficacy (or lack of efficacy) of its treatment program. At this point, I want
to limit my remarks to the ideas and attitudes of the clinical staff. When I
arrived at the center in 1986, about a year after it opened, the staff held a
variety of views about how PTSD should be treated, but over the next two
years a uniform set of opinions emerged, reflecting the ideas and policies of
the clinical director (the psychiatrist "Durocher," introduced in chapter 5):

- It is clinically useful to distinguish between the core symptoms and secondary
  symptoms of PTSD.
- The distinctive feature of PTSD is the traumatic memory and the core symptoms it produces, that is, intrusions, avoidance behavior, arousal symptoms.
- Psychodynamic therapy provides a uniquely efficacious way of working on the
  traumatic memory and, through it, on PTSD's core symptoms. This therapy
  possesses all of the treatment components—affective arousal, cognitive restructuring, and behavioral adaptation—needed to produce therapeutic
  change.
- Cognitive psychotherapies are a useful adjunct to psychodynamic therapy, but
  their effect is limited to managing PTSD's secondary symptoms.
- Antianxiety drugs are countertherapeutic, because they interfere with psychological processing by allowing a patient to disengage from his traumatic
  memory (see pp. 140–141).

The Center and Its Mission

The center occupies a two-story building on the "campus" of a large Veterans Administration Medical Center, located in the Midwest. The lower
floor contains the reception area, rooms for interviewing, outpatient services, and administrative offices. The inpatient ward is located on the upper
floor and generally consists of thirteen to sixteen patients and a full-time

staff of twenty-two, including a psychiatrist, physician's assistant, nurses experienced in psychiatric work, clinical psychologists, social workers, and rehabilitation counselors. The administrative hierarchy coincides more or less with the staff's professional qualifications, descending from the clinical director (the psychiatrist) at the top to therapists without advanced degrees (counselors and nurses) at the bottom.

The inpatient unit consists of two wings, divided by a stairway leading to the ground floor. Each wing is entered through a lockable door. The west wing contains the patients' rooms, a community room with an adjacent nursing station, several small rooms for group psychotherapy and staff meetings, bathrooms, a shower room, and a room containing a special bed and restraints for in-ward psychiatric emergencies. The east wing contains therapists' offices and several large rooms used mainly for cognitive therapy sessions. The door to the west wing is locked overnight and during the morning hours when group psychotherapy sessions are in progress.

Patients eat their meals in a communal dining hall in another building. Tables are shared with patients from other psychiatric wards, including men with psychotic disorders. The center's patients are expected to remain "on campus" during the course of the treatment program unless permission is given to leave; weekend passes are a privilege rather than a right. Each man is provided with a lockable closet for storing his clothes and personal possessions. Items that might be used as weapons against oneself or other patients are forbidden: razors are distributed every morning before wash-up and must be returned immediately after use, and patients are not permitted to take glass items, such as tumblers or bottles, onto the ward. These regulations are strictly enforced. The men's rooms and possessions are subject to surprise inspections, and if items such as knives are discovered, the patient is discharged immediately. Patients are required to sign a "patient's contract" before they are accepted in the program. This contract forbids any act or threat of violence by the patient and any use of alcohol or unprescribed drugs (aside from tobacco, which patients tend to consume in heroic quantities). Patients are routinely required to provide urine samples when they return from weekend passes and must comply with random urine tests as well.

The majority of the patients come from four sources: some men are referred to the center by staff at VA mental health clinics and VA medical centers with no inpatient treatment unit for PTSD; some are referred from drug and alcohol abuse wards located in a neighboring building at the medical center; some are guided to the center by other veterans; and some are voluntary readmissions. Many of the men who are referred to the center from other psychiatric units have been admitted with nonspecific symptoms, such as suicide attempts, generalized anxiety, depression, a history of violent behavior, or dissociative behavior. In these cases, it is often the men's military records, rather than particular symptoms, that stimulate their clinicians to send them to the center for examination.

The patients have spent an average of twelve years between the onset of symptoms diagnosable as PTSD and the point at which they first received a PTSD or PTSD-like diagnosis. The most common interim diagnoses are either disorders now said to be concurrent with their PTSD—such as chemical substance use disorders (including alcohol abuse), depression, generalized anxiety, and panic disorder—or disorders now said to have been misdiagnoses, such as schizophrenia, paranoid type. According to the diagnosticians and therapists at the center, the patients' psychiatric problems are chronic rather than episodic: embedded in maladaptive behaviors and life-styles the results of which have further demoralized the men. Before being admitted to the inpatient program, they undergo a period of diagnostic evaluation and assessment, designed to screen out those with severe personality disorders, psychoses, and factitious cases of PTSD. These procedures are described in the preceding chapter.

The treatment program is divided into three phases: a week of orientation, eight weeks of individual and group psychotherapy and training in cognitive coping skills, and a two-week reentry phase. Relatively few patients complete the second phase in eight weeks; most stay on an additional month or more. About half of the men who enter the program leave before completing it, either for personal reasons or involuntarily, because they violated regulations. Evidence of alcohol and drugs in the men's urine is the most common reason for these involuntary discharges.

During the early 1980s, veterans' organizations successfully lobbied Congress to have the Veterans Administration Medical System provide specialized treatment for PTSD. These groups claimed that large numbers of Vietnam veterans were suffering from PTSD and that the undiagnosed and untreated disorder was a significant cause of the veterans' high rates of mortality (connected with self-destructive behavior) and psychiatric morbidity and hospitalization. Congress responded, in Public Law 98–528, by mandating the VA to establish a "national center" with responsibility for developing a model treatment program. The center was established in 1985 as a result of this mandate. It was given three tasks: to provide specialized medical treatment for its patients; to accumulate and distribute to other VA treatment units clinical findings relevant to the diagnosis, assessment, and treatment of PTSD; and to conduct research relating to the etiology, symptomatology, and treatment of PTSD. With this mandate, the center's directors were given a strong incentive to develop a *distinctive* treatment program, one that would clearly distinguish the center from other VA psychiatric units treating PTSD at that time and that would justify the center's relatively large staff and budget. Given the undeveloped state of the psychiatric discourse on PTSD in 1985, the mandate from Congress had the effect of linking the center's tasks: the center would now accumulate knowledge of PTSD in the course of treating patients and would in turn use this knowledge to refine its therapeutic techniques and procedures. The drive toward producing a distinctive treatment was given direction when, soon after the center opened, the original clinical director resigned and a psychiatrist (Durocher) with strong psychoanalytical convictions was appointed in his place.

The center's program evolved over time, but two assumptions remained constant. (1) The psychodynamic core of PTSD is a repetition compulsion: the victim is psychologically compelled to reenact the behavior that precipitated his disorder, in a futile attempt to gain mastery over the circumstances that originally overwhelmed him. (2) To recover, the patient must recall his traumatic memory, disclose it to his therapists and fellow patients during group psychotherapy, and subject it to scrutiny. The memory/narrative is the Rosetta stone of his disorder. The patients' postwar histories are generally saturated with misfortune and failure: erratic employment or chronic unemployment, impulsive relocations, addiction to high-risk behavior, brushes with the law, and so on. A properly decoded traumatic memory gives the chaotic surface a coherent subtext.

Aside from a weekly film series and relaxation therapy, therapeutic activity at the center consists mainly of talking. Most of this talk takes place during group psychotherapy (daily), individual psychotherapy (twice weekly), autobiography sessions (weekly, as a group), and cognitive skills sessions (daily, as a group), on topics such as rational thinking, communication skills, values clarification, and death and dying. Even the films, which are usually commercial films and TV shows portraying themes that can be related to PTSD in some way, are followed by discussion periods led by the therapists. Talking is simultaneously a therapeutic modality and evidence of a patient's progress. From the therapist's point of view, the most important talk consists of patients' narratives of their traumatic experiences and the symptomatic events that occurred later. Therapists say that authentic narratives are evidence of the forward movement of mental processes, and acts of narration, or even efforts in this direction, are referred to as "processing [the traumatic memory]" or simply as "working."

In most psychiatric units in the VA system, pharmacotherapy is an important aspect of treatment. At the center, however, drugs play an insignificant role, despite the fact that many patients arrive with long histories of taking prescribed medicines for psychiatric problems. In principle, the center is a drug-free therapeutic milieu. Once on the ward, no patient is permitted to continue on antianxiety medication, because these drugs are said to limit the patients' ability to "process" and "work through" the psychological conflicts associated with their disorder. In the case of antidepressants, which are the most common category of drugs at the time of admission, men are either taken off the medicines entirely or they are reduced to the minimum tolerable dosages. A similar policy is followed for the minority of patients who have been taking antipsychotic medicines.

Before a patient is allowed to move from the treatment phase to the reentry phase of the program, he is expected to meet behavioral criteria that were set at the time of his initial diagnostic assessment. A man who arrived with severe sleep disturbances is expected to show improvement in his sleep patterns; a man who had problems controlling his aggression is expected to give evidence that he is now modulating his anger and effectively managing its expression; and so on. *Every* patient is expected to give three additional kinds of evidence showing that he is on the road to recovery: a coherent public account of his traumatic event, a correct interpretation of this event, and an account of his present behavior that is consonant with the repetition compulsion thesis.

In many cases, symptomatic behavior is altered at the moment that the patients arrive at the inpatient unit. For example, at the time of admission, many patients talk about being afraid of their own violent impulses. This is frequently expressed as a fear of injuring their wives and children:

> After I married my second wife, my stepdaughter used to come over to sit on my lap or maybe just stand next to me. I wanted to be nice, especially because of my wife. But whenever I looked at her, I saw the faces of Vietnamese kids we messed up and I couldn't stand to be near her. One time she surprised me and snuck up and put her hands on my eyes, and I hit her and she really got hurt. My wife got angry and started screaming, "Are you crazy?"

Other men say that they always seem to find themselves in places where people want to start fights with them and are worried that they will eventually be killed or inflict an injury on someone else and end up in prison.

Once at the center, these same men become nonviolent and remain so throughout their stay. The point that needs to be emphasized, however, is that changes of this sort are not *developments*: they do not emerge during the treatment program but are produced at or soon after admission. While the staff attributes the changes to their therapeutic practices, it is doubtful whether this is the best explanation. The changes are more plausibly attributed to the patients' adaptation to daily life in a highly structured social environment, in which prescribed behavior, including nonviolence, is a precondition for continued treatment. It would seem that, if the patients were given similar incentives, they would do equally well (with regard to these particular behaviors) in any highly structured environment devoted to treating PTSD, regardless of differences in the units' therapeutic doctrines. Evidence from other VA PTSD units supports this conclusion.

Many patients come to the center immediately after completing lengthy inpatient treatment programs for alcohol and drug abuse. Consistent with practices at other VA units, the center's assessment and evaluation section identifies alcohol and drug use among PTSD victims as a maladaptive response to the disorder's core symptoms and to concurrent anxiety and depression. Drinking and drugging among diagnosed patients are referred to as "self-dosing" and are distinguished from ordinary dependency and addiction. In practice, cause-and-effect relations between PTSD symptoms and alcohol and drug abuse are more complicated, since episodes of symptomatic violence and dissociation ("flashbacks") are often precipitated by intoxication. Here again, therapeutic changes observed at the center, especially the modulation of anger and aggression, seem to be rooted in circumstances, notably abstinence, that are not specific to the treatment program.

Therapists also credit the treatment program for the patients' good conduct on weekend passes and on those occasions when they are overtly frustrated by the center's regulations and sanctions, for example, when they are confined to the ward. But this is another instance in which behavior changes abruptly, on admission, before the treatment program has really begun.

At the same time, one *does* see certain changes gradually taking place over the course of the treatment program. These changes are in *verbal behavior*: the information that the patients communicate to their therapists about their thoughts, feelings, and memories. Many patients report feeling progressively less angry, less sad, more in control of their emotions, and more self-confident as time goes on. Once more, the source of these changes is unclear. Is it the treatment program or is it the institutional environment—structured and safe, unlike the situations the men have often left—that is producing these effects? Patients report other significant changes also, such as decreases in nightmares and disturbing intrusive thoughts and images. But the significance of these reports is not easy for an outsider to assess, since an equal proportion of patients report opposite effects—increased frequency of nightmares and intrusive images, bouts of dysphoric affect, and a rise in somatic complaints (usually headaches and gastrointestinal distress)—and these effects too are evaluated by the therapists as positive signs that the treatment program is working. The reasoning behind these assessments is that decreases in such symptoms are an index of direct improvement (the patient is metabolizing dissonant images and ideas), while increases are an index and side effect of "getting close" to the traumatic memory, a bit of pain for a long-term gain.

For the therapists, the most unequivocal evidence of progress is that most long-stay patients eventually provide satisfactory narratives and interpretations. From the therapists' point of view, these recollections are simultaneously evidence of the patient's unique pathology, proofs of cognitive processing, and vindication of the treatment program. It is this evidence, underpinned by the findings of scientific research on PTSD, that turns the otherwise ambiguous behavioral transformations that take place at the center into proof of therapeutic efficacy. (Follow-up data on discharged patients who have completed the treatment program might be a way of testing these claims, but the data do not exist.)

Ideology and Resistance

To repeat: the U.S. Congress mandated the center to develop a treatment program that is both effective and specific to PTSD. It was not difficult for the center to construct a program and an environment conducive to changing behavior. The more demanding task was to develop mechanisms for connecting changed behavior to mental structures and processes associated with PTSD: the traumatic memory, the repetition compulsion, and so on. In this sense, the center's special achievement has been its knowledge-product.

The process of making clinical knowledge at the center can be broken down into three stages:

Stage one. Therapists elicit etiological narratives from patients during group and individual psychotherapy. They renarrate the patients' accounts and use the new stories to explain (to the patients and to themselves) the meaning of the patients' current behavior at the center.

Stage two. Each therapist provides the clinical director with a double account of what has happened at stage one: an account of the patient's narrative and behavior and an account of his or her own perceptions, statements, emotions, and intentions while attempting to elicit the patient's narrative. This takes place during weekly clinical supervision and at the "debriefings" that follow each group psychotherapy session. Stage two mirrors stage one, in the sense that the therapist's narrative is renarrated by someone (the clinical director) with privileged access to the meaning of the narrator's words.

Stage three. The knowledge product of stage two is inscribed in documents for internal circulation, for a quarterly PTSD newsletter edited at the center, and for papers presented at VA conferences and annual meetings of professional groups, such as the Society for the Study of Post-Traumatic Stress Disorder.

In this process, the knowledge that is being moved across the three stages consists of meanings attributed to narratives and meanings attributed to *resistances*, that is, acts that are believed to impede the movement of the narrative meanings. At the center, resistances are identified forms of behavior—variously labeled "resistances," "stress responses," and "acting-out"—and they are attributed to both patients and therapists. The term "resistance" suggests a negativity, an act that merely prevents something else from happening. In practice, resistances are something positive, in the sense that, like the content of the narratives, they are ways of holding on to the otherwise invisible traumatic memory.

Resistance by Patients

By definition, traumatic memories are distressful. If they are not distressful, it is because they have been metabolized and are no longer able to produce symptoms. Traumatic memories at the center are especially indigestible and distressful because they are often about morally forbidden behavior, such as torturing prisoners or killing civilians or one's fellow soldiers. The job of the treatment program is to get patients to "process" these memories. They have to describe their thoughts, feelings, and perceptions before, during, and after their "acts of commission," and they have to do this in the company of their psychotherapy groups.

Some patients resist this demand by refusing to narrate a traumatic event or by refusing to provide any details concerning a traumatic event to which they have alluded. They justify their resistance on the grounds that the memories are too painful to recall and/or to relate—they cause headaches, feelings of hopelessness, nightmares, and suicidal thoughts—and that the therapists are incapable of understanding the meaning of the events. This last form of refusal, "You had to be in Vietnam to know what I'm [not] talking about," turns the knowledge production process inside out. It gives the patient, rather than the therapist, privileged access to the meaning of his traumatic memory. Patients also resist by rejecting interpretations based on the repetition compulsion, the idea that their behavior on the ward today replays events that happened to them years ago in Southeast Asia. Finally, patients resist by repudiating the idea that their anger originates in unconscious conflicts, that it represents symptomatic acting-out behavior, and that it is anything less than a justified reaction to the malpractice and humiliation they are forced to endure at the center.

Therapists believe that acting-out is a serious form of resistance and that it needs to be kept under control. If no limits were placed on acting-out, patients would walk out of therapy sessions as soon as they felt themselves becoming anxious. This would have the effect of cutting the therapeutic process off just at the point when patients are "getting close to their issues" (traumatic memory). Anxiety is the patient's response to the danger that is intrinsic to this memory. Anger is the patient's response to his inability to reconcile himself to this memory, to protect himself from the moral and psychological consequences of the act that is recorded in the memory, and to compel his therapists to stop pushing him back into the memory. According to the therapists, acting-out is a defensive move, a substitute for verbalizing his feelings and experiences. It is a form of resistance, because it is precisely this verbalization (and objectification) that is required in order for him to fully engage the therapeutic process.

Resistance by Therapists

Therapists have good reasons for not wanting to push their patients too hard. They believe that the patient's efforts to "get in touch" with his traumatic memory are painful and exhausting, and that this "work" may have the effect of exacerbating rather than reducing his level of distress during the period they are in contact with him. For example, there was a patient who had given a very fragmentary account of an event in Vietnam that might or might not have had something to do with his traumatic memory. When a therapist asked him to say "something more," he replied with bitterness,

> Last time you said "Talk," I talked. You opened me up like I was a wound. But you never closed me up. You left me like that. . . . I came here to get better, but I'm going to leave worse off than when I came. You know how to open us up, but that's all.

Likewise, patients taunt the therapists with questions about the center's track record:

> How many vets have you cured since you opened? Go on, I really want to know. I know Eddie, over there [another patient in the room]. Hey, Eddie, you been here before, right? And now he's back. And they had to put him in the crisis bed first. And I know Escobar. He was here, and now he's in detox [detoxification ward]. And I know Califano, and I know he's down on Market Street selling his ass every night. What did you do for him?

Patients make some other claims that therapists have a hard time refuting, notably, that many of their problems are rooted in their social and economic realities: the fact that they are treated like pariahs by people who were themselves too cowardly to go to war, that they were sent to Vietnam by businessmen who now call them "crazy Vietnam vets" behind their backs and won't give them decent jobs, and so on. The clinically correct response to these claims is to recognize that they are examples of "pathological blaming," and are symptomatic of PTSD. Yet many therapists are personally sympathetic to what the patients are saying and initially find it difficult to reinterpret their claims in the prescribed way.

At the same time that therapists have sympathetic feelings toward their patients, they also have punitive urges. These urges are directed against uncooperative patients: men who are suspected of concealing the memories that are the key to recovery and men who are suspected of having no traumatic memory to conceal, who are at the center for a free ride: "I felt like slapping his face [a patient] when you [a cotherapist] asked him to tell you how he felt when his buddy was killed. 'I guess I felt bad.' That's typical of him, isn't it?" Do therapists have even stronger punitive urges when the patient gives a story that describes his complicity in some atrocity? This subject occasionally comes up during clinical supervision, but it is almost always approached indirectly, through some veiled reference to "countertransference." I recall only one instance where a therapist described her own feelings of outrage, in reaction to her patient's story about the murder of a young and inexperienced American soldier by his buddies. (According to the narrator, the boy was killed because he was a liability during combat.) Is this case the tip of an iceberg of angry feelings toward men involved in atrocities? I don't know.

What I do know is that patients can be abusive and threatening toward their therapists, who in turn sometimes feel anger and punitive urges toward their patients. For example, during a cognitive skills group, a patient named Sykes took exception to something the therapist said. The group was sitting in a circle, and the two men were sitting next to one another. Sykes leaned over and began to shout in the therapist's face, drowning him out. When the therapist stopped, Sykes stopped. The therapist started talking again, and Sykes leaned over again. His face an inch or two from the therapist's nose, Sykes yelled at the top of his lungs, "Shut up, cunt. Just shut up. . . . There's nothing we want to hear from you, cunt. That's right, cunt. . . . Cunt! Just shut your cunt, cunt." There was no physical violence, but Sykes's anger and his wish to punch the therapist were palpable.

Taken by itself, the event was something the therapist would have no problem handling. But when these harangues go on for fifteen minutes, and are repeated several times a day, and alternate with periods of sullen silence during which absolutely no work is done, then the accumulated provocations begin to affect the therapist's attitude toward aggressive patients, tempting him or her to manage such men in nontherapeutic ways: that is, by avoiding the issues that might upset them (in particular, ceasing to press them for traumatic recollections) or by looking for ways to have them discharged from the center.

An Exception

In the patient contract, newcomers are informed that, while there is no denying "the painful nature of dealing with the issues of PTSD," they have the "responsibility to risk disclosure . . . and [to] continue communication despite pain and anxiety." It is generally assumed that every patient who is admitted to the treatment program has the capacity for working through the steps that lead to the disclosure of the traumatic memory. An exception, however, is made for patients who are believed to be insufficiently "psychologically minded" to make successful disclosures. These men are a minority of the center's patients, perhaps 10 to 20 percent. Some of them have suffered a cognitive or intellectual impairment since leaving military service, usually because of alcohol abuse. The others are "new standards men."

During the Vietnam War, draft-age men were required to take an examination, the Armed Forces Qualification Test, that measured their intellectual and cognitive abilities. During 1966, the Department of Defense decided that it would be necessary to lower its threshold values significantly to fill anticipated manpower needs for Vietnam. Most of the men conscripted under the new standards would previously have been rejected because of low I.Q. scores. To head off any criticism that it intended to use these men as cannon fodder, the Department of Defense gave assurances that these draftees would be employed mainly in combat support units, such as engineers and transport units, and that they would personally profit from their military service by acquiring marketable skills, enhanced self-esteem, and so on.

At the center, all prospective patients are given a battery of tests and interviews by the assessment and evaluation section, and the results are then passed on to the admissions committee (described in chapter 5). The Center's policy is that all veterans, regardless of intelligence, are vulnerable to PTSD and are equally entitled to treatment. Assessments of intelligence and psychological mindedness are sometimes made, but they are never used to screen men out. These assessments are usually introduced to explain away the vagueness or ambiguity in a man's self-reported experiences. Although there are members of the admissions committee who believe that new standards men are likely to benefit less from the treatment program than would other men, they set their personal opinions aside in the interest of equity. In contrast, therapists on the inpatient ward usually have a different set of expectations for these patients. In order for a man to move from the point where he has identified a traumatic event to the point where he can successfully disclose it, he must be able to objectify his experience and emotions in words, and this capacity is believed to be problematic for the new standards men.

Herman Barnes was a new standards man who cycled through the "treatment phase" of the program three times. At the end of each cycle, he failed to meet the criteria required for moving into the final, "reentry" phase. Barnes's medical record indicated that he had suffered a brain injury as a child, and he had a history of epileptoid seizures. In Vietnam, he had been an assistant on a truck. He may have come under fire on one or more occasions, but his narratives were fuzzy and contradictory, and his therapists were unsure what to conclude from them. During most of his time at the center, Barnes claimed that he had no especially awful experience in Vietnam, but after nearly seven months, he at last produced an event. He was riding in a truck, with another soldier sitting in the back. For a reason he is now unable to recall, the truck stopped, and this soldier hopped out. As he walked to the roadside, he stepped on a mine and was killed. Despite the efforts of his therapists, Barnes could or would provide no more details: nothing about intrusive images or thoughts concerning this event, nothing about his feelings at the time or later. Whenever he was pressed for more information, Barnes exploded and shouted that he did not remember. After his third cycle, his primary therapist suggested that it was time for Barnes to leave: "He's gone as far as he can go, at least for the time being."

Acting-Out and Limit Setting

Everyday life on the inpatient ward is intersected by two systems of control, two institutional arrangements aimed at modulating, shaping, and appropriating resistances. One system aims at setting limits on "acting-out" behavior. The second system is organized around the center's distinctive ideology.

You will recall that "acting-out" refers to the acting-out of unconscious conflicts and urges. Therapists say that acting-out needs to be controlled because it may challenge the safety and security of the unit, allow patients to evade the task of confronting their traumatic memories, and distract other patients from their therapeutic work. The following episode illustrates what they mean.

At the beginning of a "grief and loss skills group," a patient interrupted the "group process" with an emotional harangue in which he accused his primary therapist of pressuring him to talk about his traumatic experiences. According to him, his experiences had involved secret operations in Cambodia, and he had signed a commitment during the war that made it illegal for him ever to disclose any information about his activities. From this point he went on to inventory the many insults and humiliations that were daily inflicted on him at the center. All the while, he grew progressively more excited, until he rose from his chair and lifted it over his head. He held it there for maybe thirty seconds, while he poured abuse on the head of his therapist, who was seated a couple of chairs away. Then he smashed the chair on the floor and stalked out of the room. The patient was a large and powerful man, and his performance was riveting, to say the least. Once he had left, the room was silent, and the patients avoided eye contact with the two therapists. When one therapist attempted to engage a patient in talk, two other men cut him off and asked him rhetorically if he had any idea about what had just happened, about the great pain that this man was experiencing, et cetera. The two therapists then tried to persuade the patients to talk about the incident in clinical terms and prompted them by suggesting that this was an instance of "acting-out" behavior. But it was all in vain, for the therapeutic trajectory had been deflected, and the remaining hour was a void.

The center has established "limits" as a way of controlling episodes like this one. In effect, limits are rules that indicate how far a patient can go before he is obliged to appear before an MDTP panel (MDTP = multidisciplinary treatment plan). Each panel is headed by the patient's primary therapist and includes representatives from nursing and counseling services. It is formed at the time of his admission to the inpatient unit and is responsible for setting treatment goals and monitoring his progress through the stages of the treatment program. (These goals are keyed to the patient's presenting complaints at the time of his initial evaluation.) Limits are, by definition, public rules. Once they have been approved by the staff, they are binding on all patients. New or revised limits are announced at the weekly "community meetings" and are then discussed during group psychotherapy sessions.

Limits have a typical history. They start off as straightforward solutions to obvious problems, such as the counter-therapeutic effects of smashing chairs and storming out of psychotherapy sessions. Once promulgated, however, they tend to take on a life of their own, moving in unpredictable directions. An easy way to describe this process is by tracing the history of two rules.

The first rule concerns the use of headphones. Soon after I arrived, a man named James Watson was admitted. Watson had been referred to the center by the acute psychiatry unit, where he had been hospitalized after a suicide attempt. At the center, he was sleeping an average of two hours a night and was continually dozing off during the day. When he was awake, he made an effort to isolate himself from the other patients. In the dayroom, where patients usually sit together in small groups smoking, drinking coffee, and chatting, Watson would retire to an armchair in the corner, close his eyes, and listen to his radio via earphones. After a couple of weeks of this, his MDTP panel met and decided that Watson's behavior was symptomatic—"avoidance behavior"—and that a limit needed to be set. At the next community meeting, patients were told that listening to the radio or tape players on earphones would no longer be permitted in communal areas. Soon afterward, Watson was discharged on a matter unrelated to the earphone incident.

A few months later, a patient named Eddie Jackson was admitted. Although Jackson was more talkative than Watson, he too failed to develop close ties with any other patient. Shortly after beginning the treatment program, Jackson was spotted wearing earphones in the dayroom. A therapist notified him that he was violating a limit and explained the reason for the rule. Jackson removed the earphones without showing any anger. A few days later, however, he showed up at group psychotherapy wearing earphones around his neck. The sound of the radio was barely audible, even to those right next to him. At the debriefing that followed this session, his therapists concluded that Jackson was "testing the limits," a disruptive form of acting-out. According to Jackson's primary therapist, his act was probably reenactment behavior: in his intake interview, Jackson implied that he had been abandoned in Vietnam while on a patrol, and that this had something to do with his unidentified traumatic experience. If it was a reenactment, the therapist suggested, Jackson was now trying to abandon the group before it could abandon him.

Jackson was next directed to appear at a meeting of his MDTP panel, where he defended wearing the headset: "You didn't say nothing about wearing earphones on my neck. You said I was blocking out everybody [by wearing them on my ears]. I can hear everything when they're on my neck. Did I miss anything in group [psychotherapy] today? Shit, Bobby [another patient] was sleeping [during this session], and you didn't say anything to him. I hear everything, and you're going to punish me. Ain't that some shit." Jackson accepted the panel's decision to extend the limit to include wearing earphones around one's neck. A couple of days later he arrived at group psychotherapy with his earphones on his temples. When a therapist asked if the radio was on, Jackson answered, "Man, if you can't tell, what fucking difference does it make?" A second patient added, "If the man's not bothering anybody and he can hear everything, what's the problem? They're *his* earphones— You people get real pleasure from fucking with us, don't you?"

At this ridiculous point, the limit was further revised. No man would be permitted to carry a radio or tape player or head phones except during free time. During free time in communal areas, he would not be permitted to wear earphones on his ears.

A second limit originated at a staff meeting, when a therapist remarked that some patients were walking out of psychotherapy and skills groups in order to urinate. It was quickly agreed that such behavior is a countertherapeutic attempt to control the anxiety that is triggered when these men are forced to "confront their issues." A limit would have to be set and, from now on, no one would be allowed to leave the room once a session began. At the next community meeting, the patients were informed about the rule, and they reacted angrily. It was explained to them that the proscribed behavior was a way of "avoiding issues," and that they are "adults" and ought to relieve themselves before coming to groups.

A couple of days later, a patient named Tommy Spann walked out of psychotherapy to urinate. He was called before an MDTP panel and asked to explain his behavior. Spann said that the staff had placed him in a no-win situation. They required him to take an antidepressant that made him very thirsty. Consequently he had to drink large amounts of liquids, and consequently he had to urinate many times during the day. The staff were unwilling to change the limit, but the head nurse offered to provide Spann with a supply of lozenges every morning, to slake his thirst and reduce his intake of liquids. A week later, Spann was discharged following a fourth violation of limits, an incident in which he left a group without authorization, ostensibly to urinate.

The next day, patients complained angrily throughout group psychotherapy about the injustice that had been inflicted on poor Spann: he had been cast out for no fault of his own; he was a man with no family or friends to whom he could turn for help; and so on. The staff interpreted these complaints as acting-out behavior. The issue continued to be a focus of community meetings and group sessions. Spann was eventually forgotten, and talk now focused on the limit itself. During psychotherapy, several patients reported that they connected their feelings about the rule to experiences setting ambushes in Vietnam. They would lie absolutely still for long periods and, if they needed to relieve themselves on those occasions, they would have to urinate in their pants. "We come here and you treat us like a bunch of kids. I *earned* my right to piss whenever I want to."

After a couple of weeks of these complaints, the staff decided to modify the limit. Patients would now be permitted to urinate during group sessions, but they would have to stay in the room and urinate into the metal cans that are ordinarily used for wastepaper. Immediately following this change, several patients began a practice of urinating regularly during group sessions. For a while there was no more talk about the urination limit. A week later, however, patients began complaining about the *new* limit, and a patient was brought before an MDTP panel for having left group psychotherapy in order to urinate in the men's room. The patients complained that the limit was demeaning, because it obliged them to urinate in front of female therapists and to sit in rooms with cans of smelly urine: "You said it's disrupting when we leave group to piss. But you tell me, is it more [or less] disrupting for somebody to piss in a can in the middle of the room?"

Once again, the patients' objections were interpreted as acting-out behavior. Several months now passed. The situation was malodorous for both patients and therapists, but neither side would retreat. The limit remained, and patients continued to urinate during psychotherapy sessions. Eventually a new limit was set: patients could leave sessions to urinate in the men's room up to twice a week, on the condition that they took "a reasonable amount of time" and did not "make a habit" of leaving sessions twice a week. Almost immediately after the change was announced, pressure began to build for revising it. At community meetings and in psychotherapy sessions, patients posed the philosophical questions, What is a "habit"? What constitutes a "reasonable amount of time"? Why is a limit of twice a week preferable to a limit of three times? Why should arbitrary limits be grounds for forcing a man out of a program that his government created just for his benefit?

The staff (to be more precise, the clinical director) has, in general, two ways to respond to such dissidence. One response is to stand fast: set a limit and refuse to modify it. But this policy can result in serious psychomoral wear and tear on therapists. How could the center, in good conscience, prescribe antidepressants for Tommy Spann and then deprive him of access to a urinal? The alternative response is to modify limits whenever patients can demonstrate that they are unrealistic. But in each case, the patients' relentless acting-out escalated reasonable compromises to new levels of absurdity.

Over the next eighteen months, a new doctrine on limit setting emerged, promulgated by the clinical director: limits are not intended to be fair but rather to be therapeutic. The treatment program needs limits because it needs a modicum of safety and order. But, equally important, it needs limits because it needs resistances and, seen in this light, the very arbitrariness of limits becomes a therapeutic tool: it forces patients to bring their symptomatic behavior (acting-out) under self-scrutiny, enabling them to see it, for the first time, as something bounded and distinct from normal kinds of behavior.

Because the MDTP panels are the primary mechanism for enforcing limits, patients talk about them as if they were a disciplinary apparatus and as if their sanctions—warnings, confinements, discharges—were "punishments." Once again, the staff say, the patients' interpretation is symptomatic of their disorder. For instance, a patient named Mellor is called to a meeting of his MDTP panel. He is informed that he has lost his weekend pass privilege because of limit testing. Mellor is angry and says the staff has treated him with contempt (he needs the pass for a family celebration) and has done him an injustice (other patients offended similarly but were not punished). The meeting ends, and Mellor is sent away with good advice: "We are here to treat you, not punish you. One thing you should be thinking [about now] is why you think we want to punish you."

Ideology

"Ideology" is a polysemic concept, and I want to make clear what I mean by the term as used in the following pages (cf. Geuss 1981:chap. 1; Thompson 1984:chap. 3). I am focusing on a combination of ideas and practices that are particular to the center, and I am calling them "ideological" because they are deployed against points of resistance in the knowledge production process: the process of eliciting and appropriating narratives, connecting the traumatic past to the clinical present, and so forth. The job of the ideology is to coerce or convince patients and therapists to do certain things that they might not otherwise want to do: patients are to be persuaded to give up memories, therapists to inflict pain, and so on.

One can imagine a psychiatric institution that needs no ideology because it has no need to overcome resistances. It might be a place in which therapists possess a pharmacotherapy that is sufficient for controlling patients' behavior, where patients are either literally or figuratively voiceless, and where therapists and patients live in incommensurable worlds of experience. Some acute psychiatry wards are like this, and perhaps PTSD will also be treated this way one day. However, the center does *not* fit this pattern. It is an arena of contested voices and moralities, and it is populated by thoroughly unsedated patients.

The center's particular ideology springs from a combination of its distinctive mission, its commitment to psychoanalytical truths, its rejection of pharmacotherapy, and—a subject unmentioned until now—its distinctive division of labor. There is no standard table of organization for psychiatric units in the VA medical system, and plans for staffing units like the center have to be worked out on an individual basis. In the case of the center, the table of organization was worked out in advance by the regional VA chief of staff and the center's administrative director.

The VA's official division of psychiatric labor sorts workers into services. Each service is a specialty with its own career line and administration: psychiatry, psychology, nursing, social work, rehabilitation counseling, and occupational therapy. The combination of workers that composes a treatment unit's table of organization is based on the unit's clinical mission, its therapeutic orientation (units that rely on drugs are usually less labor-intensive than units oriented to psychotherapy), its annual budget (one psychiatrist costs as much as several counselors), and its ability to recruit workers from a given service (less skilled workers are generally easier to recruit than workers with advanced degrees).

Because of the center's psychotherapeutic orientation, it is a labor-intensive operation, with an average of three staff to every two patients. This proportion is economically feasible because the center recruits its therapists mainly from the less intensively trained categories. At the same time, the unit has had difficulty filling its allocated psychiatric positions, partly because younger psychiatrists are disinclined to work in a unit organized around psychodynamic principles. The center is funded for two psychiatrists: a clinical director and a head of inpatient services. Except for a brief period, the second position has remained unfilled and the clinical director has had to divide his time among the center's units: inpatient, outpatient, diagnosis and evaluation, and education.

To summarize, the center's mandate requires it to provide patients with a distinctive treatment program. The center's directors developed a suitably distinctive program, but it is also one that arouses anxiety, fear, frustration, anger, and uncertainty in its therapists. Highly trained and experienced therapists are better able to manage these emotions than are less skilled categories of therapists, who are inclined to avoid clinical practices and situations that expose them to these emotions. Forces operating outside the center's control ensure that a majority of its clinical staff will be recruited from less skilled and less experienced categories of workers.

Ideology and Narrative

The center's clinical ideology and the facts and findings of psychiatric science share the same technical language ("avoidance behavior," "symptomatic anger," etc.), the same terminus a quo (*DSM-III*'s list of criterial features), and the same stock of tacit knowledge (ideas about the traumatic memory, the movement of traumatic time from past to present, etc.). Unlike scientific discourse, a clinical ideology is a local system of knowledge, embedded in a particular institutional hierarchy and production line.

Clinical ideology and psychiatric science are interdependent practices. Science provides ideology with a distinctive object—it naturalizes PTSD, for example, giving it an existence that is independent of the clinical practices through which diagnostician and therapist encounter it—and ideology, in turn, provides science with the institutional surfaces on which its invisible object is inscribed. In the early days of the center, the unit's more experienced therapists tended to view scientific knowledge and local knowledge as parallel rather than intersecting systems. As the ideology evolved, this view was gradually displaced by the perception that the center's distinctive clinical practices and routines represent, not a separate system of knowledge, but rather the penetration of psychiatric science into clinical practice.

The center's ideology is organized around a written account of PTSD and a set of oral commentaries. The written account is inscribed in a series of memoranda that have been distributed to each member of the clinical staff. Therapists refer to this account as "the ten propositions" or "the model," and it is often mentioned in conversations between therapists concerning their patients. Every member of the daytime staff has personal responsibility for one or more patients. This means tracking the patient's progress and providing him with individual psychotherapy. Therapists are required to discuss their patients, with other staff members and with the clinical director, on four recurring occasions: daily debriefings that follow group psychotherapy sessions, weekly clinical supervision, scheduled meetings of MDTP panels, and unscheduled "limit-setting" sessions. This means that, over the course of a year, there are numerous occasions when each therapist is obliged to demonstrate his or her practical understanding of the clinical ideology before a presumably knowledgeable audience.

The written account ("the ten propositions") is a gnomic text and is represented in clinical talk mainly in the form of the following three narratives, based on Freudian notions.

The Narrative of Splitting

Our mental life is dominated by two instinctual drives: an aggressive drive, with its urge to assault and destroy, and a libidinal drive, with its urge for tenderness and longing for attachment. Normally, the drives are fused and, in this condition, can be said to neutralize one another. PTSD begins with an occasion on which the individual is faced with a situation in which both drives are mobilized, but he can find no solution that will simultaneously satisfy both. The situation is resolved when the aggressive drive, no longer coupled to the libidinal drive, satisfies itself in a violent and conscious act of commission.

The act of commission has two important consequences for the individual. First, the old self splits into two part-selves: an aggressor-self organized around feelings of anger and destructive impulses, and a victim-self organized around pathological (i.e., unfused) feelings of tenderness for victims of aggression. In the case of a soldier, pathological tenderness is often directed toward dead comrades and is experienced as a combination of grief over their deaths and guilt about his own survival. The split is marked by the peremptoriness of aggressive impulses, and this explains the individual's current tendency to explosive violence. Second, he is overcome by a compulsion to reenact this traumatic experience, in a futile attempt to achieve mastery over his actions. The split selves reproduce, rather than resolve, the conditions that evoke their responses.[2]

Splitting is simultaneously a symptom of PTSD and an obstacle to the forms of self-knowledge and self-disclosure that are required for recovery. It is only as a part-self that the man can now talk about himself and his experiences. Each part-self is imprisoned within its emotion—one part in anger, the other in pity and guilt—and is unable to see events and experiences except in terms of their personal meanings. The patient's narratives are never more than part-accounts, for he divides each event or experience into two stories, told from incommensurable points of view. His consuming anger and pathological tenderness provide his wartime narratives with such a quality of immediacy and timelessness that we feel he is actively reexperiencing the events rather than simply describing them.

"Recovery" means restoring the integral self, fusing the aggressive and libidinal drives, and *making the past dead*. To achieve this, a man must be brought back to his traumatic memory, back to his willful aggression, the moment of autonomy before the split. Authentic knowledge of the event demystifies the sources of his symptomatic behavior, demonstrates that he is responding to internal, psychic strivings rooted in the past and not, as he has been deceiving himself, to the provocations located in the here-and-now. To recover, he must, in an act of autonomous will, acknowledge that he is the author of his own traumatic act, that he chose to hurt and to destroy when he could have chosen a different course. Only through this act of public disclosure can he emerge from the timelessness of PTSD.

Recovery is possible because the libidinal drive is the source of two possibilities: *pathological tenderness*, through which the patient identifies himself with the sufferings of victims, including his own, and attempts to evade the punitive superego; and an *intent to relate*, an urge to connect and communicate his deepest feelings and thoughts. It is the second urge, the patient's intent to relate, that makes it possible for him to disclose (share) and process his traumatic memory with the therapists and patients in his psychotherapy group. While the pathological expression of the libidinal drive perpetuates the disorder's underlying conflict, the urge to relate is essential to resolving it.

The Narrative of Contagion

"Stress responses" are ways in which people attempt to manage anxiety. Anxiety is provoked by danger and, in the case of PTSD, danger is identified with the traumatic act of commission and the punitive superego. Stress responses associated with PTSD are intrinsically pathological. Thus, avoidance behavior insulates the individual from intrusive memories but also makes it impossible for him to confront and resolve the sources of his disorder. Likewise, aggressive acting-out and pathological tenderness are the patient's attempts at resolving, in the here-and-now, anxiety created by the traumatic event in the there-and-then. Because PTSD erases the line between past and present, the latter solution seems possible; in fact, it merely reproduces the traumatic situation without resolving it. More-over, like avoidance behavior, this response is itself a source of pain and impairment.

Stress responses are a recurrent theme in psychotherapy sessions, de-briefings, MDTP panels, and clinical supervision. In practice, therapists apply the term to any behavior that is overtly aggressive or disruptive, any behavior that therapists "feel" has an aggressive intent, and any behavior, such as somatic complaints, that interferes with the process of eliciting details concerning a man's traumatic event and its reenactment.

The patients' stress responses are simultaneously *forms of resistance* (both an obstacle to and an opportunity for retrieving the traumatic mem-ory) and *forms of seduction*, through which the patient attempts to incor-porate the therapist into the traumatic memory/situation. Because the patient's self is divided into two parts, seduction takes two forms. The therapist struggles with his fear of the patient's anger as well as his fear of his own anger (and his ability and intention to harm the patient). He experi-ences anxiety and conflict, and it is under these conditions that the patient's victim-self seduces him, draws him into its disorder, replacing the "thera-peutic alliance" with a pathological bond. At this point, the therapist iden-tifies himself with the suffering of his patient and is transformed into the patient's ally, in opposition to the center's clinical regime. Alternatively, the therapist may respond to the patient's anger with his own anger and lash out against him. Through this act, he confirms the patient's pathological view of life and justifies his continued aggression.

When either of form of seduction is detected, the therapist is said to be "colluding with the patient in his pathology." Collusion is a double be-trayal. The therapist betrays his patient, by abandoning him to his pathol-ogy, and he betrays his own colleagues, by undermining their confidence in the rules and rationales on which the therapeutic regime is based.

Now I want to illustrate how ideas about collusion operate at the center. My story concerns five people: three patients (Rogers, Thomas, and Preston), the clinical director (Dr. Durocher), and the inpatient unit's senior psychotherapist (Dr. Jeffries).

In the spring of 1987, a veteran named Philip Rogers was admitted to the treatment program. He was a Vietnam War combat veteran who had been diagnosed with "schizophrenia, paranoid type" in 1974. In the intervening years, he had been "maintained" on antipsychotic drugs, mainly as an out-patient. In late 1986, a VA mental health clinic referred Rogers to the cen-ter for evaluation. On the basis of the clinical interview, his test results, and his military history, the admissions committee concluded that Rogers's pri-mary diagnosis was actually PTSD and that his symptoms were serious enough to admit him to the inpatient program. His original diagnosis, schizophrenia, was assumed to have been incorrect: intrusive symptoms had been mistaken for hallucinations. Other men with similar medical histories had been previously admitted to the center and seemed to have responded to treatment.

There is a convention on the ward that, during group psychotherapy, no patient should interrupt another once he has begun to talk about his traumatic experience. Both patients and staff believe that the men have intense difficulty talking about these events. Patients expect each other to listen to the narrator with attention and respect and to show signs of support and solidarity during and after his disclosure. Disclosures are regarded as very serious moments and when it appears that someone is beginning a narrative, all peripheral movement stops. The only acceptable reason (among patients) that a man would have for breaking into someone's traumatic narrative is to "save" him, by deflecting a therapist's line of questioning if it is believed to be causing the narrator exceptional distress. Rogers was unusual in that he casually disregarded this convention. On several occasions, he broke into traumatic narratives with irrelevant remarks. (He had a reputation for also breaking into conversations in the dayroom, during free time. His comments were coherent but unrelated to what the other men were discussing. After a short while, he was shunned by other patients.)

At the end of his first month in the treatment program, Rogers' MDTP panel convened to review his progress. The panel concluded that his primary diagnosis of PTSD was incorrect and that he was, in reality, psychotic. His disorder had features of PTSD (which explained why he had been admitted), but he lacked sufficient ego integrity to develop the full-blown disorder. "There is nothing there [in his psyche] to split."

Two conclusions followed from this one: it would be unrealistic to expect Rogers to benefit from further treatment at the center, and he might be unable to continue to control his violent impulses. The clinical director was consulted and agreed with these conclusions. The case was closed a few days later when Rogers asked for an authorized absence, to attend to some family matter. His request was refused; he responded with angry threats; and an MDTP meeting was convened. At this meeting, Rogers was asked to explain his threatening behavior; the panel listened to his explanation and then discharged him on the grounds that he had violated the part of the patient contract concerning communal safety.

About four months later, a patient named Chris Thomas was readmitted to the inpatient unit. Thomas was already notorious for his strange stories and capacity for self-dramatization. He claimed to have been a member of an airborne brigade in Vietnam, but his narratives lacked any concrete information about actual sites or military units. When he was asked to describe what he thought or felt during some narrated event, he answered in a depersonalized way. If pressed for details, he would string together empty metaphors and clichés. During one session, he mentioned an ambush in which his closest buddies were killed. "What was it like for you? I mean you were there and you were looking at their dead bodies. What went through your mind? How did you feel?" he was asked. "There's no time to cry over spilled milk. You win some, you lose some," Thomas replied.

When Thomas did supply details, they were often absurd. He described an episode in which his unit was taken to an operation by helicopter. For a reason not disclosed by Thomas, the pilot would not bring the helicopter close to the ground but told the soldiers to jump out, a vertical drop of seventy feet. The therapist asked Thomas if he meant what he had just said: "You know that's like jumping off the roof of a seven-story building. Is that what you're saying?"

"Hell, yes," Thomas answered. "We just dusted ourselves off and laughed our heads off."

During the debriefing following this session, the therapist commented, "You know, you listen to Chris talk about himself and Vietnam, and he makes it sound like he was watching a cartoon."

While the content of his narratives was abstract and cartoonlike, Thomas's delivery was usually dramatic and emotional. On most of these occasions, he appeared to be overcome by such strong feelings that he was unable to complete his accounts. Consequently, it was never clear what actually happened to him in Vietnam. Sometimes his story broke off in passionate sobbing, and the other patients in the group would cluster around him and comfort him with words and soothe him with embraces. These tearful events alternated with moments of really frightening anger. Trembling with emotion, Thomas would excoriate the "politicians" who left him "swinging in the wind." And yet, in neither the tearful moments nor the angry ones did Thomas ever indicate exactly why he was so upset. When the therapists probed further, his distress would rapidly escalate to the point where it seemed counterproductive to ask anything more.

During this same period, Carl Preston, a former airborne soldier, was admitted to the unit. A serious and articulate man, who had been referred to the center from a VA medical center in a neighboring state, Preston was the scourge of his therapists. Two or three times a week, during group psychotherapy, he would fall into a state that his therapists called "paroxysmic." The intensity of his anger was terrible to see, and its onset was sudden, without obvious provocation. His message was clear and constant: once pushed beyond his limit, he would inflict grievous injury on his primary therapist, Dr. Jeffries.

It is not uncommon for patients to make threats. Sometimes it is against "society," or the "VA bureaucrats," or the staff as a collectivity. Nearly always, the threats form part of some grandiose narrative. The most common: the patient will perpetrate an act of public violence; television crews will come to cover the event; it will be broadcast on the nightly news; and viewers will learn, through his act, that "Vietnam veterans have been fucked over by the government and fucked over by the VA and fucked over by you people [therapists] here." But Preston's threats were entirely different. He had no message for the American people; he simply hated his therapists, because they were getting rich from his suffering and because they were "useless" frauds, and, when the time was right, he would not hesitate to inflict violent retribution upon them.

"When the time was right" was no abstraction. Preston had finished his tour of duty in Vietnam without any serious injury. About ten years afterward, however, he had been in a road accident, and his left foot had been amputated. (It would seem that the amputation was doubly painful because most people assumed that it was a war injury, and Preston was repeatedly put in the position of having to either explain or to deceive.) Now he moved on crutches and could rise from his seat only slowly and with difficulty. For the time being, there was no fear of an assault taking place during psychotherapy. Preston always wore short pants to these sessions, and when he harangued the therapists, he would move the naked stump up and down. While at the center, he was being fitted for a prosthetic device that would eventually allow him to walk without crutches. It was when this apparatus was finished, Preston told the therapists, that he would take care of Dr. Jeffries. He conveyed utter sincerity.

Dr. Jeffries was the primary therapist for both Chris Thomas and Carl Preston, so both patients were in the same psychotherapy group. After several weeks of working with them, Dr. Jeffries complained to his cotherapist that he was finding it impossible to "break through" to either man. No matter what strategy he pursued, he was unable to get their trust or cooperation. Although he was spending ten hours a week with each man in individual and group therapy, he knew no more about them now than when they first came to the center. Also, Preston's threats were beginning to frighten him. At the end of one debriefing, he suggested that these men were "characterological," implying that they had severe personality disorders. Gradually, Dr. Jeffries became more specific, diagnosing Thomas with "histrionic personality disorder" and Preston with "borderline personality disorder." From a clinical perspective, this is a serious claim, and Dr. Jeffries, a knowledgeable psychotherapist, recognized its significance. In Preston's case, it meant that the ego structure was incomplete: it could not support the "therapeutic alliance" on which the treatment program is based, nor could it be expected to contain Preston's violent urges. In other words, Preston, although not psychotic, shared disturbing similarities with Philip Rogers, who had been discharged earlier.

Each of the two psychotherapy groups at the center is supervised by a pair of therapists, and the daily debriefing is attended by all four. It was during these debriefings that Dr. Jeffries began to question the initial diagnoses (PTSD) given to Thomas and Preston, but his comments were politely ignored by the other therapists. This situation continued until a debriefing session attended by the clinical chief, Dr. Durocher. The visit was unscheduled, and it was unclear whether or not he had learned of Dr. Jeffries's concerns. It did not take long before Dr. Jeffries repeated his comments and remarked that he found it frustrating to work with patients with characterological problems. This observation was followed by silence, during which the other therapists stared at the table. Then Dr. Durocher spoke, slowly and with deliberation. It was obvious that he was not happy with Dr. Jeffries. He reminded the psychologist that Thomas and Preston had gone through a rigorous evaluation process, and the admissions committee had unanimously diagnosed them with PTSD. Thomas's combination of hyper-emotionality and lack of concreteness, like Rogers's rage, was characteristic of PTSD. It was ironic that Dr. Jeffries would want to discharge them for having symptoms of the disorder that he was supposed to be treating. A (good) therapist would see that these behaviors are therapeutic opportunities and that he ought to "go with the resistance, not surrender to it."

And now the coup de grace. Dr. Durocher asked Dr. Jeffries to tell him why he believed that there were important differences between Thomas and Preston and the other patients. It was now clear to everyone in the room where Dr. Durocher was headed.

"You think it's a [i.e., my] stress response," Dr. Jeffries asked.

"I think that doubting whether these patients have PTSD indicates a lack of insight," Dr. Durocher replied.

"It's possible. . . . I'll have to think about it," Dr. Jeffries responded.

The implication, as explained to me by both Durocher and Jeffries, was that the latter had been seduced into colluding with the patients' aggressor-selves. Although he would have been unaware of it at the time, Dr. Jeffries would have been assisting the men to achieve their pathological goal of being thrown out of the center. From this point on, Dr. Jeffries kept to himself whatever doubts he might have had about the patients' diagnoses.

The Narrative of Survival

The narratives of splitting and contagion form the core of the center's ideology, but they are not the only etiological accounts of PTSD that circulate on the ward. There are two other important stories. The first is the narrative of cognitive restructuring, based on a mélange of ideas and scripts culled from the "rational-emotive therapy" of Albert Ellis (1977) and the "cognitive therapy of emotions" of Aaron Beck (1976). The theme of this narrative is that the center's patients experience the world as a negative and punitive place, and this experience is perpetuated through their characteristic cognitive errors and distortions, such as overgeneralizing and "mind reading." These misinterpretations induce negative emotions, especially anger, that are acted out in ways that generate self-defeating situations. The therapy consists of didactic instruction and "take-home" exercises, in which patients are directed to test their assumptions against the objective reality. The men are not expected to connect cognitive errors directly to pathological splitting, and they tend to regard these sessions as a source of "helpful hints" for avoiding arguments and fights.

The other important narrative is the "Chap's Rap." The chaplain is an ordained Protestant clergyman who has a doctorate in clinical psychology. His primary appointment is at the center, where he serves as the senior clinical psychologist in the outpatient unit, but part of his time is spent ministering to the spiritual needs of patients who are hospitalized in the medical center's various units.

During the 1960s, when he was in his thirties and well beyond the age of conscription, the chaplain volunteered for military service and was sent to Vietnam, where he served a tour of duty as minister and counselor in an infantry division. He is well regarded by the center's patients, partly because of his reputation for being a "no-bullshit guy" who "got his ticket punched." The Chap's Rap consists of three talks that are given in the evening, after the regular sessions have finished. Attendance is voluntary, since VA patients cannot be required to participate in activities connected with religion. He is an energetic speaker, his narratives are said to make good sense, and most of the patients attend at least one cycle.

The Chap's Rap is essentially psychological and touches only incidentally on "spiritual" matters. Indeed, it is a story in which sin is displaced by guilt; traditional values take a back seat to evolutionary biology; moral authority is subordinated to reason; cognitive restructuring replaces expiation; and coping substitutes for redemption.

The narrative begins with the conundrum, How can a veteran live with the knowledge that he committed atrocities and other moral transgressions? This is a recurrent question on the ward, and the clinical ideology provides one answer, in the story of splitting. According to the ideology, splitting is a defense against the superego, whose job is to punish transgressions. In group psychotherapy, it is common to hear a patient (now speaking as the aggressor-self) describe himself as being "no better than an animal" in Vietnam. To say that I am an animal and no longer human is equivalent to placing myself outside the jurisdiction of conventional morality. At other moments, one can hear the same men (now speaking as victim-selves) describe how their comrades were tortured to death, how the comrades' bodies were mutilated, how all of them (American combat troops) were betrayed by politicians and peaceniks, and how, at this very moment, all of them (PTSD patients) are being mistreated by the VA. The message now is about powerlessness, but the moral implications are the same. The patient gets himself off the hook, this time by substituting himself (or his surrogate, the dead comrade) for the victims of his own aggression. "You're looking for victims, are you? Well, you don't have to look far. We're your victims." The part-selves resolve the patient's moral dilemma by removing him from the domain of moral judgment—but at enormous cost, since the patient is then effectively cut off from the rest of humanity. The purpose of the Chap's Rap is to suggest another solution, one that is not pathological. Instead of removing the patient from the domain of morality, the chaplain's narrative transforms the conditions under which moral judgment is to be made.

His story line is that combat soldiers put themselves under the control of a hard-wired survival circuit, located somewhere in the nervous system outside the mind's rational control. The circuit is simple: a man perceives a source of danger, and reserves of energy are mobilized to prepare him for fight or flight. The "warrior's" job is to fight, so the energy is converted into anger and aggression and expresses itself in destructive violence: enemy sighted, enemy engaged, enemy destroyed. The soldier's habituation to this neural circuit is a precondition for his survival, and it takes place behind his back. The circuit is a legacy of the time when our primitive ancestors lived in the wilderness. Over the ages, it has weakened, but it is still there, and the job of military training is to reinvigorate it.

In Vietnam, danger was pervasive and invisible: booby traps every-where, guerillas dressed as civilians, every turn in the path the site of a possible ambush. Survival required an instantaneous response to environ-mental stimuli. There was no time to filter this information through the higher mental functions, and sometimes this resulted in acts that the men now regret. But face it: the survival circuit did not develop in the course of human evolution to protect the innocent from moral transgression.

In the Chap's Rap, guilt is the key to understanding PTSD. The narrative distinguishes between "realistic guilt," which is a response to objective responsibility for acts committed, and the "pathological guilt" that is symp-tomatic of PTSD. The men's pathological guilt is traced to three sources: their ignorance of the mental transformation that took place when the locus of control moved out of the neocortex, releasing them from moral inhibi-tions; their memories of the "psychological high" that they experienced while committing transgressive acts (e.g., the pleasure of inflicting pain on one's enemy); and their symptomatic response to feeling joy at outliving their buddies (also known as "survivor guilt").

Feelings of guilt were a persistent source of anxiety even in Vietnam. Guilt-driven anxiety percolated into the soldier's consciousness as the fear of "pay back" (retribution), and his anxiety was intensified by his realistic fears of the ubiquitous enemy. Guilt, fear, and anxiety fed on one another and fueled frustration and anger. As a result, acts of aggression exploded with greater frequency and in response to weaker and weaker provocations. This explains why atrocities were often committed in "cold blood" (unac-companied by strong emotion).

The sequence of stimulus-response that originated in a hard-wired survival circuit eventually transformed itself into a self-renewing cycle. It is this entrenched cycle of guilt, anxiety, and aggression that the warrior brings home with him from Vietnam. Veterans react to their emotional distress by attempting to control anxiety rather than guilt or anger. They commonly do this by "self-dosing" with alcohol and drugs—a maladaptation that is not specific to PTSD, of course. They also employ a disorder-specific way of limiting anxiety, through efforts to control the fear of retribution (payback).

It is at this point that the Chap's Rap intersects the ideological narrative of splitting. The chaplain informs patients that "making your life a memorial to your dead buddies" is equivalent to saying that you have already been punished (killed), and therefore you have no reason to worry about retribution. Men combine this pathological identification with a pattern of self-defeating behavior, which brings them punishment in the form of poverty, beatings, broken marriages, et cetera. Through this self-inflicted pain, the patient (the ego rather than the superego, in psychoanalytic terms) becomes the agent and locus of control of his own retribution. In the ideology's narratives, this is the strategy associated with the victim-self.

The alternative strategy is to transform oneself into "the meanest motherfucker in the valley," a merciless and omnipotent destroyer of life. It is this claim to omnipotence that makes the soldier or veteran invulnerable to retribution. This strategy works no better than the first, and the men are driven deeper into their pathology. In their anger and aggression, they reach for omnipotence, but they can only grasp their guilt.

The narrative of survival now reaches its denouement. It is futile to try to break the cycle by controlling anxiety or anger. The patient's core problem is the guilt at the root of his anxiety. At this point, the distance that separates the chaplain's narrative from the center's ideology becomes clear. In the latter, guilt is an epiphenomenon of the unfused libidinal drive, and the story's trajectory leads to the therapeutic fusion of libidinal and aggressive drives, initiated by the patient's act of "disclosure." In the Chap's Rap, guilt is irreducible and is absolved through "confession" and reparation.

> You must say, "I did these bad things. There are circumstances which help explain why I did them. I went to war with an untested value system. The values were unrealistic, because what was normal in Vietnam was craziness back home. Those values didn't recognize that people are fallible. But even if I can explain what I did, I still take responsibility for it." . . .

> This is your act of confession. Your act of reparation is to recognize your own fallibility and to forgive yourself. And then move on with your life. Right now, moving on means sticking with the treatment program.

There is one other important difference between the two sets of stories. The Chap's Rap is didactic; it commands no clinical practices. Patients can talk about it, but, aside from its last bit of advice, they cannot do anything with it. Indeed, there are even limits to *where* they can talk about it. Men are free to talk about the Rap during group psychotherapy, but only so long as it leads to an intersection with the ideology of the institution as a whole. At this point, they must follow the trajectory of splitting and disclosure. The chaplain's narrative is given no place in clinical supervision, nor would any therapist introduce it into the debriefings that follow group psychotherapy or into MDTP panels. (Because he is in the outpatient service, the chaplain attends none of these meetings.)

Attitude Problems

The center's ideology is designed to do two jobs. The first is to create a language game in which the patients' traumatic memories can be connected to their behavior at the center. The second is to appropriate or overcome resistances to making these connections.

Therapists

Therapists are expected to accept both the validity and efficacy of "the model," its narratives, and the practices that are based on them. Any clinician who is perceived to be merely "going through the motions" is described as having an "attitude problem."

New staff are recruited by the center from among clinical personnel currently employed by the VA. Recruitment begins with an announcement in a VA bulletin, and applicants are interviewed by the clinical director and senior staff members. They are given a full account of the treatment program, and the difficulties ("challenges") encountered in treating the center's patients are candidly described. Consequently, no one joins the staff without having a good idea of what he or she is getting into.

The center's staff members are civil service employees. They have well-entrenched rights and an acute sense of what can or cannot be expected of them as "mental health professionals." When they are pressured to toe the ideological line, it is invariably in the language of the clinic: resistance is described in terms of the their "stress responses" and is said to work against the patients' best interests. With few exceptions, these corrective moments take place during clinical debriefings and supervision. Administrative and professional hierarchies coincide at the center, however, and when the clinical director makes a critical comment about a therapist's clinical acumen, he is giving both a professional opinion and an official evaluation. Any therapist who is unhappy with these work conditions is free to transfer out of the center into another VA treatment unit.

While I was at the center, a handful of staff members left voluntarily, mainly to work at a higher grade and salary at other VA units. During the same period, three staff members were forced to leave: a rehabilitation counselor was discharged from service with the VA because he had misrepresented his educational credentials; a psychiatrist was discharged after his state medical board revoked his license because of professional misconduct in a previous position (outside the VA); and a counselor was forced to transfer to another VA unit. The latter employee had been with the center from its inception. Her first months at the inpatient unit had passed without serious problems, but that changed after "the model" was introduced. As time went by, it became obvious to the staff and the clinical director that she was making no effort to absorb the ideas and language of the model. During clinical supervision, the director asked her to explain her lack of familiarity with the model and her failure to incorporate it into her clinical practices. The woman replied that if he wanted her "to study his ideas," the director would have to give her paid time during the workday for this purpose. She expressed a strong desire to stay at the center and, following her transfer, appealed her case to the VA grievance board, but lost.

Therapists toe the ideological line because the model is the voice of authority. This is not their only reason, however. The ideology is also valued because it is useful, in the sense that it tells therapists what to do next. It tells them how to occupy the time during sessions of psychotherapy: what to say, what to ask patients to say, how to create a semblance of momentum, and so on. It also tells them what to do when patients challenge the moral and clinical authority of the therapeutic regime: how to respond to patients' anger, for example, or cope with cases of therapy-induced suffering. In addition, the ideology produces evidence that it is correct: it accurately predicts that psychic and somatic pain will intensify when patients "get close to issues," that patients will alternate between moments of anger (aggressor-self) and guilt (victim-self), and so on.

Patients

Patients are expected to resist the ideas on which the model and treatment program are based, for this is the nature of their disorder. In practice, the patients reject only *some* of these ideas.

Without exception, patients accept the idea that PTSD is a real disorder, as real as tuberculosis. No man rejects the facticity of PTSD, and no man questions the correctness of his own PTSD diagnosis. He has been examined by a team of experts, has been through endless interviews and tests, has read descriptions of PTSD and knows that its symptoms are consistent with his own experiences. Patients have an additional reason for accepting the diagnosis: it is psychologically satisfying. It tells him he is sick but not psychotic, that he has a reversible, sociogenic disorder and not a mental disease. This is an attractive idea to someone who was once diagnosed as schizophrenic or who, until now, was sure that he was going crazy.

Patients have a further reason for accepting the disorder's facticity and their own diagnoses. Most men believe that PTSD is a curable disorder for which an effective treatment is available. This conviction gives patients the hope of a new future and a new past: an opportunity to renarrate wartime experiences and chaotic postwar careers, a chance to free themselves of guilt and pain, et cetera. The treatment program does give some patients a sense of self-efficacy, a feeling that they are gradually gaining control over their thoughts and emotions and that they are developing a new and effective style of interacting with people. This impression tends to fade over time, however, and many patients eventually conclude that the center does not possess an effective cure and that, worse still, PTSD is not a reversible condition.

Although the facticity of PTSD is accepted by everyone at the center, patients hold various ideas concerning the precise cause of the disorder. Men who lived through singularly horrible experiences are likely to accept the official explanation, that PTSD is precipitated by a discrete event. Take the case of Carl Metcalf, a patient who had been a "tunnel rat" in a marine unit in Vietnam. Metcalf's specialty was to scout the networks of tunnels that were used by Vietcong forces as command centers and field hospitals. During these missions, he was equipped with a flashlight, a knife, and a pistol, and his job was to bring back useful information. On Metcalf's final mission, he descended into a tunnel system that seemed to have been recently abandoned. Deep inside this maze, his flashlight went out. Because the tunnel systems were often very complicated and rigged with booby traps, tunnel rats developed methods of marking their routes, so that they could find their way back to the point of entry. In the darkness, Metcalf became disoriented and was unable to find the marks that he had made on the walls. He wandered around in the darkness for what seemed like hours and heard real or imagined voices, conversing in Vietnamese. His situation seemed hopeless, and he decided to kill himself. At this point, Metcalf chanced on some marks that he had cut into a wall, and he eventually found his way back to the entry tunnel. Above ground, he discovered that he had been abandoned by his unit, and he became conscious of a splitting headache. According to Metcalf, the headache has never ceased, and it is now his major symptomatic complaint. On a scale from zero (pain free) to ten (blinding pain), he says that he oscillates between seven and nine.

Metcalf had several frightening experiences in Vietnam, but this one stands out from the others. For him and his therapists, it is the ordeal that precipitated his PTSD. But there are other patients who are unable to identify any episode of this sort. Either they endured several equally awful experiences or they remember no extraordinary experiences. Some of these men accept the possibility that they are suffering from psychogenic amnesia, while others hold a neurasthenia-like theory of PTSD, tracing the disorder to discrete traumatic events (Metcalf's case) or, alternatively, to multiple subtraumatic events (their own cases).

The issue of specific causation does not have great salience for the patients, though. While this subject is occasionally mentioned in group psychotherapy, it is not a source of acting-out behavior ("stress responses"). All patients, even men who favor a neurasthenia-like theory, have a compelling motive for overtly accepting the center's official position, that a specific recallable event lies at the heart of PTSD. PTSD is a "service-connected disability." This means that diagnosed patients are eligible for compensation proportionate to the degree of impairment that they currently suffer as a result of PTSD. It is insufficient to have PTSD *plus* an impairment of some kind. A compensation rating board must be convinced that the impairment is the consequence of PTSD, not of some concurrent medical or psychiatric problem that is not service-connected.

A man who is assessed with 100 percent disability, meaning that he is too impaired to hold down a job, receives monthly compensation of $1,800, tax-free (the figure is from 1987). The declared annual income of the center's patients averages less than $1,000. Disability ratings vary from 0 percent to 100 percent, and every man is rated 100 percent disabled during periods of hospitalization. The average length of stay for men who complete the treatment program is between four and seven months, which means that by the time they leave the center they will have collected between $7,200 and $12,600 in compensation. Disability claims can also be retroactive. Take the example of someone diagnosed by the VA in 1974 for schizophrenia (which is, by definition, not service-connected) and rediagnosed in 1987 with PTSD. Let us say that he is able to substantiate the claim that his original diagnosis was incorrect; that is, he was never schizophrenic but has suffered from undiagnosed PTSD since 1974. This man will receive 100 percent compensation for the periods when he was hospitalized for "schizophrenia," as well as compensation proportionate to his service-connected disability during the same period, that is, back to 1974. Retroactive claims can be very large, and during my stay two men filed claims for $60,000 and $44,000 respectively.

At the time of discharge, each man is evaluated and a report is prepared, indicating diagnosis, severity of symptoms, and level of impairment. This becomes part of his medical record, but it does not determine his compensation rating, which is decided by a VA compensation rating board. While the boards are not bound by the center's diagnoses and evaluations, patients believe (correctly) that discharge reports may influence the boards' decisions. Typically, a man who enters the center with a low disability rating wants to leave with a higher one, and a man who enters with a high rating wants to retain it. A clinical evaluation that is positive may be a problem, since the rating board is likely to use it as grounds for disallowing or reducing compensation. This system creates opposing goals for clinicians and (some) patients. The therapist wants to reduce the patient's impairment, while the patient may want to produce the opposite effect. Although the therapists are aware of these conflicting interests, no patient is accused of exaggerating his symptoms out of economic motives. The clinically correct attitude is for a therapist to position his concerns inside a medical frame of reference, within which the patient is seen as someone who is employing his symptoms pathologically (rather than rationally), as a way of provoking the therapists' suspicions, anger, and aggression.

Disclosures

The Patient Contract stipulates that each man will make a determined effort to recall his traumatic memory and that he will disclose the contents of this memory during group psychotherapy. It is not enough for him to narrate this event to his therapist in the privacy of individual psychotherapy, nor is it enough for him to disclose it to other patients in places outside the control of his therapists—in the community room, mess hall, or dormitory. The insistence that disclosures take place in group psychotherapy rests on the idea that a patient and his therapist, by themselves, lack the psychological resources needed for recalling and processing the patient's traumatic memory. Through witnessing the honesty, strength, and open emotion that are shown by other members of the group while making disclosures, the new patient acquires the courage and confidence that he needs for revealing his own painful and dangerous memories. Further, the public quality of the disclosures—the fact that each man gives away a dangerous secret that could be turned against him—enhances the feelings of trust on which the therapeutic milieu is, in principle, based.

Group processes (the "feedback" provided by other patients) are also said to stimulate men to recall their events/memories completely and help them to objectify and process aspects that might otherwise remain unexamined and unconnected: "Narratives [given] in groups are necessarily multidimensional even when ostensibly aimed at specific others (other members can and do frequently join in). By contrast, narratives in the dyad are exclu-

sively unidirectional. The patient tells his or her story to the therapist"
(Lakin 1988:76). The group setting is believed to serve two further purposes. It allows men to symbolically reconstitute the situations in which
their traumatic experiences originated: the psychotherapy group in the
here-and-now represents the military unit in the there-and-then. Once this
substitution (the past for the present) has been brought into awareness with
the help of the psychotherapists, patients will be able to perceive a connection between their (symptomatic) behavior at the center and their etiological experience, that is, they will become conscious of the repetition compulsion that has shaped their postwar experience.

Patients spontaneously distinguish between two kinds of disclosures:
authentic disclosures, which they refer to as "events," and counterfeit narratives, which they call "war stories." Patients make this distinction overtly
on various occasions, both public and private. During psychotherapy sessions, it is made either in the abstract (with reference to no particular person) or with regard to some patient now discharged. But it is never used, in
front of therapists, to accuse someone who is actually present of fabrica-
tion. In contrast, therapists rarely make this distinction at all when they are
with patients. Clinicians believe that they are generally unable to detect
counterfeit accounts, and they regard public disclosure, in which a narrator
exposes his story to the scrutiny of knowledgeable combat veterans, as the
most effective means of encouraging truthfulness. During sessions, patients and therapists frequently mention the high "costs" of reclaiming and
"sharing" authentic narratives: the physical pain (especially headaches),
anguish ("impacted grief"), depression, nightmares, anxiety, and feelings
of hopelessness that are "stirred up" by these narratives. On the other hand,
"war stories" are painless, cost nothing, and are worthless.

Disclosures usually take place after a man has been in the treatment
program for several weeks. They are almost never spur-of-the-moment
acts. The typical patient plans ahead, to be sure that he has enough time to
complete whatever he wants to say before the session is up. Therapists
insist that group sessions begin and end precisely on schedule. The door is
closed when the session begins, and men who are late are eventually
brought before MDTP panels. The session ends when the clock reaches the
hour, the door is opened, and no further "processing" is allowed. A patient
who starts his narrative in the last minutes of a session will be cut off by his
therapist and told that he is "having a stress response." If this behavior is
repeated, the man will be called to an MDTP panel and told that he is
"avoiding his issues" (conflicts) and behaving aggressively by flaunting his
secrets. Flaunting secrets is the opposite of "sharing" and is said to interfere
with group processes by subverting the communal sense of security.

A patient who has failed to narrate an event after a couple of months is liable to be criticized by the men in his group who *have* made disclosures. Therapists and patients accuse such men of "not working"; they are shirkers who have shifted their part of the collective work load (and pain) onto other men. Eventually they are brought before MDTP panels, where they are "strongly encouraged to work." To let a man off the hook at this point, simply because he claims it is presently too painful to reclaim or disclose his traumatic memory, would be "colluding with the patient in his pathology." A man who fails to disclose after he has been given repeated encouragements will be discharged, and his departure will be described by his therapists as an instance of "limit setting." (A patient can ask to be readmitted to the center thirty days following discharge.)

Patients often respond angrily when they learn that a man has been "punished" by an MDTP panel, and the issue of limit setting is a frequent subject at the center's weekly "community meetings," when the patients' representatives discuss their collective grievances with the staff. On these occasions, and later in the day during psychotherapy sessions, the patients are reminded that the warnings, confinements, and discharges that are meted out by MDTP panels are *not* forms of discipline and that it is symptomatic to refer to these practices as "punishments." The practices are a way of supporting limit setting; limit setting is a way of managing stress responses; and stress responses are bad only in the clinical sense, that is, because they distract patients and healers from the job of recalling, disclosing, and processing traumatic memories. A stress response is not something a patient should feel sorry about, and it is not something that requires an apology, not even if it consists of vicious verbal abuse and threats. Thus, a patient who is apologetic or remorseful after an outburst of anger is told that his quest for atonement is a *continuation* of his stress response: the victim-self has taken over from the aggressor-self.

The clinically correct attitude for a patient to take following a stress response is to realize that the source of his behavior is a repetition compulsion and to acknowledge this (in his own terms) to the members of his psychotherapy group. This means telling them that, on this occasion (as on the occasion of his etiological event), he actively chose to "execute the intent to do harm." The clinically correct attitude is summed up in the phrase, "You must own your aggression." The (putative) effect of "owning" one's stress response is to objectify this piece of behavior: to know it as something different from normal reactive behavior.

During group psychotherapy, disclosures are exchanged between patients on what might be called a listen-now-and-pay-later basis. Because patients enter the treatment program individually and leave individually, the membership of each therapy group gradually changes. The men from whom a particular patient receives narratives are not necessarily the men to whom he eventually gives his own narrative. According to the therapists, the act of giving disclosures under these circumstances signifies the narra-tors' feelings of trust and safety. The free-will gifts are said to perpetuate a milieu in which future patients will be prepared to make disclosures. (This explains the concern about men who "flaunt secrets" instead of dis-closing events.) Whenever patients talk about these exchanges, the domi-nant register of meaning is economic/moral rather than psychological. The idea implicit in their talk is that a patient incurs a debt when he listens to (authentic) disclosures made by another patient. It is the urge to honor this obligation, as much as the encouragement given by therapists (disclosure is the path to recovery), that persuades many men to look for and disclose suitable memories.

While therapists can mediate these exchanges between patients, they cannot participate in them, since they themselves have nothing valuable to disclose. Occasionally a therapist makes what might be called a pseudo-confession: "While you guys were fighting in Vietnam, I was a war resister back here. I think I was right, but I don't think you were wrong." Like "war stories," the therapists' disclosures are worthless in the patients' eyes, merely risible. While exchanges of disclosures between patients and thera-pists are not possible, exchanges of acknowledgments *do* occur. This possi-bility is anticipated by the center's ideology in the idea of the narrative of contagion: the split-self is continually attempting to provoke stress re-sponses in other people, including therapists, to justify its own anger and aggression and to evade the punitive superego.

In reality, patients *do* manage to provoke the therapists' anger, and they *do* employ these occasions to shape meanings and claim equivalences be-tween themselves and the therapists. After a display of anger, patients will demand that the therapist "own" his or her aggressive reactions: "What's good for the goose is good for the gander, Lewis. Why are we [patients] the only ones who've got to apologize when we get angry?" Some therapists are prepared to acknowledge their stress responses when challenged in this way. When they do, patients respond with approval (or veiled sarcasm). At the same time, therapists are quick to scotch any suggestion that there is an equivalence between their anger (which is "normal" and originates in events in the here-and-now) and the patients' stress responses (which are symptomatic and originate in a compulsion to repeat traumatic experi-ences).

Psychomorality

The identification and management of pathological emotion—mainly anger and guilt—is a key task at the center. Anger is the more obtrusive emotion. It demands constant attention, because it disrupts clinical routines and, if left unattended, can be expected to lead to violence. Violence must be avoided at all costs, since it undermines the therapeutic alliance and the therapeutic milieu, the conditions that make it safe for patients to encounter and process their traumatic memories.

Therapists

Clinicians have several ways of talking about anger at the center. In the language of cognitive therapy, anger is a product of cognitive errors and the mistaken view that the world is a hostile place; in the narrative of the split-self, anger is the voice of uninhibited and pathological emotion; and in the ritualized exchanges of "acknowledgments" that occur during psychotherapy sessions, anger is a "stress response."

On these occasions, anger is pictured as something both negative and positive. Anger is negative because it distorts perception and cognition, undermines rational self-interest, and, when it is turned into verbal abuse and threats, drives a wedge between therapists and their patients. This is the version of anger encoded into the narratives of cognitive therapy and the split-self. Anger assumes a positive nature in the exchange of acknowledgments, however. Here it is a bridge rather than a barrier, something therapeutic, a psychomoral bond that connects the anger of the therapist, expressed in the urge to punish patients because they are abusive and frustrating, with the anger of the patient, the anger that precipitated the traumatic act and that is now compulsively repeated on the ward. The therapists cross the moral chasm separating them from their patients when they say (to their colleagues) that, in their urge to inflict clinical pain, they are "being the same as the patients." In these moments, their anger is transformed into a professional asset—"clinical insight." This, I believe, is how anger is experienced by the therapists.

The Moral Emotions

Patients and therapists can tell when a man is "angry" by reading his signals: what he says, the tone of voice in which he says it, his gestures, et cetera. They decode these signals in more or less the same way and have no problem in agreeing when someone is angry. It is a bit more difficult for them to know when a man is feeling "guilt," since they must rely mainly on the content of his narratives and remarks, and these can be interpreted in a variety of ways. But here again, patients and therapists share a stock of cultural knowledge and tend to agree on what sorts of statements signal feelings of guilt.

While patients and therapists can generally agree on whether a man is expressing anger or guilt, they often disagree on the *meaning* that should be given to "anger" and "guilt" on these occasions. In the case of the therapists, the meaning of these emotions is fixed by clinical practices, viz., cognitive therapy, the narrative of splitting, acknowledgment of stress responses. Although these meanings also penetrate into the consciousness of many patients, there are two significant differences: patients in general have fewer motives for accepting these meanings, and they operate with a concurrent set of meanings of their own.

Before I describe this alternative set of meanings, I want to contrast how most Americans *talk* about "emotions" and how they *experience* them:

The English language has a large number of emotion terms, and it is commonly assumed that these terms/states have something essential in common. However, efforts to include terms such as "anger," "boredom," "love," "happiness," "stubbornness," "hope," "disgust," "pride," "scorn," "distress," and so on within a unitary framework have proven fruitless (Ekman et al. 1972:39–55). In practice, these terms are connected by family resemblances rather than necessary and sufficient features. Further, most of these emotion terms are polyvocalic. That is, a single term is given different meanings in different situations, and different people may give the same term different meanings in what seem to be similar situations (Ortony and Turner 1990:323).

In both everyday talk and psychiatric discourse, a contrast (opposition) is drawn between emotion and cognition, that is, the faculty of knowing, perceiving, conceiving. In these conventional accounts, "emotion is to thought as energy is to information, heart is to head, the irrational is to the rational, preference is to inference, impulse is to intention, vulnerability is to control, and chaos is to order" (Lutz 1988:56–57). It is assumed that emotions, including anger and guilt, are the result of thoughts and meanings rather than their cause. While this is also how the center's patients and staff talk about emotions, it is not how anger and guilt are evidenced on the ward, where they are associated with three interconnected components: a sensate element (affect), a behavioral element (action scripts, narrative scripts), and a distinctive cognitive element.[3]

In reality, then, emotion is neither opposed to cognition nor produced by it. Rather, emotion is a structure that both incorporates and organizes cognition. Emotion performs its cognitive work (and meaning making) by foregrounding elements in the perceptual field via selective attention and inattention, mobilizing memories, and connecting the foregrounded perceptions and memory traces into constellations of meanings. In Michelle Rosaldo's phrase, these meanings are experienced as "embodied thoughts." Because they are multidimensional and complex, they are difficult to reduce to a coherent verbal form (Rosaldo 1984:143; Ciompi 1991:99).

Emotions such as anger and guilt are organized around conceptions of the self. The Western self is simultaneously a *psychological* construct, identified with ideas about the "mind," and a *moral* construct, associated with ideas about obligations (to other people, to nonhuman authority, to oneself). This culturally constituted self possesses three basic features: it is the subject ("I" and "me") of the individual's experiences and self-narratives; it is the initiator of his purposive actions; and it is the locus of responsibility for these actions. Moral responsibility further implies a single internal center of control.

Among the various emotion terms/situations expressed at the center, "guilt," "anger," and "shame" constitute a distinctive system of affects, scripts, and meanings that persists despite determined efforts, by therapists, to assimilate it to the clinical ideology. To describe this system and explain its tenacity, I will focus on certain meanings that are associated with these three emotions, setting aside for the moment the associated affects and behavioral scripts. The system of meanings is schematized in figure 4.

Shame is associated with three elements: an *act* performed (or not performed) by the person who feels the shame; a self-directed adverse *judgment*, tied to the idea that this individual now feels he is not the kind of person he assumed himself to be, hoped to be, or ought to be; and an *audience* before which he now feels degraded. This audience is a moral community, not simply an aggregate of people. Just as the existence of the moral community is a precondition for feeling shame, so is the capacity to feel shame a precondition for membership in this community. It is not shame that removes one from this community. On the contrary, it is shamelessness, described at the center as "a lack of self-respect," that puts one beyond the pale. In this sense, shame is different from embarrassment, which is "not an adverse judgement on the person as a whole, but an adverse judgement only on the person in a given situation" (Taylor 1985:75).

The patients constitute a moral community through a shared sense of shame, centering on their wartime experiences. It is within this community that their confessions are valorized and exchanged as gifts, each gift creating a moral debt, as described above. The efforts of therapists to give such gifts are ridiculous gestures precisely because they cannot belong to this community.

Anger is associated with three elements: an *act* performed (or not performed) by the person who is the object of the anger; the perception (by the person experiencing the anger) that this act has caused him an *injury or loss*; and the further perception that the injury is *unjustified* (Myers 1988; Lakoff and Kovecses 1987:209–211). Anger is a response to an assault on the self. The assault has the effect of taking away a right—the right to determine one's own actions or the right of ownership, for example—and it signifies a deficit or incompleteness. The way to recoup this deficit and to repair the self is by retribution or revenge, by inflicting a corresponding

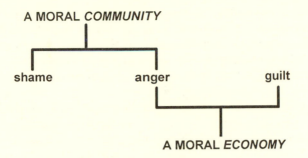

Figure 4. The deployment of the moral emotions at the center

injury or loss on the source of the anger. At the center, this is expressed as "getting even" and "paying back."

Shame and anger are connected by ideas and feelings about *autonomy*. Autonomy, the freedom to make one's own choices, becomes possible only when a person can both determine or initiate his actions (and therefore take responsibility for their consequences) and correctly identify his own best interests. When a man lacks either capacity, it may be morally justifiable to place him in the hands of a beneficent authority, either to protect him from himself (from "self-destructive" behavior) or to protect other people from him. Autonomy is seen as a natural right and a signifier of manhood. Thus, attempts to deprive a man of his autonomy (the ability to determine his actions) or his grounds for claiming autonomy (for example, his claim to know his own best interests) is an injury, actual or threatened, to the self and therefore provokes anger. The loss of autonomy is *shameful*, and the absence of anger under these conditions signifies the acceptance of this loss and it is *shameless*.

The patients' righteous anger over threats to autonomy focuses on the repetition compulsion, the splitting of the self, and exploitation (the claim is that politicians, war profiteers, and therapists make or made money from the sorrows of soldiers and patients). This anger erupts during disciplinary meetings of MDTP panels, since the main business of these panels, according to patients, is to deprive men of their rights. But the most frequent site of righteous anger is in psychotherapy. Day in, day out, therapists labor to transmute the patients' collective memory, a thing that is located in historical events (to which the patients have privileged access), into individual memories, a thing situated in mental events (to which the therapists have privileged access). Memory is the proof as well as the record of the self's existence, and the struggle over memory is the struggle over the self's most valued possessions: "Who should know better what these things [events, experiences] mean, you or me? Hey, they happened to *me*, not you!" The therapists' claim to possess the authentic meaning of these memories is equivalent to diminishing the self, or, to be more precise, to discovering a true but subversive self hidden beneath the surface of consciousness.

Guilt is a counterpart of anger and is constituted of three elements: the guilty party knows he has inflicted an unjustified loss or injury; the loss or injury represents a gain of some sort for him, as in an increment of pleasure; and because of this act and his obscure ideas about cosmic payback—"what goes around comes around"—he knows he that he has put himself in a position where punishment is due. When feelings of guilt are so intense that the guilty party is unable to ignore them, he has five options: (1) He can try to balance the books, by reparation (restore the injured party to status quo ante) or expiation (inflict an appropriate amount of suffering on himself). (2) He can try to have his debt forgiven (by his victim) or to have himself forgiven (by a spiritual authority). (3) He can try to "adjust himself to the alteration in himself by now continuing in a way consistent with it, by making the disfigurement disappear by disfiguring himself still further" (Taylor 1985:93). (4) He can accept the center's solution and pathologize his guilt. "If he can neither restore himself to his unblemished self nor adjust himself to the altered one, there seems no alternative for him but to see himself as two distinguishable selves" (Taylor 1985:96). (5) He can continue to suffer his guilt.

Option 1 is useless because most patients possess no suitable moral currency, such as "good deeds" or acts of mortification, with which to balance the account. Many patients say that their moral debts are simply too great to pay back in one lifetime or are owed to people who now are beyond compensation, because they are dead, their families have been destroyed, and their lives have been ruined. Option 2 is out because most of the men would appear to lack the kinds of religious convictions that make rituals of atonement and forgiveness meaningful. Option 3, mentioned in the Chap's Rap, is likewise unattractive, because most patients would argue that the costs are too high. A man might successfully handle his guilt in Vietnam by becoming "the meanest motherfucker in the valley," but this solution means incarceration once he is back in the United States. Option 4 is possible, but it means surrendering autonomy and, as the patients see it, becoming shameless. All roads, therefore, seem to lead to option 5: the persistence of guilt, the righteousness of anger, and the community of suffering warriors.

The Rhetoric of Emotion

I want to underline that I am writing about the *rhetoric* of emotions, not the totality of the patients' experiences of shame, anger, and guilt at the center. The "anger" that the men experience is a universe of meanings, feelings, and memories, and the righteous anger that I have just described is simply a sector within this universe. The same can be said about their shame and guilt. The sense of shame that centers on the community of warriors is not the whole of their experience of shame. There are other, invisible audiences—parents, children, mentors, friends—to which these men hold themselves accountable.

What gives *this* rhetoric of emotions its salience and coherence is the clinical ideology. Life is complicated, and consciousness floats over a confusion of meanings. Ideologies of the sort found at the center make truths by eliminating other truths: such ideologies are great simplifiers of experience, and this is the source of their power and seductiveness. The most conspicuous effect of the center's ideology is to create a compelling language game, in which the rhetoric of emotion plays a double role: it makes selves (for the patients) and it makes resistances (for the therapists). This effect is achieved through clinical practices that are aimed at divorcing anger from guilt, by depositing each of the emotions into an autonomous segment of the mind (a part-self). When anger is separated from guilt, it loses its moral and self-signifying power; it passes into pathology. The clinical ideology breathes life into the center's treatment program by attacking the moral foundations of the self, but its effect is to mobilize precisely the network of meanings, affects, and scripts that it is intended to displace. It is in this irony, and not in some ghostly movie hidden in the recesses of the mind, that we can find the traumatic memory.

Seven _____

Talking about PTSD

IN ALFRED HITCHCOCK'S FILM, *Spellbound* (1944), Gregory Peck plays the victim of a traumatic neurosis similar to PTSD. He is tormented by a memory he cannot recall but is certain that it concerns an act of terrible violence. He also suffers from a mysterious phobic horror of objects decorated with parallel lines. By the film's end, it is discovered that the phobia mirrors a visual element of his traumatic experience, which involved attempted homicide and accidental death on a downhill ski slope. Like many of the patients who speak on the following pages, the Gregory Peck character is chronically angry and deeply disturbed by his own barely controlled aggressive impulses. With the help of a psychoanalyst, Ingrid Bergman, he reclaims his repressed memory and experiences a complete remission of symptoms. This is the film's climactic moment, and it takes the form of an epiphany, in which the audience is allowed to watch on the screen the same event that Gregory Peck's inner eye is watching inside the theater of his mind.

Hitchcock was not the first person to compare the act of recovering a traumatic memory with the act of watching a film. Abram Kardiner (1941) made a similar connection, comparing the structure of the recalled memory to a slapstick comedy. Judith Herman makes the same connection in *Trauma and Recovery*. In this book, the ordinary memory of the ordinary adult is pictured as information encoded into a verbal, linear narrative, and assimilated into an ongoing life story. The traumatic memory is different: it is dominated by imagery and bodily sensation, and, in these respects, is similar to the memories of young children (Herman 1992:37–38). The therapist's job is to help the patient to reconstruct the traumatic memory from "the fragmented components of frozen imagery and sensation" and to "slowly assemble an organized, detailed, verbal account, oriented in time and historical context" (Herman 1992:177). Herman's account ends with a cinematic vision, narrated by a psychotherapist (Jessica Wolfe) whose specialty is treating combat veterans diagnosed with PTSD: "We have them reel it off in great detail, as though they were watching a movie, and with all the senses included. We ask them what they are seeing, what they are feeling, and what they are thinking" (Herman 1992: 177).

Herman traces this idea back to Breuer and Freud and their description of the abreactive therapy that they practiced a century ago. The patient was put into a hypnotic state and then induced to retell her memory and to reexperience the attached affect. The account given by Breuer and Freud is precinematic, but the parallel between what happens in the abreactive moment and what happens when we watch a film reel off would seem obvious to most people today. But Freud did not stick with abreactive therapy. By 1914, he had replaced it with a distinctively un-cinematic form of treatment (Freud 1953b [1904]:250; 1958 [1914]. Two decades later, Freud reflected on his change of method: "It was true that the disappearance of the symptoms went hand-in-hand with the catharsis [abreaction], but total success turned out to be entirely dependent upon the patient's relation to the physician and thus resembled the effect of 'suggestion'" (Freud 1955d [1923]:237). Throughout his professional life, Freud was sensitive to charges that his clinical successes might be products of suggestion, the implantation of an idea or intention in someone's mind so that she believes it is her own and proceeds to act on it. As early as 1888, Freud had discussed this possibility in connection with Charcot's work, writing that Charcot's clinical observations would be "worthless" if his patients' symptoms were shown to be products of suggestion (Freud 1966a [1888]:77; also Grünbaum 1984:127–139, 157–158, 250, 282–284). Freud resolved this problem by adopting a new technology, psychoanalysis, that shifted clinical attention away from the traumatic memory per se and onto the obstacles (resistances) that patients place in the way of reaching these memories:

> In psycho-analysis the suggestive influence which is inevitably exercised by the physician is diverted on to the task assigned to the patient of overcoming his resistances. . . . Any danger of falsifying the products of a patient's memory by suggestion can be avoided by prudent handling of the technique; but in general *the arousing of resistances is a guarantee against the misleading effects of suggestive influence* (Freud 1955b [1923]:126; my emphasis).

In other words, if heterosuggestion did occur, it would be expected to work in just the opposite direction, inducing recollections and encouraging patients to accept the therapist's interpretations.

In abreactive treatment, the process of remembering took "a very simple form." The patient was merely led back to "an earlier situation, which he seemed never to confuse with the present one." With the shift to resistance, however, the therapist was no longer concerned with "bringing into focus the moment at which the symptom was formed." Now he concentrated his efforts on *"studying whatever is present for the time being on the surface of the patient's mind"* (Freud 1958 [1914]:147, 148; my emphasis). Freud reported that, in some psychoanalytic cases, clinical events would at first proceed as they might have under hypnotic technique: a steady movement back to the etiological event, accompanied by a willingness to adopt the therapist's views. However, resistance would eventually be encountered and in some cases was present from the start. Moreover, it always took the same form, namely, a compulsion to repeat:

> [T]he patient does not *remember* anything of what he has forgotten and repressed, but *acts* it out. He reproduces it not as a memory but as an action; he *repeats* it, without, of course, knowing that he is repeating it. . . .
>
> Remembering, as it was induced in hypnosis, could but give the impression of an experiment carried out in the laboratory. Repeating, as it is induced in analytic treatment . . . implies conjuring up a piece of real life. (Freud 1958 [1914]:150, 152)

A new division of labor had been created. The physician's job would be to uncover resistances—defensive operations of the ego in the analytical process—and to acquaint the patient with them. But this would be only the starting point of the process, Freud wrote, since the patient can be expected to continue to resist even after the resistance has been identified. Indeed, it is only when the resistance is at its height that the analyst and his patient can expect to discover the repressed impulses that are feeding it.

> I have often been asked to advise upon cases in which the doctor complained that he had pointed out his resistance to the patient and that nevertheless no change had set in. . . . The treatment seemed to make no headway. This gloomy foreboding always proved mistaken. . . . [G]iving the resistance a name could not result in its immediate cessation. One must allow the patient time to become more conversant with this resistance . . . , to *work through* it, to overcome it. (Freud 1958 [1914]:155)

It is the experience of successfully working through resistance that "convinces the patient of the existence and power of such impulses" (Freud 1958 [1914]:147, 155).

After 1914, the cinematic metaphor is no longer consistent with Freud's conception of the traumatic memory. Nor does this way of thinking have any place in the center. The center is not, of course, a Freudian institution, but under the tutelage of Dr. Durocher, it has come to be organized around orthodox Freudian notions of remembering, and its conception of the traumatic memory is realized (made visible) through notions of resistance and repetition.

Therapists are continually bumping into traumatic memories at the center, but there are no epiphanies of the sort depicted by Alfred Hitchcock and Jessica Wolfe. There are occasional "disclosures" of course, since this is what the clinical ideology demands, and some of these narratives are vivid and charged with emotion. But there are no real climaxes; there is no point at which everything—narrative, affect, and remission—seems to come together.

In fact, life at the center is rather monotonous, consisting of unending hours of talk, punctuated by the incessant drip-drop of tiny signifying moments. Yet for most of the therapists and a handful of patients, it is a special kind of monotony, filled with narrative expectations. Paul Ricoeur has distinguished between two kinds of narrative expectations. When people hear or read a narrative for the first time, say, a new novel, they are "moved on by a thousand contingencies" and anticipate that the narrative's ending will prove to have "required these sorts of events and this chain of actions." In other words, time is pulled forward toward the ending. But when a story is so familiar that it can be only reheard or reread, time is liberated, free now to flow back and forth. Being able to follow the actions and events to an unpredictable end becomes meaningless. What is important now is the ability to apprehend the end that is implied in the beginning and the beginning that is implied in the end. It is *this* knowledge, typified by the psychoanalytic notion of "reenactment," that distinguishes time on the ward from "the open-endedness of mere succession" or empty seriality (Ricoeur 1981:170, 175).

The clinical ideology presumes that every patient conceals a narrative, his pathogenic secret. In practice, the patient carries a double narrative. There is a first-told narrative of a singular life: a story about this particular soldier, his particular family, his particular wartime event, his particular postwar life. But inside each of these novel stories, there is a single familiar one, a story first told in 1980 in *DSM-III* and now expanded into the narratives of splitting described in chapter 6. It is the therapists' ability to apprehend the double narrative that gives a coherence to the talk that we are about to chronicle.

This chapter will replay a series of clinical sessions that took place over the course of two years. Because I was not permitted to audiotape these exchanges, my accounts are based on notes taken during the meetings. I have changed people's names and sometimes deleted or altered details in order to protect the speakers' anonymity. Most of my accounts cover only a part of the actual sessions. The date at the head of each entry is nominal and is included to indicate the interval that has elapsed since the previous session.

The first set of sessions were recorded in 1986, a few months after the clinical director, Dr. Durocher, introduced the center's therapeutic ideology. During this period, patients were often restive, and sessions tended to meander along in a desultory way. The second set of sessions dates from 1987, by which time the staff and patients have been incorporated, to different degrees, into the center's official language game. Most of the action takes place in one of the two psychotherapy groups on the inpatient ward.

Three staff people appear throughout these accounts, Dr. Durocher and the group's cotherapists, Carol Thompson and Dr. Lewis Jeffries. It may be helpful to know something about them before we begin.

At the time, Dr. Durocher was in his mid-fifties. After completing medical school, he served as an army doctor. This was during the Korean War, but Dr. Durocher was not posted overseas. Afterward, he completed a residency in psychiatry and a teaching analysis and over the next three decades worked as a practicing psychiatrist. Before coming to the center, he lived in another state, where he had a private practice, a faculty appointment at a university medical school, and a position as a forensic psychiatrist in his state's prison system. Dr. Durocher now lives in a suburb of Petersburgh, a metropolis located about thirty miles from the center. He holds an adjunct faculty appointment in psychiatry at the Petersburgh University School of Medicine.

Dr. Durocher is an urbane and personable man, but he does not welcome behavior he perceives as boundary testing. His relations with the staff are generally amiable and positive, and his intellectual authority is unquestioned, at least in institutional spaces. He likes to run a tight ship and does not overlook behavior that needs correcting. On these occasions, correction is framed as a moment of supervision and an opportunity for the errant staff member to acquire deeper insight into his/her real motives and emotions. His commitment to Freudian psychoanalytic views is complete. Although these views are alien or confusing to a majority of the people on his staff, Dr. Durocher is generally respected for his honesty, intelligence, and clinical experience.

Dr. Durocher monitors the status of each inpatient, but his clinical contact with them tends to be intermittent and indirect, usually mediated through reports and conferences with his staff. Most patients perceive him as tough-minded and as someone who must be shown respect, for it is he who decides whether they stay or go and who determines their disability evaluations at the time of discharge.

Lewis Jeffries is about fifteen years younger than Dr. Durocher. After college, he completed law school but found life as a lawyer extremely stressful and suffered a near-fatal heart attack that he connects with his work. He underwent complicated cardiac surgery that was, in his words, a traumatic experience. After recovering from the operation, he enrolled in a graduate program in clinical psychology at a prestigious midwestern university. He earned a Ph.D. and returned to Petersburgh and a clinical position in a Veterans Administration spinal cord unit, where, he says, he found the work interesting and fulfilling. Dr. Jeffries learned about the center from a VA memorandum that solicited applications for staff positions. He was attracted to the new program by its psychodynamic orientation and distinctive patient population. In addition to his appointment at the center, he has an active private practice in a suburb of Petersburgh.

Dr. Jeffries is a knowledgeable student of psychodynamic psychology and sees his relationship with Dr. Durocher as a form of memtorship. On the surface, Dr. Durocher is the patient teacher and Dr. Lewis is the bright student, eager to learn but not inclined to hide his uncertainties. Relations between these men are respectful, but hardly relaxed.

Carol Thompson is in her late twenties and is rated as a psychological counselor, a position several grades lower than Dr. Jeffries's. Like Dr. Jeffries, she has a strong intellectual interest in psychodynamic psychology and sees her work at the center as a kind of internship. Carol projects a sense of calm (indifference?) that sometimes upsets her patients and also Dr. Jeffries, whose basal metabolism is set four or five stops higher than hers. Like Dr. Jeffries, Carol grew up in Petersburgh but in less affluent surroundings. She was orphaned as a teenager when her parents were killed in an accident, and she occasionally describes this to patients as her traumatic event.

The National Center for the Treatment of War-Related PTSD, 1986

Four patients are introduced in the first session: Flip, Eddie, Chris, and Ray. According to Flip, he was a non-commissioned officer in the air force during the Vietnam War. He was seconded to the CIA to assist in secret operations over Laos and Cambodia but, because of the unorthodox nature of these operations, his military records do not include this information. After leaving military service, he obtained a university degree, married, and earned a good living as a sales representative. He gives the impression of being reasonably well off, and he periodically informs the staff that he has no interest in obtaining a disability pension.

Flip's first two months at the center were a period of unremitting resistance, punctuated by bouts of highly visible mental and physical suffering. His primary symptom was pain, usually in the form of headaches that followed his attempts to process his traumatic experience. He mastered the language of the treatment program, but his attitude remained generally negative. After four months, he disclosed a confusing story about shooting Vietnamese civilians during an evacuation. Soon afterward, he claimed that the program was putting everything together for him. Following his voluntary discharge from the center a month later, another patient produced evidence suggesting that Flip's military exploits were fabrications and that he had probably never even served in Vietnam. The following sessions take place during the period before he narrated his traumatic experience.

Eddie Jackson is an African American. He was born in rural Alabama but grew up in Petersburgh. He is married and claims to be a regular churchgoer. He is a former paratrooper and gives numerous accounts of his wartime exploits; most of them are surreal, and none of them involves a traumatic event. He has a passion for drawing Lewis into tedious debates about his rights as a patient, as a veteran, as a man, and as a spiritual being, and how these rights are being systematically violated by Lewis, Carol, Dr. Durocher, the VA, and the U.S. government. Eddie had been a patient at the center previously, and the inpatient staff did not welcome the news of his readmission. He has a particular animosity toward Carol, for something she was supposed to have done during his earlier stay. Eddie is not well liked by other patients, but he seems unconcerned. (The affair of Eddie's earphones was described in chapter 6.)

Chris is large, dim, and histrionic. He was a marine and claims to have participated in classified missions. He says that he is bound by an offical oath of secrecy and therefore cannot reveal any details of the dirty work in Vietnam that traumatized him. Chris was a patient at the center in its early days, prior to Dr. Durocher's arrival, and participated in a televised mutiny. He moved to the Petersburgh area to be with old war buddies.

Ray is an army veteran in his late thirties. He was raised in rural Appalachia, a member of a family with deep fundamentalist convictions. In Vietnam, he served in a reconnaissance unit and claims to have participated in several atrocities involving women. He left the army with a less-than-honorable discharge that was subsequently upgraded. He has been previously hospitalized in VA hospitals with a diagnosis of schizophrenia. This is his second stay at the center. Sometimes Ray is an amusing cynic, and sometimes he plays, with great conviction, the part of a psychologically tortured victim-perpetrator. Although he usually sticks to one persona per session, he occasionally transforms himself in the middle of group psychotherapy, shifting from a contrite and hopeless sinner to a cheerful and unrepentant son of Satan. According to the therapists, here is a classic case of traumatogenic splitting: the division of the psyche into a victim-self, filled with pathological tenderness, and an aggressor-self, threatening and intimidating, the personification of evil.

7 April 1986. Group psychotherapy.

EDDIE: This place is designed to make people submissive—not to bring things out of them—to make them into children. Isn't this true, Carol? Why don't you deal with us like men, not children?

CAROL: We do. It all depends on what your definition of "men" is.

EDDIE: Seems like a kindergarten here. We're not even treated like adolescents. If you don't do "this," or you do do "this," then you got to go before "the team" [MDTP panel]. We have no rights; our beliefs and feelings are trampled on over and over. Basic reason I'm still here now is that I don't like to believe that someone has beat me [forced him out]. Your "program" is below my level of thinking!

CAROL: I hear two things here, Eddie. I hear Eddie's feelings of being treated like a child, and I hear his feeling that there is a contest of wills. What you said yesterday about feeling like a warrior, that implied a contest of wills.

EDDIE: I don't care how other people see me.

CAROL: Eddie, you're putting yourself in a warrior position . . . reliving your life . . . using your *feelings* to make sense of the world.

EDDIE: Why do you call it a "treatment team," Carol, when it's really a *discipline team*? Bullshit. [Eddie is going before an MDTP panel this afternoon.] Another one of your words: "clients." My definition of a "client" is someone you're supposed to serve. This definition does not apply here—we're your "patients."

CAROL: Patients have things done *to* them: they are given pills and medicines. You're "clients" because you have responsibilities on your part; you're actively working towards goals.

EDDIE: You're saying I was a patient on 41A [drug rehabilitation ward], but I'm a client here. What's the difference? When I was on 41A, I talked to therapists, nurses, social workers—even to the chaplain.

CAROL: It involves a different mind-set . . . the entire system you're in . . . whether its medication-oriented.

EDDIE: I was also a patient in psychotherapy groups in 41A. What's the difference now?

CAROL: Sounds like a struggle with identity—with your conflict.

EDDIE: Carol, I'm showing how you play with words, how you play with people's minds. You want to break us. If we don't see things your way, you throw us out. A client's got no choices. Miss a meeting and you're restricted. You have guidelines here. We have a choice: either we go along with what they say or get out. Personally, I'd prefer that you tell things the way they are here. When you call the clients, you should say, "Hey, fool!" Just don't try to say it to me. The program would be better if we dealt with problems and not all this petty bullshit: wearing earphones, holding back our passes. When I came here, I thought I'd be in a place where you sit with doctors and nurses and I'd talk about my problems—not listen to all this petty bullshit.

CAROL [to Eddie]: How do you *feel*? What's going on inside you?

RAY: *You* tell me, Carol. You have the power; I don't. I don't know how I feel; you're the one who knows. At the VA, you [patients] have no rights. About six years ago, I was beaten up while I was hallucinating [in another VA hospital]. I brought a case against the VA. You know what I got out of it? Nothing—a lot of bullshit. Same thing here. You call them "treatment teams," but they're here to discipline us. It's the same thing with electroshock. [Ray has been treated with ECT in other VA units.] They don't call it that any more, but they're still using it, and it's the same old thing. You people are *punitive* here.

CHRIS: The only difference between this place and a prison is that there are no bars here, and you can walk out. We're just pissing in the wind here, wasting our time. I'll listen to what you guys want to talk about, but it's just pissing in the wind. Hey Carol, I want to change your chain of thought. You motherfuckers lied to us. You said we weren't going to have sessions on Monday and Friday afternoons.

FLIP: We're moving right along, aren't we? What do you [therapists] want to talk about now?

RAY: How's the shower curtain situation coming along? How much longer are we going to have to wade into a fucking swamp every fucking time we want to take a shit or brush our teeth?

CAROL: Chris, how are you feeling? [Chris is bent forward, his hands covering his face.]

CHRIS: Just leave it alone. I don't want to talk about it.

RAY: I'd like to talk about Vietnam . . . but I remember only the beginning and the end.

EDDIE: It'll take three months more until we get back to Vietnam in group. On Tuesday, they'll say we can't piss in the waste can any longer and that we have to piss in our pockets now.

FLIP [to Lewis]: How come you never answer us directly? Whenever we ask you a question, you always answer our questions with a question.

8 April 1986. Group psychotherapy.

CHRIS: In Vietnam, we used to talk about "the real world"—the world we left behind. But when we returned, it was gone.

LEWIS: Couldn't it be that it is you who changed?

RAY [to Lewis]: Did you change when you were a war protester? Fucking turncoat. I came home for thirty days, and I was shunned by everybody: by my wife, my family, my old friends. I couldn't wait to get back.

LEWIS: It was you who changed. Take the example of a woman who is raped in an elevator—

FLIP: I hate that fucking comparison!

LEWIS: —she is afraid to get back on elevators.

CHRIS: Take that same woman and imagine that she gets raped over and over, every time she gets into an elevator. That's us.

LEWIS: All I'm saying is that the combat experience—even just being in the combat zone—is a traumatic experience. You've got to be changed after this: your priorities are changed, your brain is reordered. This isn't abnormal, pathological . . . but you get hung up in this process.

FLIP: Your fucking orientation film preaches "freedom of choice." But now you're forcing us to go to the Wall against our will. [According to the orientation videotape, every client is free to choose to get better, to change his old way of life and start anew. The Wall is a traveling replica of the Vietnam Veterans Memorial, in Washington, D.C.]

CHRIS: You train us like a dog: pat us on the back if we do what's right, punish us if we don't. And you expect us to automatically trust everything that you—

LEWIS: The therapists in this program are no different than any other health professionals, even a dentist. You go to the one you trust.

FLIP: Except that this is the only dentist in town! I build up some trust here . . . but every time I do, then you throw an obstacle in my path . . . a fucking toy Wall.

LEWIS: The job of psychotherapy is to hold you in place and look at your problems. We put you in a box in order to cut off the avenues of acting out. We say, "Talk about it." Being boxed in is painful. If you don't trust us, then the pain isn't worthwhile. The trip to the Wall is part of this. It's not punishment or harassment.

Three more men make appearances in the next sessions: Stan, Carl, and Gary. At this point, Stan has been at the center about five months and is about to move into the reentry phase, prior to "graduation." Carl was introduced in chapter 6, where Lewis Jeffries and Dr. Durocher gave their contrasting appraisals of his death threats and paroxysmic behavior. Gary is an African American in his mid-thirties. He was born in Petersburgh and served as an airborne soldier toward the end of the Vietnam War. He used drugs before military service but became a serious doper in Vietnam. At the time of his admission to the center, he was involved in litigation involving criminal activities; the case is still unresolved, and it is understood that his diagnosis and treatment will somehow affect what happens next in court. Gary arrived at the center after drug detoxification. His right arm is in a sling, broken cleanly in two places—a message sent to him by business associates, according to his intake interview. He is a highly energized and engaging man, very clever and very verbal. Gary claims to have done terrible things in Vietnam and says that he is beyond making "meaningless gestures." Talk is cheap; the dead cannot be raised from the grave; and the mutilated cannot be made whole again. Gary also has neurological symptoms for which he is being tested. It is suspected that they result from his heavy doping and drinking.

20 April 1986. Group psychotherapy.

EDDIE: Since you brought up my MDTP [a meeting of his multidiscipli-nary treatment plan panel], I'll tell you what it was about. [An inventory of Eddie's grievances follows.] And they also told me I got to talk. And so I'm going to talk about whatever you bring up—so that I don't lose my pass—that will enable me to spend time with my wife—and maybe with someone else, too.

LEWIS: The issue is not just "talk," Eddie. It's the need to share [verbal-ize] your *feelings* with the group.

GARY [to Lewis]: Is it your *intention* to make people angry with your stupid remarks?

LEWIS: My remarks did not make the anger. The anger comes from an-other source—inside the individual.

GARY: You talk like Mr. Spock on "Star Trek." You always talk about logic. You're missing something—you're like a clone. You're like a book that's talking. You don't talk like a human being; you talk like a robot. She's the same way. You've got no compassion about learning how people really feel. There's no emotion about you two. You sit there and analyze. People ask you a question and you answer with a question. I don't see the need for you two to be here. What you want to find here, you can find in a bookstore. Oh, the hell with you! You sit like a dummy, saying nothing. You know, if you're not part of the solution, then you're part of the prob-lem. If someone says something to you that's not in the book, then it's not "logical." Personally, I like you, Lewis—with your preppy look—watering flowers, with flies buzzing around your head. But you have no purpose—don't give us any feedback. I don't think that you could even cast a shadow. You're like vampires. We want information—but you don't have the tools to work with. Now tell me, Lewis: what am I thinking . . . feeling right now?

LEWIS: You tell me.

GARY: But you can *see* what I'm thinking! I don't have to *tell* you!

LEWIS: You sound frustrated, annoyed, disappointed. But I can't guess what's going on inside you.

GARY: What's your next step, then?

LEWIS: Did other people see things that I didn't? Are other people able to make some assumptions about Gary's behavior?

STAN: From what I've seen in my time here, Lewis and Carol want us to examine our behavior, so we can get along with my life—

GARY: But it's *his* way of doing it. It's got nothing to do with Vietnam. A load of bullshit.

LEWIS: We disagree on that point.

GARY: You can say what you want to say. You're talking on levels beyond me. Don't tell me bullshit: break it down so that it's less complicated.

22 April 1986. Group psychotherapy.

GARY: What's your success rate?

CAROL: It's normal to doubt, but—

GARY: I'm not doubting, I just want to know the name of one graduate: someone who has no flashbacks, bad dreams, et cetera.

CARL: You can go to the alcohol ward, and they'll give you the name of a reformed alcoholic they've treated. Will this program allow us to get back on the street? I came here all the way from North Carolina, and I spent forty-one days on the alcohol ward—that was the condition you people set for me to get into this program. I put a lot into getting here; now I want to get some answers.

FLIP: The program won't change anything. We have to change ourselves. But I'm in a state of confusion now. My mood changes from minute to minute. You can get something from this program, Carl, and you can just reject the rest.

CARL: There's a difference between me and you, Flip. I've been going to the VA for years—long before PTSD. They browbeat you at the VA, and then they send you out on the street. [To the therapists] When I ask if you have answers, don't fuck with me. Everywhere in the VA, it's the same. Patients against the staff. We don't need their rules. They've got to treat us like men. We don't need to take bullshit anymore. All over the country you get the same bullshit.

25 April 1986. Group Psychotherapy.

GARY: I need this program because I need to get a certificate to get into the Disabled Veterans Disability Program. So far as I can see, this program isn't doing anything for me. I'll have to pretend I'm doing okay so that I can get your recommendation to get into the other program.

LEWIS: You can't change unless you change inside.

GARY: If the rewards are big enough, I'll change for sixteen weeks!

CARL: You know there's a big difference between changing your behavior in this controlled environment and then going on the streets. If you don't conform to the ideas and demands of the staff, you're thrown out. [To Lewis] What makes one psychologist qualified to treat PTSD and another not qualified? Did you go to a special school and get special training?

LEWIS: Training in clinical psychology . . . internship . . . treated many patients . . . private patients . . . paraplegics.

CARL: It shouldn't be the center that chooses the staff, anyway. It should be us—because *we* have PTSD. I don't see how someone who doesn't have PTSD could treat it. What it comes down to is that this program is *experimental*! You don't know what PTSD is. Flip's been here for ten weeks, and you [Flip] don't think it's helped you much. He went out for a weekend to see his parents and his sister and her kids, and he couldn't handle it.

The following clinical debriefing takes place about a week later. As you will see, Lewis is expressing his uneasiness about aspects of the clinical ideology and his doubts about the extent to which his fellow therapists understand its content.

28 April 1986. Clinical debriefing.

LEWIS: Everything's "reenactment" now! Five months ago, we had never heard of "reenactment" and now—

CAROL: Did you hear what Eddie said in communications [group] today? "My family never listens to me." *This* is a reenactment. It's his experience in Vietnam. At one point, he was abandoned by his outfit. Then later he was its radioman, its "ears."

LEWIS: Listen, Carol, some of the things that they say have *nothing* to do with Vietnam.

CAROL: The experiences they describe don't have to be from Vietnam. They can be experiences that occurred before Vietnam. But these experiences can be screen memories for their Vietnam events. And this stuff [disagreement] between you and me, this is reenactment too.

LEWIS: Carol, you're just stringing together buzzwords—warmed-over Freud. Anyway, how many people [staff] understand what we're talking about? How many could understand these four pages from Kernberg on splitting? Only three of us: you, me, and Peter [the clinical director]. Do you think that even Malcolm [like Lewis, a staff member with a doctorate in clinical psychology] really understands these ideas? I'm not so sure. And the nurses—they don't have the foggiest idea about any of this. They're struggling just to understand the cognitive therapy stuff. And they're working in one-on-ones! So how can you believe that the clients know what we're talking about? You keep telling me that they're "getting it," that they're "absorbing it." Really? How can they?

CAROL: I can't believe that they're moving James Church [a patient in the other psychotherapy group] into reentry—and on Cindy's evaluation. [Cindy is a clinical psychology intern.] Doesn't his MDTP understand anything? Our criterion says that a client has to resolve his traumatic event. And all Church has done is to *mention* it!

LEWIS: You're really incredibly naive, Carol. Of course they don't understand. Look, you and I have a lot of disagreements. And I don't understand some of the things you say. But at least we're in the same ballpark.

The big difference between the two of us is that I believe in the model in an instrumental way—like I believe in any theory. I judge it according to its usefulness, and I decided to adopt it because it feels right to me. You, on the other hand, believe in it to the exclusion of every other way of understanding PTSD. This habit of yours is one of the reasons that the staff respond to you negatively: it's your rigidity.

CAROL: Have you noticed that Flip's pains are an inventory of the deaths that he carries with him? You know the way that he's always putting his finger in the corner of his mouth? That's where his friend Steve was shot. It's where the bullet entered that caused his head wound. And according to Flip, Steve was also shot in the shoulder, and Flip also has chronic shoulder pain. [Steve's death is one of several traumatic events reported by Flip.]

The National Center for the Treatment of War-Related PTSD, 1987

It is now about one year later and Dr. Durocher's "model" has evolved into its final form. The staff and patients have been incorporated, to various degrees, into an official language game, organized around the notion of "stress responses." The subject of each vignette is transparent, and captions are superfluous. The patients who were encountered on the previous pages have all departed.

The initial sessions center around three men, Marion, Roger, and Alvin. Marion was referred to the center by a VA unit in another state. At the center, he has formed a close friendship with Roger, a former airborne soldier. Marion served in the marines and trained to lead an attack dog. Shortly after he arrived in Vietnam, his dog was intentionally killed by other marines. Marion now became an ordinary infantryman and joined a rifle company as a replacement. After leaving the marines, he continued to train dogs and that is his occupation today. He is a star patient: quick to learn the rules and language, quick to put the clinical ideology into practice.

Group psychotherapy is supposed to run on talk exchanged between patients. This is the ideal, but it is not often achieved at the center, where clinically meaningful talk depends on the continual intervention of the therapists. The group that includes Marion and Roger is unusual in this respect, since these two men are eager and able to keep the group in motion.

The third patient, Alvin, is a sad and friendless little man. He looks as if he were sixty but is actually forty. Since leaving military service, he has lived a marginal and alcoholic existence. His current residence is a shack that he and another veteran have built in a forest reserve near the center.

There are several other patients who speak during these sessions, but it is unnecessary to give them any additional introduction at this point.

My account begins with Marion's disclosure of a traumatic experience. This is an important moment for Marion, because it makes him eligible to move into the reentry phase. The event he recounts here matches the program's profile of a traumatogenic experience, and it contrasts with some of the much nastier incidents that Marion has related to other patients.

19 February 1987. Group psychotherapy.

MARION: Our squad was cut off. NVA [North Vietnamese Army] was close enough to lob grenades into our position. Perimeter fire kept them back, but when we stopped firing, they lobbed grenades in. The grunt [marine] next to me was on the ground, curled into a ball and crying. Whenever I attended to him, more grenades were lobbed in. I returned to firing, and this guy jumped up and ran away from our position. I shot him in the back. For years, I felt justified. Now I feel guilty. He was only eighteen. It was his first fire fight, and he couldn't handle it.

24 February 1987. Group psychotherapy.

MARION: I feel real aggression towards Alvin now. He's been here two weeks longer than me, but there's still nothing about his trauma complex [traumatic memory] and nothing about his reenactment. He gives back just enough of the program to be left alone. I want to help Alvin. I don't want to feel aggression towards him. I see a lot of myself in Alvin.

CAROL: What aspect of *your* reenactment is being triggered by Alvin?

MARION: Withdrawal—it's the withdrawal that I see. It's the same pattern I see in my relation with Sheryl [his current companion]. I hate it because it's the way my father was to me, completely nonresponsive whenever he was confronted. . . .

CAROL: What did you feel inside when your father didn't respond to you?

MARION: Anger. I wanted to ask: Why do you do this to me? Tell me how you're feeling? What's going through your mind? Alvin here is putting me right back home.

CAROL: Also back to Vietnam?

MARION: Well, there weren't many times when I felt I was being cut out there—except possibly in my event, when I saw the grunt wasn't firing, and then I saw him run. I felt frustration and anger. It's funny, because Alvin even looks like the guy, now. The way he's hugging himself, the way he's curled over in his seat.

ROGER: It's the same reason why the guy got shot. Isn't it? I mean because he was a "nonparticipant."

CAROL: In Vietnam, you felt as if you had been abandoned. You felt you were helpless. The enemy were pushing in on you, and you couldn't rely on the people you thought you could. You felt anger and you—

MARION: I'll tell you, I felt lots of anger against officers. They were supposed to be our leaders, and they put our squad out in front. It was obvious to anyone, from the beginning, that they were putting us where it would take nothing for the NVA to cut us off.

CAROL: How does this feeling get personalized here, at the center? In your feelings towards Alvin?

MARION: You know, we've all tried to help him in group. We tried to help him to process, but his response is aggressive. He doesn't give anything.

[A long silence]

ALVIN: I think that this is where I get off. I'm leaving the program. I'm going to come back in September—after I get some things worked out.

ROGER: That's great, Alvin. It's been nice knowing you. This is just another piece of his aggression. He's trying to make us feel that we've been picking on him and that he's leaving because of us.

CAROL: Alvin, are you aware of what you're responding to?

ROGER: Hey, Alvin, are you coming back in September to get six more guys angry at you? Your strategy is to make people feel sorry for you—and we found you out. I know I'm being aggressive now, and I understand this. But, you see, he doesn't respond to anything we say. The rest of us do all the work, and Alvin doesn't do a damn thing. He's cheating us. It's like all of us are moving a house, and one guy is sitting in the back of the house, drinking booze. And that guy's Alvin. And it's not just the patients who feel this way. I've heard staff members express frustration about Alvin.

CAROL: What are we doing? Are we responding to Alvin's reenactment? Do we cooperate with Alvin, to make sure that he will continue to reenact Vietnam? Do we collude with him? Leonard, what are you thinking?

LEONARD: Alvin doesn't exist for me anymore. He's gone. He's disappeared. I've killed him in my mind.

ROGER: If Alvin stays, he should be put in a corner of the room, away from the rest of us. We shouldn't have to work around him. His presence threatens to break up the group.

MARION: With Alvin in the group, I got to feel that I didn't want to disclose. Not in front of him.

CAROL: We've got to watch our own stress responses and work with them. We have to remember this—always. We know that each individual, including Alvin, is ultimately responsible for himself, for making his own choices. He's making his own choices. But we still have to watch our stress responses.

MARION: There are important things we've shared here. There are things that we haven't shared with anyone else. But Alvin never shared.

For the outsider, it is unpleasant to see Marion and Roger at work, to see the triangulation of tenacity, sanctimony, and totalizing right-think. Carol is not insensitive to what is taking place and, predictably, she translates her uneasiness into the rhetoric of the treatment program: "We've got to watch our own stress responses. . . ." After the session, Marion respectfully suggests that it may be Carol who is acting out stress responses, displacing her unresolved conflicts onto himself and Roger. If this is true, she owes them an acknowledgment.

The next session starts off with Carol's reply. The meeting is also notable for James's efforts to carry on the work of the group by psychologizing Leonard.

25 February 1987. Group psychotherapy.

CAROL: I want to begin today by acknowledging my own stress response in yesterday's group.

MARION: I was glad to see it, Carol. I couldn't go on in group with Alvin's behavior. Alvin told Redman [another patient] two days ago that he was planning to leave. I felt good that I got a chance to tell him about his behavior before he left.

CAROL: It's important to be able to continue to monitor your stress responses—to modulate your aggression.

MARION: I feel good. I'm at peace with myself. I feel that nothing's left over. Yesterday, in group, I felt a little uncomfortable. But a few hours afterwards, I decided that I did everything I could. The man had to be told that he was clamming up—not sharing. He had to know that his rejection of us was giving rise to our aggression. I dealt with my aggression, and I have no sadness.

CAROL: Stress responses are going on all the time. They take constant monitoring.

MARION: I couldn't do that when I first came here. I would have said, "Fuck you," and I would have choked him. The important thing for me is that I dealt with my aggression. It's more important than Alvin's leaving and not disclosing. I don't feel sad at all about that. He had an opportunity here and didn't take it.

ROGER: A week ago I had the desire to smack the piss out of him. And so I opened up [disclosed an event], and I thought that maybe he would too. But he didn't. I got some of my frustration off yesterday. Before I came here, I would have smacked the shit out of him—or anyone else who did that to me.

MARION: Carol, I *felt* you were being aggressive yesterday. When I came to the program, I couldn't recognize my own aggression. But now I can recognize it in me.

CAROL: Well, I acknowledge my aggression. Yesterday, we were all displacing anger and aggression.

MARION: Alvin was a constant reenactment. The guy never got outside of it. It's a relief that he decided to go. The only sadness that I have is that he didn't take his opportunity. [A pause.] He's really down, depressed. I think if he's this far down, he may try suicide again.

CAROL: Henry, you're being quiet. What's your response to Alvin?

HENRY: Me? I'm only surprised that the staff is so incompetent. The man's been here for five weeks, and during that time, he's doing nothing. And you don't get him to do something. I've been in other PTSD programs and—

CAROL: This is a common response at times like this. Henry's remarks have to be examined in the light of everything that we know about stress responses.

HENRY: Well, I see a lot of me in Alvin. If I have a choice, I'm going to be quiet too.

MARION: I hear a lot of different things in what Henry's saying. I hear stress response, aggression towards the staff. I hear him doing splitting—into "us" versus "them." Maybe this is the result of Henry's own experiences, in Vietnam.

CAROL: Leonard, you said yesterday that you killed Alvin in your mind. How are you responding to your aggression? "Killing" is a pretty extreme word to use.

LEONARD: Right now, I'm responding to a whole lot of shit. I'm responding to my aggression against myself. I'm depressed again. I need space to get myself together. And my physical shit is overriding the program. My headache is back; I can't sleep anymore. And I'm blaming myself more and more for my wanting to kill. I'm pretty close to the point I was at when I was in the hospital [after a suicide attempt]. Before I came here, I blamed other people for my problems. Now I recognize that I am the one who is responsible. I recognize that, and it's driving me down even more.

CAROL: The question you should be asking, Leonard, is, Why now? Why did the headache return now? Why am I having this desire to kill now? Could it be your response to Alvin? Your need to do harm? When you talk about your total "responsibility" and about blaming yourself for your thoughts, it's a sign of your belief in your complete and total power, and it's a sign of your wanting to make the rest of the world powerless.

LEONARD: No, all I know is that it's morally wrong to want to kill.

MARION: It's the vicious circle, Leonard. You want to kill and you get a stress response. Then you feel guilt, and then you want to kill. But it's okay just to *think* that you want to kill.

LEONARD: To me, *thinking is doing*. That's the way I've been brought up, and I can't help believing it.

JAMES: Your problem, Leonard, is that you just don't want to give up your belief. You just want to go on feeling sorry for yourself.

CAROL: James is saying that you want to stay inside your stress responses, Leonard. You don't want to give them up. And this is being very aggressive to other people. You're feeding on your images, and that's an aggressive response too.

JOHNNY [to Leonard]: Are you angry at James for saying this to you?

JAMES: Leonard, choose for once not to punish yourself. If you want to say to me, "James, you're a motherfucker," then go ahead and say it. Say it; James can take it.

LEONARD: I can't. These are my feelings: blaming myself for wanting to kill.

JAMES: Leonard, they're your feelings because you *want* them to be your feelings.

CAROL: You must decide that you don't want to be a victim of your own aggression and that you don't want to make other people your victim. You have to acknowledge that you have the intent to do harm but that you also have the intent to relate to people. You have them at the same time.

JOHNNY: This is the main thing: to be able to have both of these feelings at the same time. Realizing that is the way to recover, Leonard.

JAMES: Go ahead: Leonard, tell me to shut my fucking mouth. I know what you're thinking. You're thinking, "James is my friend." But at the same time you want to shut me up, to leave you alone. You also want to hurt me—but can't. Go ahead: tell me, Leonard, to shut up.

CAROL: You're smiling, Leonard. Are we colluding with you in your reenactment? Remember, Leonard, you have the choice to stay with your stress response and to keep blocking off. You can keep splitting between the intent to do harm and the intent to relate. But remember that the splitting keeps you reenacting. If you stay in it, you'll stay there for years. Instead of dealing with the conflict, you'll keeping it going, on and on.

JOHNNY: I feel your aggression towards us, Leonard. But you're still my friend.

LEONARD: For twenty years, ever since Vietnam, it's been my fault.

CAROL: This word, "fault," it's a piece of your stress response. It's a decision you're making, to feel this way—to blame yourself.

HENRY: It's easy for me to recognize myself in Leonard. I carry this same shit around with me. It's my way to punish myself, and I won't give it up. On my second tour, I did things—awful, unnecessary things. I was brought up to be a Christian. I went to church twice a week. But after four days in Vietnam, I realized that these beliefs would get me killed. My first tour was for my country, but my second tour was for my buddies who were wasted. I got carried away—to the point where my commanding officer told me that it's time for me to leave Vietnam. I know I'm punishing myself now for what I did, but I can't get rid of this shit. I didn't always feel this way. It's only in the last four or five years, now that I've gotten treatment for PTSD, that I've come to realize the awful things I've done.

1 March 1987. Group psychotherapy

JOHNNY: I still feel bitterness towards Alvin: not knowing more about him at the end than I did in the beginning. I think he may have been here for the monetary reward. Maybe he was setting things up for his next compensation physical.

MALCOLM [a psychotherapist who is substituting for Carol and Lewis]: We need to be aware of our own aggression, monitor it and keep track of it. When clients decide to leave—

JAMES: Are you suggesting that we were a major influence on Alvin leaving?

MALCOLM: I'm not saying major or minor. And I have to acknowledge that some staff were also glad to see Alvin go.

JAMES: The fact is that Alvin sat in group and he set us up. He acted as if he was going to disclose, and everybody would shift around in their chairs, with an expectation. And then: zilch!

MALCOLM: Okay. When Alvin did this, he adopted a powerful position. It was his way of trying to control and manipulate the other clients.

The next segments center on two patients, Martin and Jack. Martin grew up in Petersburgh, in a working-class neighborhood. After high school, he enlisted in the marines and served two tours in Vietnam, as a sergeant. He left the marines after twelve years of service and returned to Petersburgh. Later, he moved to nearby Roxboro, where he married and worked for a meat packer. He has been unemployed since the packing house closed down three years ago. Martin currently lives in a Veterans Administration domiciliary in Roxboro.

Jack was a patient at the center about eighteen months ago. He grew up in North Carolina. Before completing college, he joined the navy, where he was trained as a SEAL (naval commando). Jack was sent to Vietnam, where he served two tours of duty in a long-distance reconnaissance unit. During this period, he also participated in special operations outside of Vietnam. During his second tour of duty, Jack suffered a psychotic breakdown and was evacuated. Following a long hospitalization, he completed a degree in psychology. He subsequently joined a motorcycle club that supported itself through unorthodox forms of commerce. Three years ago, he retired from the club.

27 March 1'987. Group psychotherapy.

LEWIS: Forgetting traumatic events originates in a conflict. You don't want to remember. Your conflict is always driving you back to the original event, but you don't want to go back to it. This is why you "forget." Stress responses do two things for you. First, you don't have to face your conflict, and second, you punish yourself. They allow you to have your cake and eat it. It keeps you from working out the conflict of your trauma complex. We know that things work this way. Because of this, we can assume that everything you do here is a reenactment. At least, that's what you can assume until you have good evidence otherwise. Of course, we know that *not* everything you do is reenactment. And that's one of the reasons why you're here: to find out just *which behavior is reenactment*. One of the jobs of combat training is to remove some of the conflict over aggression, so that you can be aggressive. In wars like World War II, society helped remove this conflict. In these wars, society tells soldiers that their aggressiveness is good. When the war's over, society welcomes them home. But that's not what happened with the Vietnam War.

MARTIN: Well, I can tell you that we had no problem being aggressive in Vietnam.

LEWIS: Okay, Martin. Give us an example, but use the word "I" instead of "we."

MARTIN: Well, you get orders to burn a village, and a gook tries to put the fire out while you're trying to burn his hootch. He fucks with you, and you show him that you can fuck with him. You can push him away, or you can kick his ass, or you can do what we usually did: you can shoot him.

LEWIS: The word "gook" is a good example of how we depersonalize people, turn them into objects. It's how we make it easier to—

MARTIN: I wasn't even conscious that I was using this word.

HENRY: My aggression is against Americans: against the smug, sanctimonious, hypocritical, silent majority of Americans. My fantasy is to release the black plague on them. If I could do that, then maybe I'd have some satisfaction.

LEWIS: Okay, Henry, what you need to do now is to examine what you've said and to ask yourself why you said it *when* you did. It was what we call a "displacement." The real target of your aggression is somewhere else. You may have a legitimate complaint about these people, the silent majority. But you're using them as a target for your anger. The real source of the conflict is hidden in these feelings against the silent majority. If you take the time to think about it, you'll discover two things at the bottom of it. First, there's the intent to do harm, and second, there's the intent to punish yourself. You need to stop trying to see what makes sense "logically"—whether or not the majority of Americans earned the negative feelings you have for them. You need to see what makes sense emotionally and psychologically.

HENRY: The staff are always referring back to *the* model. You make it sound as if all the patients here are on the same par and have the same problems.

LEWIS: But that's exactly right! You *do*. You have the same psychological processes. And so do I. Listen, stomach ulcers are all the same, even though the subjective experience of the pain may be different, different from person to person. And the length of time that they take to heal isn't the same for everyone. Our model says that you're afraid of your aggression, and the stress responses hide this from you. But no matter how horrible reality is, no matter how horrible your event is, it's more horrible if you don't come to terms with it. Confronting reality is always more painful than denying it. But it's there anyway, and denying it will only make you feel better in the short run.

4 April 1987. Group psychotherapy.

CAROL: Say to yourself, I've been punishing myself and people around me for twenty years. Say, Jack, you *can* choose to stop!

JACK: Listen, Carol. On some nights, I feel anxiety going through my body like it's electricity. It started in Vietnam. It wasn't just a feeling. It was anxiety together with terrible chest pains and difficulty breathing. Just like having a heart attack. They sent me over to a field hospital to get an EKG. The doctors told me that there's nothing wrong. They said I was just hyperventilating. They told me to breathe into a paper bag when I got these feelings, and they gave me a supply of Valium to take back. But I got these attacks again anyway. And I'm still getting them.

CAROL: What would you call it?

JACK: Well, I know that it's called a "panic attack." But I didn't know it then.

CAROL: No, I mean what would you call it using the terms of the model—the model that you learned about during orientation phase?

JACK: I don't really know, Carol. My mind is confused right now.

CAROL: The model says that we're dominated by two drives, aggression and sex, and that—

JACK: Listen, Carol. When I got these attacks, I sure didn't want to get fucked, and I can't believe it was my aggression.

CAROL: We've got to think of these events, your difficulty breathing, we've got to think of them in terms of *guilt*, of your wanting to *punish* yourself. We need to get in touch with your conflict—

JACK: Wanting to punish myself? Fuck, no! It's the most awful thing. You can't breathe—your heart is—

CAROL: Do you know someone who died from a chest wound? Can you go with this question?

[Jack suddenly stands.]

LEWIS: What's happening, Jack? What are your feelings, your thoughts?

JACK: Christ! I feel a load of anxiety right now. I have lots of thoughts. I have thoughts of coffins lined up like when I was going home, when I was leaving Vietnam the second time. I don't like to get like this. [Jack is visibly upset.] I don't want to talk about it. Hopeless feelings. Listen, I want to stop now.

CAROL: Put your thoughts out. You're safe here.

JACK: No, I'm not.

[During the next thirty minutes, Jack does not participate but gradually calms down.]

JACK: Towards the end of my second tour, I smashed my rifle and some other equipment at the base camp. I was in pretty bad shape, and they decided to send me to a hospital. The doctor says to me, "Talk to me about your childhood." It was irrelevant, to say the least. In fact, it was like a lot of what goes on here. After they talked to me, their solution was to give me Thorazine, put me on a plane, and pack me off to a naval hospital in the U.S. In this hospital, they put me in a room with a guy I knew from Vietnam. Both of us were in restraints, and they kept on injecting us with Thorazine. I was there for almost two years. I have a lot of anger. When I watch Lewis talking now, I want to slap his face. When I stand next to strangers, I want to bust them in the face. For no reason. And then I'm filled with anxiety.

LEWIS: Exactly. You experience the intent to do harm. Then, when you don't act on it, the result is anxiety. This anxiety is the way in which you punish yourself.

JACK: Well, I can take this anxiety only so long. Then I blow up. Next thing that happens is that the cops take me in. I tear up the police station. And then, the next thing I remember is waking up in a hospital, beaten up. In Vietnam, when I'd have anxiety attacks, they'd be followed by depression. And then I'd think to myself, You're thousands of miles from where you want to be, and all you want right now is to see your mother, and all that you want to hear is her saying that you're really all right.

LEWIS: Why have you hung onto this all these years?

JACK: Hung onto it! I've taken every fucking drug for it. I've driven to a VA hospital three in the morning for Valium injections. I've hung onto it for no reason. Look, let's stop. We've been at this for an hour. Let's go on to something else.

[Carol and Lewis continue to look into Jack's face. Silence. Jack turns his chair away from them.]

JACK: Forget it. You're not here.

LEWIS [to Jack]: Put it into words, Jack. Put into words what you're *feeling* now.

CAROL: You've shut down, Jack. That's okay. We can always come back to it. Henry, you've also told us that you turn away from people because of your fear of hurting them.

LEWIS: Why do you limit yourself to these options, Henry? Why must it be either/or, either hurt or love? Why must there be this splitting?

HENRY: You're right. The first thing that I do in relations with people is to split this way.

LEWIS: But you can always go *backward*. You always have the option of going back and discovering the source of your aggression.

CAROL: Yesterday, when we thought that there might be a thief here [among the patients], you became very angry towards the group. You said that you wanted to "gut" the man. Can you go back to something in Vietnam?

HENRY: Well, the word "gutting" has a special meaning for me. It makes me think of a time when a guy in my outfit threw some ears on the lieutenant's table. The ears were still wet with blood, and the lieutenant got pissed off, because he had to begin his report over again: his paper was bloody. In our outfit, it was policy to bring in ears as proof of confirmed kills, for body counts. And when we'd do this, it reminded me of hunting. You know, you go out, track your deer, shoot it, and then you gut the carcass.

LEWIS: But yesterday you used this word towards the group.

HENRY: Yes, I know. I don't like feeling that way. I don't feel that I'm too tightly wrapped. Maybe I am crazy. I was raised a Baptist. I went to church every Wednesday and Sunday, and I went to Bible camp. By the time I was eighteen, I was in Vietnam. After I was in country for only three days, I killed a sixteen-year-old boy. I began questioning my religion. I was asking myself what kind of a god would put me in a position like this and let me do this. At first, I felt sad. But then people started telling me, "Way to go," and the captain and the sergeant congratulated me for having a kill after such a short time. After a while, I fell into the program. I'd see guys lose legs and other shit happen, and it didn't bother me anymore. I began to enjoy it and went back on a second tour. I wanted to get revenge, and I wanted do as much destruction as possible.

LEWIS: Henry, when you talk about yourself, you keep giving us the image of someone who was being "singled out." You were singled out to kill after only three days. You were singled out for congratulations. And you've already told about the incident when you were singled out to go down into the [Vietcong] tunnel, and then you were separated from [abandoned by] everyone. And yesterday you were talking about wanting to hurt someone who had singled you out here— to steal from you. There's a pattern here, a repeating over and over.

6 April 1987. Group psychotherapy.

LEWIS: We are not here to *learn* the model. I am not your teacher, and you are not my students. You don't have to know the model to the extent that I do. Your job is to be able to apply it to your own situation. It's not possible in psychotherapy to sit back and be quiet. You must talk in order to get better.

JACK: Sometimes, I don't want to talk, because I feel that I shouldn't stir people up. When I talk, I make them feel my rage. I know I'm wrong when I do it, because I can feel my intention to make them feel my rage.

LEWIS: Stop right there, Jack. No talk about "shoulds." That belongs to "morality": thou shalt do this or not do this. And no talk about "wrong." Your job is to look at your behavior on these occasions—when you stir people up, when you try to make them feel your rage—look at your behavior as a stress response. It's not a question of whether it's right or wrong, but a question of what does it go back to.

7 April 1987. Group psychotherapy.

JACK: I'm jumpy today. I had a bad night. Trouble breathing, like Vietnam when my anxiety attacks began.

LEWIS: What caused your anxiety in Vietnam? And what caused it last night? What did they share?

JACK: I don't know. I tried to trace it last night. I felt that I was afraid of something. What? Dying? A heart attack? Everything mimics a heart problem. I couldn't get my breath. But I decided that it was caused by the heat and high humidity.

LEWIS: What was happening in Vietnam when you had this feeling for the first time?

JACK: Nixon announced that he would pull troops out. We reduced the number of missions going out. I spent more time at base camp, working out with weights and drinking beer. But not on the day it happened. I can't remember anything special; just everyday bullshit.

LEWIS: Well, we know that last night there was a major stress response and reenactment. Go back further.

JACK: You know, I have just one big fear: that it's going to happen all over again: that they'll pump me full of Thorazine and send me to an active-duty hospital. I know it's not my heart because I've been examined for it so many times.

LEWIS: You keep going outside of yourself for your explanation. Go with the model. It's a tool. It gives you something to look for. It's a map, a hypothesis to check out. It helps to focus you.

JACK: I started sweating—my head was pounding—I got out of bed—talked to the night nurse—played solitaire—got back in bed. Everything seemed back to normal. But then it started all over again. I'd like to know what's under this motherfucker.

LEWIS: You *do* know, Jack. It's conscious. When you recognize it, it will be painful, and that's why you don't uncover it.

JACK: You know the motherfuckers gave me shock treatments in the naval hospital. I was awake: the worst pain I ever felt.

LEWIS: You're choosing to avoid—you're continuing the aggression by talking about shock treatments.

[The session moves to other patients. Fifteen minutes pass.]

JACK: This mind-racing hurts. I wish I could say, Fuck it.

LEWIS: But the mind-racing is enabling you to avoid a worse pain. Did you ever have a bad pain? You want to relieve it? Then cause another pain. It distracts you. Anxiety is important to you. It motivates you. Too little anxiety leaves you unmotivated.

WILLIAM [to Henry]: Do you think the program has helped you in the last eight weeks?

HENRY [very softly]: I don't see any improvement. I'm not doing anything different. I'm not fucking myself up, but I haven't done that for the last three years anyway—after I got off of drugs and alcohol. When I gave these up, I thought I'd be better, but it's not true. I've had to stay away from my old friends and contacts and relationships. In 1981, they sent me to hospital. They juiced me [electroshock] to hell, turned me into a vegetable. An attorney went on his own initiative to the judge who committed me and got them to reexamine me. The doctors who examined me were the same doctors who examined the Iran hostages, and they said I had PTSD and didn't belong in this hospital. I don't blame the judge for what he did. I was a violent person back then. Even in the hospital, I spit on people. They moved me to the Salemville VA—a good place, and no Thorazine. I got active in AA and NA [Alcoholics and Narcotics Anonymous] and made the choice to give up alcohol and drugs. I felt good about it, and I accepted the fact that I am a lifetime cross-addicted addict. But now—I look back, and I see that I've got nothing. I used to be in the Vandals Motorcycle Club. And I used to go on beer runs with the Mongols. I had a lot of friends in the club, but I gave them up. Now I'm thinking about what I want to go on to.

JACK: I was in a motorcycle club for twelve years. I used to think that you go through life only once, and I'll do it my way. But I got tired of waking up fucked up. I was drinking a lot, and I knew that if I kept this up, I'd die. I had forty of those fucking shock treatments. And not just three second jobs. It all hurts: the treatments hurt; waking up sick hurts; being here hurts; anxiety attacks hurt. But you've got a choice. You don't need two extra burdens, alcohol and drugs.

LEWIS: Well, life's not a bowl of cherries. When you get stress and anger in the present, you don't have to jack it up with the shit of the past. You don't have to add to your level three anger in the present your level seven anger from the past. You've got to ask, Why am I doing this to myself? So far as shock treatment is concerned, it's a legitimate treatment for certain problems. I knew a patient who suffered from periods of acute depression that would last eight months at a time. He kept requesting shock treatment for his problem.

JACK: Then he really was crazy! [To Henry] Were you awake during your treatment?

LEWIS: Jack, you're working hard at avoiding, escaping. And you're engaged in aggression again. Why don't you talk about it, instead of acting it out? What were you thinking at this point?

JACK: Only a blank. When you're mind's racing, you're not thinking about anything.

LEWIS: See. Jack *knows* he's being aggressive. Look at the smile on his face.

JACK: Today's my birthday, Lewis.

LEWIS: You must be eighty-two going on twelve. Oops—that was very aggressive!

12 April 1987. Group psychotherapy.

JACK [to Paul, another patient]: What do you think are the greatest weaknesses of the program?

CAROL: Jack, what do you see your intention is here, by asking Paul this question?

JACK: In yesterday's clear thinking [a cognitive skills session], Paul said he hasn't been angry for over three months. Margaret and Janine [the nurses managing this clear thinking session] did their best to get Paul angry, but he had normal responses, even though the staff aggression was very high—very high, Carol. Listen, if he can give normal responses in these conditions, maybe he can help me to understand the causes of my own aggression, what's keeping my aggression so high. Now that he's decided to leave the program, I want to get his impressions.

PAUL: I know that the anger is there and always will be. But my anger is lower than it was before. I don't have to walk away anymore.

MARTIN: Well, it's not the same for everyone. Clayton [Martin's closest friend] was here for a visit last night, and he kept asking me, "Why are you exploding? You're not the same guy you were in Molton. If this is what the program's done to you, then I don't want any part of it." [At the Molton VA hospital, Martin was receiving antianxiety medication.]

JACK: I don't think that you staff really know the weaknesses of your program—

PAUL: Basically, it's a great program. I don't have such a high level of aggression day in and day out like I did six months ago. Now I can think—I don't have to say "fuck it" to every frustration. I know that strong feelings of aggression will return to me one day, but I hope I'll be able to say "Whoa!" When I first came here, I didn't want to look inside of me, because I knew what was there. But I did look inside: I faced up to it and acknowledged it is part of me. But now I like me. I recognize that I'm a pretty nice guy. I have some bad parts, and I can be a miserable son of a bitch because of this. But I can control that part now.

CAROL: You can say to yourself, I have an aggressive part of me, but I have a choice. I don't have to condemn myself because of this bad behavior. It's not necessary to go on either acting out or reenacting. It's not necessary to focus on externals, like the stress responses of the staff during clear thinking [sessions].

HENRY: Well, I can't recognize any reenactment in me—not since Day One here.

CAROL: Yesterday you were wearing all black, Henry, a black shirt, black jeans. And what was Paul's response? He said that the clients are like troops being shot at by people [Vietnamese] in black pajamas.

HENRY: Listen, the people who were shooting at me in Vietnam weren't wearing black uniforms. Black clothes don't mean anything to me. There was no aggressive intent. After the session, when I realized that my clothes produced a stress response, I went to my room and changed, and I came to lunch in different clothes.

CAROL: Even if it is not your conscious experience right now, its meaning will come to realization. Same thing about the meaning of your remark yesterday, about the Vietcong not being your enemy.

HENRY: Well, that's right. The Vietcong were acting the same way we would have acted in their place. I had no anger against them. My anger's against the people who sponsored the war.

CAROL: Which is equivalent to saying, "My anger is against all Americans."

PAUL: Why is it "reenactment" every single time? The staff still hasn't convinced me that this is true.

JACK: Where I can relate my behavior to some act in the past, I accept it. But I really can't just accept this idea without my own evidence. And I can't do it when someone who isn't really a therapist tells me that my behavior's a reenactment. Are the nurses here psychiatric nurses, Carol? Did they come here from other psychiatric units, or did they learn it all here at the center?

CAROL: Everyone on the staff is a therapist, Jack.

JACK: Tell me, Carol, am I splitting the staff? A patient can never be right here. If a person behaves perfectly normally and the staff says, "That was a stress response," but I know that it's really a normal reaction, then—

CAROL: What was it like in Vietnam when you didn't do it right?

JACK: In Vietnam, I did my job: look at my quarterly reports. No one ever told me that I was doing something incompetently. In our operational unit, we had a minimum of supervision and—

PAUL: Only one person was perfect, and he got nailed to a cross for it.

JACK: The aggression towards me is so high that— Listen, Carol, I don't mean *your* aggression. I'm talking about my own aggression now. It's so high when I came in here—

CAROL: But tell me about your anger *towards me*, Jack. If your aggression gets too high, we can handle it.

JACK: Carol, you process everything I say in a way that is different than how I actually feel. Everything I say is constantly being judged and then turned against me. I feel as if I'm in a fish tank. It's almost like persecution. The staff looks down on me because of my anger. I feel as if you feel that I can't do *anything* right. Even if you thought I was right, you'd pick at it until you found an ulterior motive. Because I have PTSD. You intimidate me, because I know that everything I say will be analyzed. I'm always off-balance. I feel vulnerable, unable to control my environment.

CAROL: What thoughts come to mind, right now?

JACK: The thought of slapping the bitch [Carol] off the chair. I'm trying to verbalize! I've tried to dislike you—both of you. But I don't. Sometimes I feel hopeless when I speak to you. I can go through fifteen emotions in three seconds, so that I can't remember each one. I feel as if my head is full of anger, and that my other emotions are deadened. I'll tell you, I'd like to backhand some people here. It's a good thing that there are rules. Right now? My anger is filling my ears. I can hear it.

CAROL: But you seem very calm, Jack. What do others think?

PAUL: He's repressing like hell!

CAROL: How about you, Johnny, what do you think?

JOHNNY: I don't know. When Jack started talking about wanting "to slap the bitch out of the chair," I got distracted. I drifted off. I started listening to the sounds outside . . . the birds, the tractor. . .

[Jack pushes his chair against the wall, out of the circle. Carol looks in his direction.]

JACK: Look, I'm *not* angry at the group. I'm not going to transfer my anger from the people who are responsible onto these guys. I know who's responsible. These guys here are just as uncomfortable as I am. I do accept responsibility for being aggressive against Paul, especially during yesterday's clear thinking. But I don't feel personally responsible for his decision to leave. That's not something I have to feel guilt for. It's not something that you can ask me to connect with Vietnam. It's the staff that took Paul through the eggbeater. The guy held up better than Oliver North.

PAUL: It did feel like an attack on me by the staff. But there were some patients who reenforced my self-confidence. And Jack gave me support when he said that my behavior was a "normal response" to what they were dishing out. But I'm not finished with the work I started at the center. I intend to continue as an outpatient.

JACK: I'd like to get kisses from a couple of the staff . . . I like a little affection when I get fucked.

CAROL: Thank you for your added aggression.

JACK: Thank *you*.

13 April 1987. Group psychotherapy.

JOHNNY: I remember putting a guy on a helicopter after a firefight. We lifted him up, and before we got him in, his body broke in half. When I first got to Vietnam, I spoke out against some of the awful things I saw going on. But then I got wrapped up in the killing scene—and got caught up in payback. I didn't like mutilations—

CAROL: When did you get caught up in payback?

JOHNNY: It happened gradually. There wasn't any one point that I could point to. There'd be memorial services, and the helmets of the dead guys would be set out, and they'd play taps, and then, afterward, we'd go into town and get boozed up, and then you'd go back to the bush with a pounding head.

JACK: Whenever I saw U.S. dead in Vietnam, I always thought, I'm glad it's not me.

MARTIN: The battalion next to us got a new colonel who was hot to get a reputation. He led them into an NVA ambush, and there were lots of dead marines. My battalion chased these NVA for two days. On the third day, we cleaned up—stacking dead marines. There were too many bodies to fit into the helicopters, and we had to tie some of them to the tracks. I spent the whole day crying and puking. At Khe San, General Walt [the marine commander] said he doesn't want any B52s. "This is the marines' playground," he said. The estimate was that there'd be 70 percent casualties—but Walt wanted to make a name for himself. He didn't want any army, air force; he threw out the Special Forces who were there before us. Hill 881 [Khe San] took two thousand rounds a day—rockets, mortars. NVA patrols were probing our lines all the time. All the marines did at Khe San was to take incoming fire, run patrols to clear the NVA off the barbed wire, and roll them down the hill every morning. The NVA didn't give a shit about Khe San. They could've taken it anytime they wanted. It was a fucking diversion for them. Their real business was running supplies around through Cambodia.

CAROL: The group should look at the issues that have been brought up so far: the sense of loss, hopelessness, futility, anger, the link between the here-and-now and the there-and-then.

JACK: I helped clean up after the An Shau mess—three thousand U.S. dead! The smell—guys crying, puking. When it was over, I had an empty feeling, like something had gone. I stopped going to memorial services. They only made things worse, emptier. I arrived in Vietnam with two close friends who had gone through training with me. They were both dead by the end of the first month. Patricks just never returned after a helicopter assault. The other guy was shot in the neck, and the bullet came out the top of his head. You know, I'm a thirty-nine-year-old man, and the last time I remember being happy was when I was eighteen.

CAROL: Henry, how about you?

HENRY: I lost my optimism—

JACK: When you come in here, it exposes your weaknesses to yourself and other people. I realize I couldn't do anything good since Vietnam; everything I touch turns to shit. I couldn't find anyone to relate with. A gunnery sergeant was my closest friend, and he drank himself to death. I joined a motorcycle club. [During this period,] I could fall into my own puke, and it was okay. I felt fear just like in Vietnam, and it was okay. I don't feel that I had to be here, like this. It's because of what the mother-fuckers did to me when they brought me out of Vietnam—the Thorazine and electroshocks. I didn't have to be like this.

PETER: In Vietnam, we didn't have an objective. We weren't allowed to accomplish anything. They just sent people there to fart around and to die.

CAROL: And the center—is it like this too?

PETER: Well, I wonder if there *is* a cure for PTSD. Lewis says that what we're into here is "a recovery on the way to a cure."

HENRY: Everything I've read before coming here says that PTSD can't be cured.

CAROL: We don't talk about it in terms of a cure, because PTSD isn't a "disease." PTSD is a "disorder," and this implies that we have to work together in the process. And we're not talking about just *coping*. We're talking about getting the intent to relate to dominate over the intent to aggression. This is the work that you've come here to do.

MARTIN: You know what makes me want to puke? All this shit about how "they" are welcoming us home now. I mean this rally they're planning for Washington, to "welcome home" the Vietnam vets. If I could arrange it, I'd go down there for the rally and shit in the middle of the street.

19 April 1987. Group psychotherapy.

JACK [sad and subdued]: And at the end of the patrol, I felt so fucking tired. All I wanted to do was sit down and go to sleep. I felt incredibly old. I felt old then, and I feel pissed now.

CAROL: Who are you pissed at, Jack? Holding that pissed feeling in is what is keeping it alive.

JACK: I have no problem keeping anger alive, Carol. During my second tour, I was in an assault on a large village. It was ringed around two-thirds by dug-in army. Our helicopter landed outside the village. I was leaning against it, smoking, when firing begins: a Cobra [helicopter gunship] slants down and pours high-volume fire into the village. You can't see anything because of the smoke—awful smell—all kinds of shit burning. We entered the village, and there were bodies all over the place. Near me, there was a dead old woman and a young girl—but the girl was alive. She'd lost one leg and was going around in a circle on the ground, crying out but not making any sound. The marine next to me takes out a hand gun and shoots her in the head. I was completely pissed by the whole fucking thing, all of it. All I wanted to do was trash people and that's what I did. I didn't care who they were, and I'd just as well have killed U.S. Back in the helicopter, I could hardly control myself from killing the pilot. I just pissed out the door and emptied my M16 into a pig. Now I hate everyone. I don't have a special target for my anger.

21 April 1987. Group psychotherapy.

LEWIS: Martin, you're still wearing your [tinted] glasses. Are you ready to take them off yet?

MARTIN: Not unless you turn the [ultrabright fluorescent] lights off. I told you that the VA said I'm photosensitive. I need these lenses to prevent headaches. The VA prescribed them.

LEWIS: Would you be angry if you had to take them off?

MARTIN: Yes.

LEWIS: Would you talk about this anger?

MARTIN: No. I'd just get angry.

LEWIS: But you know this is your task, to get under your anger. The glasses are a stress response, a way of shutting things out, a symbol covering up your psychic trauma.

MARTIN: You turned the lights down in the afternoon groups.

LEWIS: Well, we can't turn them down in this room. We can only turn them off. And that would cut down communication. Your choice of wearing the glasses is not morally bad, Martin. But it does hurt you, even if they're protecting you from short-term pain.

[Henry now redescribes a traumatic event, in which he killed civilians. At one point, he uses the word "frenzy."]

MARTIN: You know what frenzy is? It's how we were during action: no thinking; you only want to react. We assaulted a village and were pinned down for eight hours by NVA. Finally, they called in gunships, and the NVA started to run. Then we attacked the village. It was like something from the Civil War: us running in—yells—shooting everything in sight that wasn't a marine. Villagers were running around everywhere, trying to get out of the way of us and the NVA. After we finished with the NVA, our colonel told us to burn the village. I was torching a hut and a boy, maybe twelve years old, runs out. I grabbed him by the neck and threw him on the ground and told him to stay there. But he jumped back up, and I did the same thing again. When I turned my back on him, he reached for a machete, to cut my leg. But my buddy George saw all this and stabbed the kid in the neck. He's dead—like an NVA—and I emptied my magazine into the body. When it was all over, there are maybe eight hundred dead: marines, NVA, villagers, dogs and cats. We got the order to stack the U.S. bodies for the copters. [Martin removes his sun glasses.] George and me see two guys new to our outfit and they're looting the bodies of dead marines. We wanted to shoot them, but then we decided to take them to the platoon sergeant. The next day, we have another assault and S-2 [field intelligence] predicted 60 percent casualties. They made these two guys point men—and they didn't last out the day.

DONALD [a black veteran]: Were these two guys black?

MARTIN: Yes, but that didn't have any effect—didn't mean anything, or influence anyone's feelings.

DONALD: I had this "race" feeling from my Day One here, and I'm glad it's brought out in the open now. I don't like it when people think in terms of all-or-nothing about other people. Not all blacks are thieves. Some blacks are good, and some blacks are bad. Same thing with whites. My first day in Vietnam, while we're still at the airport, waiting for transportation, a white guy near me said to another white guy, "I'm going to get me [kill] a Vietnamese nigger as soon as we get in the bush."

CAROL: Racism and racial prejudice are ready-made splits for aggression to use.

MARTIN: Well, I felt pretty bad when the staff showed that movie, *Bloods* [a film about black soldiers serving in Vietnam]. I was in Vietnam in the first wave of U.S. Relations between the races were good. I had a black buddy, Spunk, who was blown away and . . . [Martin is having a hard time speaking, is finally overcome, and cries.]

CAROL: Martin—what thoughts are you holding inside?

MARTIN: I don't know. I think about partying together, talking about going home. Guys would talk and drink together, talk about their old ladies and cry. And you'd cry with them. [Martin breaks down again.] And then you'd say, Fuck it, and you'd smoke another joint. Vietnam was a crazy place. Lots of anger and revenge. You'd be glad to have a corpsman [medic] to take the edge off of things [with tranquilizers].

LEWIS: How do you other people feel?

DONALD: I'm glad Martin shared his feelings. I hope it made him feel better. Made me feel better. Maybe this is my problem too—holding things in.

JACK: I got a whole lot out of what you [Martin] said.

HENRY: Thanks for sharing it, Martin.

MARTIN: I've been holding it in since the night of the movie. What made me mad about the whole fucking movie was that there wasn't any time for race riots [among American troops] like in the film. [Martin is tearful again.]

[Fifteen minutes pass, during which other patients speak.]

MARTIN: This morning, around three 3 A.M., there was a garbage truck outside my window. It picked up a dumpster and slammed it down twice. I wanted to tear the bed apart. I got images of the fucking village in my mind, and I couldn't fall asleep. I kept thinking about it until it was time to get up.

LEWIS: Why did you take your glasses off, Martin? Why at that point?

MARTIN: Because I knew that I wanted to cry.

LEWIS: Go back to that point, Martin: look at what you were feeling, your thoughts. You've got to look inside. There was a surge in the intent to relate—

CAROL: And to recover.

PETER: You said the kid in the village was "just like NVA." Was that your feeling then, or is it your feeling now?

MARTIN: That's how I felt then.

PETER: Has the feeling changed for you since then?

MARTIN: No, not at all. I wanted to hold on to the kid to keep him for the ARVN [South Vietnamese Army] interrogators. I knew what they'd do to him: torture him. Not one doubt in my mind. They'd just start out with torture, no fucking around. Maybe he was lucky he was killed when he got it. Even if he didn't talk, the ARVN would kill him. He just went a couple of hours early. I knew he'd die one way or the other. I just wanted the information; he wanted to cut my legs off. When the assault was over, we felt great—running the NVA out of the village.

CAROL: You made a good start today.

JACK: A gunnery sergeant once told me, "Little cockroaches grow up to be big cockroaches."

25 April 1987. Clinical debriefing.

LEWIS: Aren't we doing limit setting anymore? I mean, how much longer is Martin going to wear his glasses?

CAROL: We're working on our stress response.

LEWIS: That's not an answer. What is "our stress response" supposed to mean?

CAROL: Yes, Lewis, we're still limit setting. And if we fail to set limits, then it's our stress response. But your question seems aggressive.

LEWIS: I feel as if I'm being left out in the cold. I take him up to this point and, then—

CAROL: Listen, your question *is* very aggressive. It's like you're saying to me, "You bitch, I'm working with the program, but what about you?" I don't know where you're going. You're implying, "Carol, you should have told Martin to take his glasses off."

LEWIS: Sometime last week, Peter [the clinical director and author of the model] said to us that we may be reaching the point where limit setting may be unnecessary for confronting clients with their reenactments and evasions. Is limit setting no longer allowed within the alliance? Is there some kind of understanding now that the model doesn't need limit setting? Is there something you know that I don't? This morning, I was left not knowing *what* to do.

CAROL: I heard your anger and—

LEWIS: Well, my anger was about not knowing what to do.

CAROL: On Friday, when Martin wore his glasses to group, you gave him a choice. And now you're angry. I sense your anger at Martin.

LEWIS: Yes. I am.

CAROL: And you feel trapped about whether or not you're making a proper intervention. On Friday, it seemed like a good intervention. Does it feel like Martin's set up [entrapped] now? You give him a choice and then—

LEWIS: It's damaging to him to leave things this way. Either he works, or he goes to an MDTP. Listen, I've got to go.

MALCOLM: What would you say to him at the MDTP?

LEWIS: That he stop wearing his glasses to group.

MALCOLM: And you'd explain that—

LEWIS: We've already done this. He recognizes that wearing his glasses may be a stress response, but he doesn't yet understand the aggression, the intent to do harm, in this. At this point, he has only a primitive understanding of the model.

CAROL: I still have questions about exactly what we're supposed to be doing in debrief. Why are we hooked on content—the ins and outs of Martin's glasses? We should be asking, Why are we reenacting now? We should be looking into process.

LEWIS: Well, we don't have any time or place to discuss content other than here. The fact is I feel foolish saying the same thing—about the glasses—day after day.

MALCOLM: You should be processing him in group, at the beginning of the sessions. If he doesn't respond, then—

CAROL: Yesterday, Martin had a splitting headache after his one-on-one.

LEWIS: If I know that a client is going to do something like this—wearing his glasses—it seems awfully aggressive of me to ask him not to do it. Now I really do have to go.

CAROL: You know, Lewis, what we really should be processing right now is why, each time we reach a point, you say "I want to go." Two times in ten minutes.

LEWIS: Listen, we could go on forever, analyzing. I'm going to miss lunch.

CAROL: Okay. Then this is it: if Martin comes in with glasses, we'll tell him no. And if he persists, we'll have an MDTP.

I close this chapter with a clinical supervision. These sessions are held weekly and are attended by the entire clinical staff. Dr. Durocher often focuses on a staff member whose work or comments have attracted his attention during the previous week. The following session revolves around Fiona, a nurse in her early forties. Fiona's initial duties at the center were medical and administrative. Subsequently, she was asked to conduct cognitive skills sessions and to act as a therapist in one-on-one sessions. On this occasion, she discusses one of her patient's, James Tyrell, an African-American army veteran.

1 May 1987. Clinical supervision.

DR. DUROCHER: During yesterday's debriefing, the image of patients as "war criminals"—people involved in torture, raping women, killing children, and so on—was brought up.

FIONA: It was in connection with James Tyrell's disclosure, that he executed a GI, that he set this man up and then killed him. James says the man was a fuck-up, that he would eventually get some of his own people killed. James said that on one occasion, the man had thrown a grenade at the enemy, but it landed near his own people—that just before this happened, he had been told *not* to throw grenades. James went to his commanding officer and told him what happened. According to James, the officer told him "to take care of it." James went out and beat the man up, and afterward the man said to him, "What did I do? I don't know what I did wrong." James describes the man as being "a rich kid without street smarts" and says that he shouldn't have been sent to fight a war. After he said this, James just stared me in the eye. I held his stare, and, after maybe thirty seconds, I asked him to continue. I said, "You can tell me, James." And then he said, straight out, "I killed the fuck-up. I was glad I did it, and I'm glad now that I did it. We were moving out, and I told him to move ahead toward the tree line. When he got in front of me, I shot him. Then I moved up and shot him in the head with an AK-47, so there wouldn't be any problems when his body was sent back. And I saw the guy's eye fall out of its socket, onto his face." And James rubbed his own eye when he got to this part.

DR. DUROCHER: And what was the effect of his narrative on you?

[Fiona is tearful but stops crying after a moment]

FIONA: At first, I felt very aggressive, but then I fumbled my way through the rest of the session. [She is tearful again.] I vacillated between the image of this kid saying to James that he didn't even understand what he'd done wrong and the image of him throwing the grenade and endangering his own people—and between the fact that things like this happen during any war and the idea that James is a war criminal. When the session was over and James left, I thought to myself, Okay, James is a genuine motherfucker. I knew this was my aggression, my stress response. I had heard other staff talk about hearing disclosures like this one—but I never thought I'd hear one.

DR. DUROCHER: You're *still* vacillating between the victim and the aggressor. Is it your stress response or is it your moral judgment that we're seeing?

FIONA: It's a stress response, but I also have my beliefs about what is right and what is wrong.

DR. DUROCHER: We all do. The important question is, How does it affect your clinical behavior?

FIONA: Well, when this happened, I mobilized myself. And he became very aggressive.

DR. DUROCHER: Mobilized yourself? What do you mean?

FIONA: I reminded myself that this is something that happened to James in Vietnam and that he's been carrying it around with him. I told myself that I was colluding with him in his aggression.

DR. DUROCHER: You mean that you mobilized yourself to his aggression. The urge that motivated you was your intention to deal with his aggression. Yesterday, you identified with him as a war criminal through your aggression. Today, you identify with him as a victim, through your crying and tenderness. What's your reaction now to what I'm saying?

FIONA: Confusion. . . .

DR. DUROCHER: How did James evaluate your behavior?

FIONA: At first, he was confused by my reaction. He couldn't tell what was going on in my head. Then he became angry at me.

DR. DUROCHER: We're not evaluating you. You know that?

FIONA: No. I don't know it. I feel lost, confused.

DR. DUROCHER: Perhaps you're not confused. Perhaps you're angry *now*—the way you were angry yesterday, with James. But now you're angry at me. Why? What comes to your mind? Tell me about your anger towards me.

[Silence]

DR. DUROCHER: You're not stuck—are you? It's not easy. Take your time.

[Silence]

DR. DUROCHER: What did you say when James was silent?

FIONA: I just maintained eye contact with him.

DR. DUROCHER: And I'm now maintaining good eye contact with *you*. Am I aggressive? Where are we, Fiona? Are you struggling with your anger towards me?

FIONA: Yes. James said to me, "I admire your persistence: you don't let me off the track." That's what's happening here.

DR. DUROCHER: James is able to like you, and you are able to like me. This is what allows us to go on—so that we don't get stuck. What name would you call me? What would you like to say to me?

FIONA: I don't know. I guess I would just say, "Listen up, mister!"

DR. DUROCHER: Go on.

FIONA: Well, this is how I feel. I'm very angry. Oh, shit! I do believe in the model. The thing is, my moral judgment gets in the way.

DR. DUROCHER: Maybe you're judging the model.

FIONA: I'm not that skilled. I don't know enough. How far do you take someone into the model? How far do you lead someone like James? You don't want to hurt him. If you're incompetent and confused. . . . [Fiona is tearful again.]

DR. DUROCHER: When you talk about the model, you're talking about me—about me hurting people.

FIONA: I'm talking about *me*.

DR. DUROCHER: Fiona, you're talking about me. If I hurt people, what would you call me? A criminal?

FIONA: Maybe.

DR. DUROCHER: There is a psychological connection between the term you would use for me and the term you would use for a patient like James. What are we doing here?

FIONA: We're working out my stress response—so that I can continue to work here. If I can't do this, I can't work with patients. And I can't stay.

DR. DUROCHER: Do you want to go? If you're struggling with people who are criminals—

FIONA: What's touchy is that this thing with James is touching the core of my own aggression.

DR. DUROCHER: Do you have a concern that *you* are a criminal? Do you ever think this? Have *you* done things in the past that you're ashamed of, things that say something that you don't like in yourself? Sometimes we realize that it is only because of chance that we ourselves are *not* in jail—that we've committed some culpable act. When we're bothered by this, we sometimes project these feelings onto other people—and call them criminals. How does this relate to yesterday? You didn't want to be victim, so you became aggressor. Then you projected this onto James—and then onto me. What do you think about this?

FIONA: It makes some sense. It gives some order—

DR. DUROCHER: Order? What do you mean?

FIONA: Control.

DR. DUROCHER: Over what?

[Silence]

DR. DUROCHER: Over your understanding of how we judge our patients?

FIONA: Yes.

DR. DUROCHER: But why couldn't *you* have said this?

FIONA: Because it touches my own dark side . . .

DR. DUROCHER: It's no different for the rest of us. [To the others] Some comments? We have fifteen minutes left for processing.

DONNA [like Fiona, a nurse]: My prior one-on-one could have been labeled a war criminal. But when he left, I could put these thoughts away. But I know that I have to look at my contribution to him deciding to leave.

DR. DUROCHER [to Donna]: Should Fiona process this with James? Should she try to learn what his reaction is to her behavior?

DONNA: No. It was probably this same thing also that made me say no. I mean my strong feelings towards my client—my stress response.

DR. DUROCHER: Is it possible that your stress response is your attempt to protect some memory of your own behavior in the past? Or am I putting words in your mouth?

DONNA: Oh, no! I could be doing this.

DR. DUROCHER: How did you react to Fiona now?

ARLENE [another nurse]: As both a victim and a perpetrator. I cried; my eyes welled up. But I don't know if I was identifying with Fiona and with James's victim or with my own thoughts.

DR. DUROCHER: You felt as if you were my victim—through your identification with Fiona?

DONNA: Well . . . I walled that one off.

VINCENT [a psychotherapist]: Fiona said that she feels that she might have been pushing James too hard for a disclosure. And now you're pushing her in pursuit of making her disclose. Fiona says she told James, "You can tell me, James." And now you're saying to her, "Go ahead, tell me."

DR. DUROCHER: What was your experience during this?

VINCENT: I was irritated with you: for goading Fiona, for pressuring her.

DR. DUROCHER: *How* angry were you with me?

VINCENT: Oh, I would say that I was just irritated. It's not as if I felt like running you over and tearing your lips off!

[General laughter]

DR. DUROCHER: But you do now.

CAROL: I've struggled with identification, as a victim *and* as a criminal—I mean, as being exposed for being a criminal.

DR. DUROCHER: Working with veterans who have been 100 percent involved in atrocities raises your own criminal tendencies and recalls your own experiences, whether or not we can admit this. [To Fiona] How do you feel?

FIONA: I'm glad they took the [traffic] cones down on I-580 [the interstate highway outside the center]. Yesterday, after I left the center, I hit three of them. I ran one over, and it got stuck under my car—and I was pulled over by a highway officer.

DR. DUROCHER: Here the identification with the criminal is on the surface. You were ready to commit atrocities.

FIONA: I guess so.

DR. DUROCHER: And you still want to knock a few cones over. . . . Thank you very much.

[The session ends.]

Eight

The Biology of Traumatic Memory

There are many signs and symptoms observable
in some severe "shell-shock" cases which point
to an affection of the endocrine glands and vege-
tative nervous system. . . .

Crile, in his work on Shock, asserts that there is
an interrelationship of function of the medullary
adrenal gland, the thyroid gland and the brain:
"Environmental stimuli reach the brain and
cause it to liberate energy which in turn directly
or indirectly activates certain other organs and
tissues, among which are the thyroid and adrenal
glands."
    (Report of the British War Office Committee
of Enquiry into "Shell-Shock" *[1922]*)

THE BIRTH OF PTSD followed a historical transformation in psychiatric knowledge making. Out of these changes emerged an invigorated psychiatric science that identified progress with the accumulation of facts
by means of testable hypotheses: "Hypotheses prove themselves superior
. . . by surviving strenuous attempts at disconfirmation. Science advances
by the replacement of falsified theories by yet to be falsified ones" (Wallace 1988:140). Psychiatric writers associate "testability" with Karl Popper's epistemology of falsificationism (Faust and Miner 1986). There
are important differences between Popperian falsificationism and knowledge making in psychiatric science, though, and I will use the term "fallibilism" when I refer to the style of reasoning now practiced in psychiatric
science.

In psychiatric science, hypothesis testing means following standards and
procedures established by other sciences, most especially by medical science; for example, the use of blinded randomized trials (Klerman 1986:25).
These procedures nearly always entail statistical techniques and probabilistic reasoning: a desire to determine and control levels of uncertainty and
contingency and a wish to generalize from observations based on samples

to the populations that they are intended to represent. Statistical techniques are easily transferrable from one science and domain of nature to another, because they are assumed to be a purified form of rationality—reasoning purged of subjectivity and indifferent to the context and content of its subject matter:

> [T]he function of rules is to eliminate ambiguity, and mathematical rules are the least ambiguous of all. Since it was precisely in those areas where uncertainty was greatest that the burden of judgement was heaviest, statistical rules seemed ideally suited to the task of ridding . . . the sciences . . . of the arbitrary, the idiosyncratic, and the subjective. Our contemporary notion of objectivity, defined largely by the absence of these elements, owes a great deal to the dream of mechanized inference. It is therefore not surprising that the statistical techniques that aspire to mechanize inference should have taken on a normative character. Whereas probability theory once aimed to describe judgement, statistical inference now aims to replace it, in the name of objectivity. (Gigerenzer et al. 1988:288)

As in other sciences, hypothesis testing in psychiatry is based on calculating probabilities. The calculations enable researchers to say whether an observed outcome has confirmed (justified) the experimental hypothesis. For the process to work, researchers require standards to identify the point at which confirmation can be claimed. A variety of statistical techniques are available, but psychiatric science strongly favors measures of significance. (Alternative techniques, such as confidence intervals and Bayesian inference, are employed far less often.) Using this technique, a researcher calculates the probability of his outcome and then chooses between his own experimental hypothesis and "the null hypothesis." The null hypothesis is that the outcome is just as likely to have occurred by chance. Before the choice between hypotheses can be made, though, researchers have to decide what will count as a significant difference. One way to do this is to set an "alpha level": the null hypothesis is accepted when chance probability exceeds this level. The alpha level is set entirely by convention—a point that is made routinely in statistics and epidemiology textbooks. However, convention has decreed that in no scientific field today does the alpha level rise above 1 in 20 ($P < .05$). A researcher who rejects the null hypothesis (and thus accepts his own hypothesis) when the outcome exceeds this level is said to have committed a "type 1 error." The editor of a psychiatric journal who set the limit of the alpha level higher than 1 in 20 (e.g., at 1 in 15) or who left the level to the discretion of his authors would violate no theorem of statistical science nor would he offend any principle of logic (Gigerenzer et al. 1988:78). Yet no editor would be likely to take this step, nor would any competent referee when evaluating a manuscript submitted for publication. Deviations of this sort would be particularly perilous for psychiatry, given its relatively low status in the hierarchy of medical sciences.

Textbooks on statistics also warn readers about "type 2 errors." The message here is that, in some circumstances, the alpha level may be too stringent and it would therefore be a mistake to accept the null hypothesis. This happens when the predicted effect (outcome) is present but can be detected only weakly ($P > .05$), because of interference from unanticipated attributes of the experimental groups, for example, medical or social characteristics that dampen the differences between the groups.

If we look at the way in which calculations of probability are actually employed in psychiatric discourse, type 1 and type 2 errors mirror different conceptions of facticity. (As in other sciences, probability functions as a surrogate for "truth" in psychiatric science. I use the term "facticity" for knowledge claims based on probabilistic reasoning.) Type 1 errors are conceived in such a way as to make facticity a *dichotomous* variable, in the sense that positive and negative findings are distinguished by means of a determinate threshold (cf. Faust and Miner 1986). Type 2 errors turn facticity into a *continuous* variable: chance probability of 1 in 20 is a stronger indication of facticity than 1 in 15, but 1 in 15 might be stronger than 1 in 10 (Riegelman and Hirsch 1989:34–37).

Both standards, dichotomous and continuous, are employed in psychiatric discourse. In journal articles and conference papers, they are distinguished from one another by a variety of linguistic markers. The most conspicuous of these is the term "statistically significant," which is used to identify results that meet the dichotomous standard. Logically, there is no reason to prefer a dichotomous standard of facticity over a continuous measure. Pragmatically, though, the standards are *unequal*, in the sense that decisions to publish the results of experimental or epidemiological research as "regular articles" in mainstream journals strongly favor the dichotomous standard (Bailar and Mosteller 1992:316–317; MacRae 1992:94; Ware et al. 1992:185–186, 188–189, 196–197). The superiority of the dichotomous standard is based on its usefulness for aggregating results published by different authors, and the desire of psychiatric researchers, editors, et cetera. to match the standards employed by the "harder" sciences.

It is on this basis that psychiatric discourse distinguishes three kinds of experimental results:

1. Results that measure up to standards of testability and facticity. These can be called *facts*.

2. Results that do not measure up to such standards but are preserved nevertheless. (See Wallace 1988:140 on "partial truths.") These can be called *findings*. They are either results of nonstandard procedures, such as drug trials that lack control groups, or results that failed to reach statistical significance. Because facticity exists in fallibilist science as a continuous

property (as well as a dichotomous one), researchers are justified in pre-serving their findings and connecting them into networks of knowledge.

3. Results that do not measure up to the standards and are abandoned by the researchers. These are *discarded results*.

The evolution of a facts-versus-findings distinction can be traced back to the origins of significance testing in the 1930s and the writings of R. A. Fisher. According to Fisher, the existence of a natural phenomenon

> cannot be established on the basis of one single experiment, but requires obtain-ing more significant results when the experiment, or an improvement of it, is repeated at other laboratories or under other conditions. Therefore, not only sig-nificant, but also non-significant results should be published in order to let the literature correctly reflect the frequency with which a certain experiment has led to significant results. . . .

> [Thus] Fisher distinguished *significance testing* from *the demonstration of a nat-ural phenomenon*. Careless writing on Fisher's part, combined with selective reading of his early writings has led to the identification of the two, and has encouraged the practice of demonstrating a phenomenon on the basis of a single statistically significant result. (Gigerenzer et al. 1988:96)

While "facts," "findings," and "discarded results" are my own terms, they map categories of statements that are identified (but not named) in the psychiatric research literature. For example, look at the following state-ments taken from a recent article on PTSD in the *American Journal of Psychiatry* (Pitman et al. 1990). In each instance, the initial citation comes from the section of the text titled "Results," in which the authors describe the outcome of their experiment.

Facts: "As predicted, the subjects with PTSD showed an analgesic re-sponse in the placebo condition. . . . Furthermore, naloxone reversed this response." The text indicates that probability levels are .03 or less. These results are restated in the paper's concluding section, again without any qualifiers: "The data in this study demonstrate a significant naloxone-reversible analgesic response" (542, 543).

Findings: "These increases [in heart rate and skin conductance] tended to be larger in the subjects with PTSD than in the control subjects; but for heart rate, this *just missed statistical significance*" (my emphasis). These results are carried into the concluding section where they are said to "fall short" of distinguishing the PTSD subsample. They are positioned with other findings that, in retrospect, are found to be consistent with the re-searchers' thesis (542, 544).

Discarded results: "The subjects with PTSD tended to have lower pain intensity and unpleasantness ratings than did the controls after the first vid-eotape . . . , but these differences were *not statistically significant*" (542; my emphasis). These results appear only in the "Results" section.

Meaningful Time

My point so far is that psychiatric scientists have two ways of talking about facticity, in terms of dichotomy or in terms of continuity, and these two ways are mirrored in the tacit distinction between facts and findings. As we shall see now, facts and findings are connected in another way: through conceptions of time.

Scientific discourses grow because researchers continue to encounter points at which reality appears to be pushing back, by refusing to yield any facts. Endangered theories are the sites of this kind of resistance and, for this reason, are potential sources of new knowledge. Seen in this light, effort spent on rescuing an endangered theory is not necessarily a bad thing. No theory should be rejected "before it . . . [is] able to make its contributions to the growth of science" (Popper 1972:30). But, practically speaking, how do we determine when rules and conventions have been stretched too far? At what point does the desire to preserve a theory or hypothesis become unreasonable? When is an experiment's outcome too feeble to be carried forward as findings? "[In practice,] assessments of conformity to Popper's rule of scientific method hinges on scientists' interpretations of 'falsification'; and the meaning of 'falsification' depends *entirely* on researchers' technical and scientific judgements" (Mulkay and Gilbert 1981:398). Within any scientific field, the line between results worth preserving and results that ought to be discarded is fluid and usually contested—unless the theory or hypothesis has attracted no attention. Where the line is finally drawn by a particular network of knowledge producers depends on the play of contingencies: the urgings of interests, the ability of individuals and groups to control access to the means needed for producing knowledge (money, institutional support, patients, etc.), the skill and energy with which participants deploy their rhetorical resources, and so on (Mulkay and Gilbert 1981:403; Latour 1987).

Gilbert and Mulkay's 1984 study, *Opening Pandora's Box*, describes how scientists narrate these circumstances. The book's informants are biological scientists, whose discourse is divided into two contexts: a "formal context," consisting of articles they have written for refereed journals and papers they have delivered at professional meetings and conferences, and an "informal context," consisting of their conversations with Gilbert and Mulkay. The formal context is said to be dominated by an "empiricist program," based on the informants' belief that scientific knowledge is the product of falsifiable hypotheses, competently performed experiments, and full disclosure of research methods. Their readers and auditors are given the impression that scientific knowledge is the product of highly routinized procedures, that equivalent results can be easily obtained by any competent scientist who follows the same rules, and that scientists' actions and beliefs flow "unproblematically and inescapably from the empirical characteristics of an impersonal natural world."

When located in the informal context, however, informants switch back and forth between the empiricist program and what Gilbert and Mulkay call the "contingent program," where scientists describe research practices that clearly depart from rule-following. Informants report that key bits of information are regularly omitted when experiments are described in journal articles; that even if this were not the case, no article can include all of the information that is needed for repeating its experiment precisely; and that even if it were possible to provide this information, researchers would have compelling economic and professional disincentives for not attempting to repeat other people's research in a precise way (Gilbert and Mulkay 1984:52–56).

When informants were asked to explain the inconsistencies between statements made in the two contexts, their response was to give an overarching narrative in which they nested their previous remarks about contingency inside a second argument that pictured knowledge production in science as ultimately self-correcting and cumulative. Gilbert and Mulkay call this move the scientist's "truth will out device": "Experimental evidence is depicted as becoming increasingly clear and conclusive over time and enabling scientists to recognize, discount, and eventually eliminate the influence of contingent factors" (Gilbert and Mulkay 1984:56). Through this device, knowledge production is represented as a teleological process. On the one hand, compromises and deviations from rule-following are inevitable and produce truths (knowledge claims) that are defective or contingent. On the other hand, such truths are temporary and local, a means to an end. (This is the fallibilist notion that findings, *because of their defects*, lead to useful hypotheses and facts.) Science's enduring truths are emergent and transcend the contexts and compromises in which they were born. When the informants switched back to the empiricist program, the distinction between a present time in which truth is a regulative idea and a future time in which truth is a durable quality disappeared. At this point, the researchers' "scientific assertions become indistinguishable from the empirical world under observation" (Gilbert and Mulkay 1984:109–110).

(Re)biologizing Psychiatry

The style of reasoning that Gilbert and Mulkay observed among biological scientists is common to psychiatric researchers. What is distinctive to the psychiatric scientists is that ideas about time and emergent truth (the presumption that "the truth will out") are connected to assumptions about the unity of the sciences. "Unity" has a double meaning in this context. It means that the sciences are joined by a shared methodology, "the scientific method." It also means that the sciences are joined into a dual hierarchy: a *hierarchy of nature*, whose levels ascend from the simplest (subatomic matter) to the most complex (aspects of human consciousness, such as memory), and a corresponding *hierarchy of scientific fields*, with the physical sciences at the base, the biological sciences at the middle level, and the psychological and behavioral sciences at the top. Fallibilist researchers find truths at each level of the hierarchy—biological truths, psychological truths, and so on—but the truths that are found at underlying levels are progressively denser and more dependable. Further, relations between levels are asymmetrical. The highly complex and often poorly understood phenomena that are typically found at higher levels are explained by (caused by, reducible to) the less complex and better understood phenomena located at underlying levels (cf. Charlton 1990).

This is the context in which one has to understand the claim, made by various commentators, that American psychiatry is going through a process of biologizing itself: "Biological psychiatry is seen as the 'new psychiatry' and as the means by which psychiatry will raise its status within medicine and become part of 'mainstream medicine'" (Gaines 1992:190). Guze describes these changes as a return to reality. Psychiatry is "a branch of medicine, which in turn is a form of biology," and psychological aspects of psychopathology are epiphenomena of "disordered processes in various brain systems" (1989:317). While there are psychiatric writers who would prefer to resist this verdict, it is significant that they picture themselves as members of a beleaguered minority (Eisenberg 1986; Reiser 1988; Charlton 1990).

Important changes are taking place in American psychiatry, but it might be more accurate to describe them as representing its *re*biologization. Into the 1920s and 1930s, the majority of psychiatric researchers and clinicians tended to look for and find somatic explanations for mental disorders. It was during the following three decades that the mainstream shifted in the direction of theories based on personality and intrapsychic conflict, associated either directly or indirectly with the work of Sigmund Freud, and the psychosocially oriented mental hygiene movement, associated with the ideas of Adolph Meyer, William Menninger, and others:

> There are [today] some psychiatrists, described as "organically minded," who believe that many types of mental illness do have a physical or chemical cause. . . . A minority of psychiatrists of an older vintage believe that heredity is the major causative factor in mental illness. There has always been a majority who . . . accept the clinical picture as they see it and describe it in terms of the

symptoms. In contrast to these, there is an increasing number of psychiatrists, described as "dynamically oriented," who analyze the case on the basis of defective or misdirected psychological forces within the personality. (Menninger 1948:256–257)

By the 1950s, more than half of the chairmen of departments of psychiatry in the United States were members of psychoanalytic societies. Within another decade, however, the mainstream had shifted direction once again, redefining mental disorders in terms of disease and renewing psychiatry's historical connections with biological medicine. Today, psychodynamic and environmental approaches to mental illness continue to drift from the center, "outgunned by nosologists and neurobiologists in an era of scientific research" (Eisenberg 1986:498). The move away from biology was less obvious in the case of the traumatic neuroses of war, however. Biological accounts remained continuously important from World War I onward (e.g., Kardiner 1941: chaps. 3, 4, 5; Grinker and Spiegel 1945: chap. 7).

Psychiatric science has changed in two ways since the earlier period of biological ascendancy. The previous emphasis on hereditary taints is diminished, and researchers as well as clinicians have acquired powerful technologies for actualizing biological etiologies. Psychiatry is now reconnected to biology through a variety of linkages: through psychoactive drugs that connect diagnostic classifications to specific biochemical processes (phenothiazines for schizophrenia; tricyclics, monoamine oxidase inhibitors, and selective serotonin reuptake inhibitors for depression; lithium carbonate for bipolar affective disorder; etc.); through imaging technologies, such as positron emission tomography, that give pictures of the brain's pathoanatomy and pathophysiology; through a standardized nosological system that models psychiatric diagnosis after medical diagnosis; and through a division of labor that shifts the psychiatrist's gaze away from the psyche and psychotherapy (increasingly the responsibilities of clinical psychologists and social workers) onto the medical management of psychoactive drugs (Klerman 1986, 1990; Sabshin 1990; Scadding 1990; Sharfstein and Goldman 1989).

In a fallibilist perspective, the continuity of nature (signifying a hierarchy of causes) is mirrored in the unity of the sciences (signified by the "scientific method"). This perspective has a double significance for psychiatric science. First, it provides psychiatry with a legitimate place among the sciences, next to biological medicine. Second, it justifies a strategy of looking for etiologies of mental disorders in underlying levels of nature belonging to other sciences and incorporating somatizing technologies into research, technologies that make biological causes visible (operational) and measurable.

For many years there appeared to be no . . . independent means by which the claims of various schools concerning the causation of mental illness and the efficacy of their treatment could be established as valid or dismissed as unproven and unsubstantiated by empirical evidence. . . .

In the past two decades, however, this situation has changed dramatically, particularly in the United States. . . .

The empirical approaches within psychopathologic research have been related mainly to the emergence of a new [Kuhnian] paradigm. Its methods have been applied primarily to biological research and to the evaluation of drug and behavioral therapy. . . .

This new paradigm in psychopathology emphasizes two aspects of the modern philosophy of science: (a) that the essence of modern science is testing hypotheses through experimental and quasi-experimental methods in the laboratory and clinic; and (b) that quantification of phenomena . . . is necessary. (Klerman 1986:24–25)

It is this hierarchy of science and nature, and psychiatry's claim to a place within it, that obligates researchers to the fact/finding distinction, an obligation to maintain respectable standards of statistical significance and experimental procedure (equivalent to the "nuts and bolts" of the scientific method). The obligation has created many fact-poor discourses in psychiatric science; on the other hand, it has permitted psychiatric researchers and writers to buy time for their findings—time to work with them, time to mature them into facts—through connecting them (via research technologies) to fact-rich sciences, such as neuroendocrinology and molecular biology.

The Neurobiology of Traumatic Memory

In the remainder of this chapter, I examine two biological research narratives of PTSD: "Naloxone-Reversible Analgesic Response to Combat-Related Stimuli in Post-Traumatic Stress Disorder," by R. K. Pitman, B. A. van der Kolk, S. P. Orr, and M. S. Greenberg (*Archives of General Psychiatry* 47[1990]:541–544); and "Urinary Free-Cortisol Levels in Post-Traumatic Stress Disorder Patients," by J. W. Mason, E. L. Giller, T. R. Kosten, R. Ostroff, and L. Podd (*Journal of Nervous and Mental Disease* 174[1986]:145–149). Before I turn to these articles, I am going to outline the background knowledge that competent readers bring to such texts.[1] Readers who are familiar with the neuroendocrine model of stress can safely skip to page 276.

The prevailing biological account of stress reactions begins with the nervous system. The system is pictured as a functionally divided communications center whose parts variously (1) process messages arriving *from* the external environment, other body regions, and other parts of the nervous system, and (2) originate and transmit messages *to* other body regions and parts of the nervous system.

Within the nervous system, messages are transmitted in the form of electrical impulses. The impulses pass over synapses between adjacent nerve cells by means of neurotransmitters, chemical substances usually secreted in the nerve endings from which they are later released. It is estimated that there are more than fifty different neurotransmitters, but only a few have been closely studied. These few include the catecholamines: norepinephrine (NOR), epinephrine (EPI), and dopamine. Within psychiatry, NOR and EPI are widely believed to play a part in the etiology and symptomatology of several mental disorders, including depression and the anxiety disorders (which include PTSD).

NOR and EPI function both as neurotransmitters and hormones, that is, chemical substances produced by endocrine glands and carried by the circulatory system to receptor sites on target tissues. As neurotransmitters, they are synthesized in the central and sympathetic nervous systems; as hormones, they are produced in the medulla of the adrenal glands. In contrast with NOR, only relatively small amounts of EPI (also called adrenaline) are found in the nervous system, but they constitute about 80 percent of the catecholamines produced in the adrenal medulla.

PTSD researchers and writers are interested mainly in the roles played by NOR and EPI in the sympathetic-adrenal system and the limbic system, and with respect to stress reactions (see figure 5).

The hypothalamus (a division of the midbrain) is the principal locus of integration of the sympathetic and parasympathetic nervous systems (divisions of the autonomic or "involuntary" nervous system). PTSD writers and researchers are interested in the sympathetic and parasympathetic nervous systems because of the central role they play in stress reactions. These systems—a network of adrenergic neurons (NOR is the neurotransmitter) and cholinergic neurons (acetylcholine is the neurotransmitter)—connect the brain to various organs, including the circulatory system and the adrenal glands. Among other things, the network regulates the secretion of adrenal hormones, heart rate, blood pressure, cardiac output, circulation, and respiration.

The hypothalamus is connected to the sympathetic system by adrenergic neurons and regulates adrenal secretion along two routes. One route leads through the pituitary, a gland adjacent to the hypothalamus. The hypothalamus secretes a substance, CRF (corticotropin-releasing factor), that stimulates the pituitary to release another substance, ACTH (adrenocorticotropic hormone), that in turn stimulates the adrenal cortex to produce corticoste-

A PTSD-SCALE TRAUMATIC EVENT impacts on ⟶

HIGHER BRAIN CENTERS AND COGNITION and initiates ⟶

A NEURAL HORMONAL RESPONSE that is equivalent to the following:

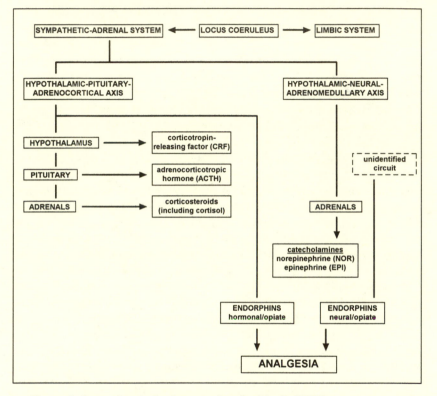

Figure 5. Some chemical substances involved in the PTSD stress response

roids. The second route leads to the adrenals along sympathetic nerve impulses and bypasses the pituitary: the hypothalamus stimulates the adrenal
medulla to synthesize and release NOR and EPI. (A third route, connecting
the hypothalamus to the thyroid through the pituitary, also affects catecholamine activity but has not yet found a significant place in the literature on
stress responses.)

Perceptions of threat and danger ("stressors") stimulate the sympathetic-
adrenal system to discharge as a unit, along both routes. Large amounts of
medullary catecholamines are released (especially EPI), and large amounts
of CRF, then ACTH, and then corticosteroids are also released. (Of the
corticosteroids, it is cortisol that plays the most important part in stress

reactions.) In this way, the individual is mobilized for "fight or flight": the central nervous system is alerted and aroused; reaction time is speeded up; heart rate and cardiac output are increased; blood pressure rises; blood flow is shifted from the viscera to the brain and the skeletal muscles; plasma glucose rises; bronchioles dilate, and the rate and depth of breathing increase.

The term "stress" is used in two senses in this connection. Sometimes it is used to identify any instance of the fight-or-flight response, and sometimes it is used to describe those situations when stressors, because they are massive or inescapable, clearly exceed or exhaust the individual's capacity to successfully respond to them. It is in this second sense that "stress" is described as "trauma" and enters the PTSD literature.

When individuals are reexposed to initially provocative stimuli (stressors), the hypothalamus-pituitary-adrenocortex route loses its responsiveness, and the release of corticosteroids returns to normal levels. This contrasts with the hypothalamus-neural-adrenomedullary route, where the adrenals continue to release catecholamines for a longer period. This difference is traced to the influence of higher brain centers and cognitive mediation on each of the two routes. Researchers have suggested that the release of EPI can be traced to situations that are novel and unpredictable or where individuals are anticipating an encounter with stressful stimuli; while the release of NOR can be traced to situations that demand continued attention, vigilance, or effort.

There is an additional way in which higher brain centers and psychological processes are said to shape the neurophysiology of stress reactions. There is abundant evidence, both experimental and epidemiological, that people exposed to the same stressful environment do not necessarily respond in the same way. Some respond with rapid increases in cortisol and catecholamine levels, while others show little or no hormonal response. The differences in responsiveness are said to reflect differences in how people perceive the same physical environment, and their different perceptions are traced to their different experiences, developmental histories, and models of reality.

The Limbic System

In the brain stem there is a nucleus known as the locus coeruleus. A system of NOR neurons ramify out from the locus coeruleus, reaching most regions of the brain, including the cerebral cortex. The major neuronal projection of the locus coeruleus is into the limbic system, the "visceral brain" located beneath the cortical mantle. This system, which includes the hypothalamus, hippocampus, and amygdala, regulates the individual's emotional response to his external environment and integrates his emotional states with his motor and visceral activities.

The limbic system has the greatest concentration of adrenergic (NOR) cells in the brain. A single adrenergic cell, rooted in the locus coeruleus, can innervate the hippocampus, amygdala, and cerebral cortex. By making cortical neurons "aware" of changes in the organism's feeling states, the limbic system gives emotional coloring to the individual's perceptions and memories. It infuses them with fear, anger, disgust, pleasure, sadness, and surprise. In this way, the system signals which elements in the environment and memory need attention and demand action.

It is at this point, where the axis linking the locus coeruleus to the limbic system intersects the axis connecting the hypothalamus to the adrenal glands, that the stress reaction is produced.

Endorphins

The (anterior) hypothalamus, amygdala, and hippocampus also have high densities of opioid receptor cells. Exposure to certain stressors is said to activate these receptors and to stimulate the secretion of endorphins (endogenous opioids) in the brain. These receptor-endorphin responses are said to explain the phenomenon of "stress-induced analgesia," a condition in which individuals become either insensitive or less sensitive to pain during or after stress reactions (and analogous to the kind of analgesia produced by acupuncture). This form of analgesia is also said to explain the passivity of laboratory animals when they are exposed to inescapable shock. (See Bloom 1985; Baldessarini 1985; Rose and Sachar 1981; Snyder 1986; van der Kolk et al. 1985; Weiner and Taylor 1985.)

The Neurobiology of PTSD

Biological accounts of PTSD are based on three ideas: (1) The biological processes that explain "ordinary" stress responses also explain the etiology, symptoms, and chronicity of PTSD. (2) While the biological effects of ordinary stress responses are transitory, PTSD is characterized by enduring neurophysiological changes. (3) To explain the disorder's symptoms and chronicity, psychological processes also need to be taken into account.

The disorder's traumatic events are said to stimulate the release of large amounts of NOR into the neuronal synapses. This leads to a "long-term augmentation" or "potentiation" of locus coeruleus pathways leading into the limbic system according to van der Kolk and his associates (1985), and to changes in "synaptic structures . . . in the temporal-amygdaloid complex concerned with agonistic behavior" according to Kolb (1987). It also produces neuronal changes in the sympathetic system, making it hypersensitive to autonomic arousal (see also Mason et al. 1990 and Watson 1988).

There is no direct evidence that these neuronal changes occur in PTSD, but researchers know how they might occur. A neuron's sensitivity to a neurotransmitter is partly determined by the numbers and molecular properties of its receptors for this neurotransmitter. These receptor features are known to alter in response to changes in the neuron's biochemical environment, such as the presence of persistently high levels of endogenous opiates. In the case of PTSD, the flood of NOR would, then, precipitate the changes that "augment" the movement of impulses along (adrenergic) stress response pathways. This effect is described by van der Kolk and his associates as the "neurophysiological analogue of memory" (1985:318)—the somatic memory whose origins were described in chapter 1. Kolb writes about the somatic memory in terms of "subtle neurochemical changes . . . that currently defy detection by available methods" (1987:993).[2]

The psychological dimension of the explanation is based on learning theory and the distinction between unconditioned and conditioned aversive stimuli. In PTSD, the unconditioned stimulus is the traumatic event. Exposure to this stimulus induces extreme levels of arousal. During these events, there are nonthreatening elements in the environment—smells or sounds for example—that become psychologically fused to the unconditioned stimuli. These nonthreatening elements are called conditioned stimuli, and individuals react to them in the way that they originally responded to the unconditioned stimuli (Keane et al. 1985; van der Kolk et al. 1985).

These biological and psychological accounts are said to explain PTSD symptomatology in the following ways:

1. The "best available biological model for PTSD involves the exposure of animals to inescapable shock (IS)" (van der Kolk et al. 1985:316). In both cases, individuals are exposed to overwhelming stressors. They are unable to predict the onset or termination of the stressors, unable to effectively fight against them or to flee from them. Experiments indicate that IS initially stimulates the release of large amounts of NOR in the central nervous system. (This is the point at which the augmentation of pathways would occur.) The NOR is eventually depleted, because demand for it exceeds supply. In PTSD, the depletion effect is maintained by feelings of "chronic helplessness." Chronic NOR depletion explains the PTSD patient's "diminished motivation, decline in occupational functioning, and global constriction" (numbing, anhedonia, etc.), as well as his hypersensi-

tivity to transient stimulation, especially conditioned stimuli. A vicious circle is created: NOR depletion → diminished motivation, et cetera. → helplessness → NOR depletion, and so on (van der Kolk et al. 1985:317). ("Chronic helplessness" in PTSD is analogous to "learned helplessness" in animals traumatized by IS. When these animals are reexposed to IS, they surrender to their situation and make no attempt to separate themselves from the source of pain even when they are given the opportunity to do so.)

2. The behavioral counterpart of autonomic hypersensitivity is hyper-reactivity. In PTSD symptomatology, this is evidenced in startle responses, intrusive images and memories, nightmares, a tendency to overreact to (conditioned) stimuli, and difficulty modulating emotions, especially anger.

3. The release of large amounts of NOR during the traumatic experience produces hypermnesia, a state in which details of the etiological event are locked into the person's memory (see Pitman et al. 1987). Hypermnesia is facilitated by the fact that a single neuron located in the locus coeruleus can simultaneously branch into the limbic system and the temporal neocortex, the place where long-term memories are stored. The combination of hypermnesia and hyperreactivity accounts for the flashbacks and the unusually vivid and detailed (eidetic) nightmares that sometimes occur after individuals are exposed to conditioned aversive stimuli.

4. Changes in the synaptic pathways in the locus coeruleus and hypothalamus impair the cerebral cortex's inhibitory control over the parts of the brain concerned with the expression of aggression and the patterning of sleep cycles. These changes, together with autonomic hypersensitivity, account for the explosive anger and the difficulties falling asleep and staying asleep that often accompany PTSD.

5. Endorphins are released when PTSD victims are exposed to stimuli that resemble their traumatic experiences. The endorphins reduce the feelings of rage, inadequacy, paranoia, and depression that often accompany PTSD, and they enhance the person's sense of being in control. People with PTSD often oscillate between states of quiescence and agitation. According to van der Kolk and his colleagues, this is similar to what happens with exogenous opioids. Morphine inhibits neuronal firing rates in the locus coeruleus and creates a state of quiescence; opiate withdrawal syndrome is connected to the hyperactivity of these (NOR) neurons and is characterized by increases in anxiety, irritability, outbursts of anger, insomnia, and hyperalertness. "The parallels between the symptoms of opiate withdrawal and the hyperreactivity in PTSD are striking" (1985:320). Endorphin-releasing practices provide temporary relief but also create a cycle of self-addiction: anxiety → endorphin-releasing practices → relief → endorphin depletion → hyperreactivity → anxiety, and so on.

Self-addiction to endorphins also explains why PTSD victims would put themselves in situations that resemble their traumatic experiences, as when traumatized veterans who become policemen or join motorcycle gangs, or traumatized soldiers continue to volunteer for dangerous missions.

Duhem's Thesis

The neural-hormonal theory connects the criterial symptoms of PTSD into an explanatory system in which time and causality always run in the correct direction. (See chapter 4 for a discussion of this problem.) In addition, the theory affirms the intrinsic unity of the PTSD classification by providing it with a distinctive biological mechanism (e.g., van der Kolk et al. 1985). The theory is also a blueprint for biological research on PTSD, and in the next sections, I scrutinize two examples of this research.

The latter articles are representative of biological research on PTSD in at least two ways: the researchers produce many findings but few facts, and their narratives follow a pattern described by Pierre Duhem. A philosopher of science, Duhem observed that when researchers predict that a particular outcome will occur under specified circumstances, but the outcome does not occur, they will usually rescue their original hypothesis with an auxiliary hypothesis that finds fault with their methods: "if the stars aren't where theory predicts, blame the telescope not the heavens" (Hacking 1985:115, 187, 251; Duhem's intellectualist version of this thesis is contrasted with Hacking's materialist version [Hacking 1992b:30–31, 52–55]). PTSD researchers use the same strategy to explain why a *predicted* outcome did not happen (Pitman). A complementary strategy is to add intervening variables to the original hypothesis, thus explaining why the *observed* outcome did happen (both Mason and Pitman). The first strategy is defensive, allowing researchers to preserve what they had prior to the research. The second strategy is more productive, in that it is simultaneously a defense and a technique for entrenching new findings in established networks of facts and citations. I will return to this point after describing the following studies.

Urinary Free-Cortisol in Post-Traumatic Stress Disorder Patients (*Mason et al. 1986*)

Mason and his associates compared cortisol levels in the urine of psychiatric inpatients with five different diagnoses: PTSD, paranoid schizophrenia, major depressive disorder, undifferentiated schizophrenia, and bipolar disorder. The main findings are that PTSD cortisol levels fall at the lower end of the clinical endocrinological normal range and that PTSD levels are similar to that of paranoid schizophrenia but significantly different from the other groups (Mason et al. 1986:145).

When NOR and EPI levels were measured, it was found that "the PTSD patients have low, stable cortisol levels at the same time that they show markedly increased urinary norepinephrine and epinephrine levels." This is an unexpected finding, since it "has been a common experience in experimental studies of acute stress to find that levels of both cortisol and norepinephrine often *rise together*," consistent with the neural-hormonal theory (Mason et al. 1986:147, my emphasis; also Kosten et al. 1987).

A "possible explanation" for the unexpected results is that a psychological mechanism characteristic of PTSD patients "is exerting a selective inhibitory or suppressive influence upon the pituitary-adrenal cortical system. It has been previously established in basic psychoendocrine research that the use of certain psychological defenses, especially denial, can exert a strong suppressive effect upon urinary corticosteroid levels on a chronic basis." In support of this auxiliary hypothesis, the authors cite a study of parents of children with leukemia. The study is said to have "confirmed a powerful inverse relationship between the effectiveness of psychological defenses and on-going pituitary-adrenal cortical activity." Additional research on the "general relationship between effectiveness of psychological defenses and adrenal cortical activity" is cited: studies of women awaiting breast tumor biopsy and recruits undergoing basic training (Mason 1986:147).

Mason and his associates propose that similar defenses occur among veterans with PTSD, citing "the common use of denial and splitting . . . [and] characteristic numbing of responsiveness and feelings of detachment from others" that characterize the disorder. (See chapter 6 for an account of "splitting.") This makes low levels of cortisol among men with PTSD, consistent with the neural-hormonal theory. The similarity between PTSD patients and paranoid schizophrenia patients is brought under the general theory by a reference to a study of schizophrenia patients which found that "while corticosteroid levels were very high and unstable during the initial phase of acute psychotic disorganization and turmoil, levels subsequently dropped and became low and remarkably fixed as a paranoid delusional system became established as a prevailing organizing framework for the patient." These findings point to a "paranoid system of adaptation," oriented to "maintaining relatively low psychoendocrine tonicity on the pituitary-adrenal cortical system on a chronic basis." According to Mason and his colleagues, a similar adaptive system is found among men diagnosed with PTSD. Four studies are cited as indicating the "prominent use of projection and themes of mistrust of others, especially of those in authority" among individuals diagnosed with PTSD (Mason et al. 1986:147–148).

But why would psychological mechanisms that affect pituitary-adrenal activity (depressing cortisol levels) have no concomitant effect on sympathetic-adrenal activity (NOR and EPI)? The auxiliary hypothesis is now extended: a "certain cognitive organization may selectively suppress cortisol while allowing catecholamine levels to rise." A study is cited that is said to show this effect "in response to achievement demands during a choice-reaction task in coronary-prone type A subjects" (Mason 1986:148).

Naloxone-Reversible Analgesic Response to Combat-Related Stimuli in Post-Traumatic Stress Disorder (*Pitman et al. 1990*)

Pitman and his colleagues started with two hypotheses. When veterans with PTSD are exposed to a combat-related stimulus, endorphins are released and analgesia is produced. The analgesia will be reversed by naloxone, a drug known to block interaction between neural receptors and the opiate class of endorphins (Pitman et al. 1990:541).

The experimental group consisted of eight Vietnam War veterans with PTSD (plus concurrent psychiatric diagnoses). The eight veterans in the control group had no current mental disorders and no history of PTSD. The two groups were matched for age and "combat severity."

In the experiment, all of the men viewed a fifteen-minute "neutral" videotape, followed by a fifteen-minute videotape segment of the movie *Platoon* (dramatizing combat in Vietnam), then a thirty-minute neutral videotape. Thirty minutes before viewing, an intravenous line was inserted into each man. Immediately before the first videotape, he was injected with either a dose of naloxone or saline placebo. Booster doses were injected immediately before the second (combat) and third (neutral) videotapes. "No subjects reported detectable physical or psychological effects of the naloxone injections" (Pitman et al. 1990:542).

Researchers measured a variety of physiological variables: pain intensity and unpleasantness; autonomic arousal (via heart rate and skin conductance); and hormonal levels (NOR, EPI, corticotropin, cortisol, and two opiate endorphins known to be naloxone-reversible). Pain assessment entailed applications of a hot thermode stimulator to each man's forearm. After each application, the man rated the intensity and unpleasantness of his pain, using a standardized technique.

In addition, researchers measured self-reported "emotion states" (happiness, sadness, fear, surprise, anger, disgust, guilt) and "emotion dimensions" (arousal, pleasantness, and sense of control).

Data on physiological arousal were collected *during* each videotape. Emotion self-reports, pain ratings, and hormone levels (via blood samples) were collected *after* each videotape. Two weeks later, each man went through an identical session under the alternate drug condition: the men who had been given the placebo were retested with naloxone and vice versa.

The neural-hormonal theory predicts that exposure to the combat videotape will produce a marked hormonal response in combat veterans with PTSD. Indeed, it is this hormonal response, rather than the release of endorphins, that is the core element of the stress reaction. However, the researchers found no significant differences in hormonal responses between the PTSD and control groups, nor were significant differences found in endorphin levels (beta-endorphin and met-enkephalin). What they did find was a significant analgesia effect in the PTSD-placebo group. These men reported significant decreases in pain intensity after watching *Platoon*, while the other groups reported increases (Pitman et al. 1990:543).

In other words, the PTSD men were different from the other combat veterans in one respect: they experienced an analgesia effect following the stressful event (viewing *Platoon*). When naloxone was administered to them, pain ratings were similar to those reported by the control group, indicating that the analgesia effect was produced by naloxone-reversible endorphins. The anomaly is that the analgesia effect was accompanied by neither the predicted hormonal response nor increases in the two endorphins. The authors respond with two alternative hypotheses.

The first is that the PTSD men really did experience the predicted hormonal response, but it was not detected because of certain technical problems. ("More frequent hormonal sampling, or sampling during afternoon hours when the hypothalamic-pituitary-adrenal-cortical axis would be expected to be more responsive, might have been more revealing" [Pitman et al. 1990:544].)

The alternative hypothesis is that the findings are correct as they now stand. There were *no* significant hormonal differences between PTSD and control groups, but this does not challenge the general theory. The authors cite a "review of neuroendocrinologic research on anxiety disorders . . . [which] noted that experimental stimuli that produce behavioral evidence of intense distress in human subjects often result in *small, inconsistent, or even absent peripheral hormonal changes*" (Pitman et al. 1990:544; my emphasis). To explain how this might accommodate the observed analgesia effect, a second review article is cited as pointing to the existence of four kinds of endorphins: opiate/neural, opiate/hormonal, non-opiate/neural, and non-opiate/hormonal, where the terms "neural" and "hormonal" identify releasing mechanisms.[3] Pitman and his associates argue that, in their experiment, they were incorrectly looking for endorphins belonging to the opiate/hormonal group. In reality, the naloxone-reversible analgesia effect could have been produced by opiates released by a neural mechanism, operating through a yet-to-be-identified circuit (Pitman et al. 1990:544).[4]

Wordless Memories

In each of these studies in the predicted outcome failed to occur. They produced few facts, but many findings. This they achieved by embedding substandard results in networks of facts and citations, connecting PTSD to the neural-hormonal theory of stress. Neuroendocrinology is rich with facts and is based on the relatively privileged epistemology of experimental biology. In addition, visible tokens of neural-hormonal processes are easy to obtain. Standardized technologies and measurements are already available for identifying and measuring endocrine secretions and autonomic arousal. Even single tokens, such as urinary cortisol, possess great explanatory power, signifying multiple connections between the analogy's source (the animal model of inescapable shock) and its targets (diagnosed veterans).

The neural-hormonal theory offers another advantage to PTSD researchers, for it provides a solution to the problem examined in chapter 4 in connection with veterans' verbal accounts of their traumatic memories: the problem of getting time to run consistently in the right direction. The neural-hormonal theory solves the problem by shifting the locus of inquiry downward, from words and meanings to biological states and substances. To obtain facts and findings, researchers now interrogate blood and urine rather than men.

This move is justified through the analogy that researchers make between patients' Vietnam War experiences and experiments involving inescapable shock inflicted on laboratory animals. For this analogy to work, it must either ignore or accommodate the heterogeneous quality of the veterans' etiological events. Even among the fraction of diagnosed veterans who traced their present difficulties back to life-threatening combat situations, many (perhaps most) of the events centered around the successful execution of fight or flight impulses. But it is precisely the impossibility of either fight or flight that defines inescapable stress. (World War I provides a better fit: the image of soldiers huddled in trenches and bunkers over days of unremitting shelling and the image of soldiers buried alive following explosions.) The model of inescapable shock seems likewise problematic in those cases where men were the authors of their etiological events—the perpetrators of violence rather than its victims.

There is one more possible objection to the analogy, namely, that the inescapable-shock model is based on a very simple kind of memory:

> In rats, even 1 year after extinction (i.e., more than a third of the lifetime of the animal), the aversive memory can be restored to its original magnitude by a single training trial. This indicates the essentially permanent nature of conditioned fear and the apparent fragility of extinction. This phenomenon may help to explain the common clinical observation that traumatic memories may remain dormant for many years, only to be elicited by a subsequent stressor or unexpectedly by a stimulus long ago associated with the original trauma. (Charney et al. 1993:296)

Does the word "memory" as applied to laboratory rats have the same meaning that it has in connection with the traumatic memories narrated by people diagnosed with PTSD? Is this another instance of the family resemblances described in chapter 4?

Analogical reasoning—matching features between sources and targets—is not determined by rules. Analogical reasoning in science proceeds through technologies and social relations—the negotiation of meanings within networks of knowledge producers (see chapter 4.) Within these networks, inescapable shock is connected to its target through an essential element, "stress." Inescapable shock purifies this element: strips it of words, purges it of subjectivity, and engraves its physical existence on neural pathways. It is less edifying to ask whether PTSD writers and researchers *ought* to be using this analogy than to ask *how* the analogy is made persuasive.

From the nineteenth century on, it has been observed that people do not respond uniformly when exposed to the same potentially traumatizing event. The neural-hormonal model accommodates this finding by making the brain's neocortex the locus of individuation and a key element in the stress response circuit. Sited between the environmental stressors (input) and the rest of the central nervous system, the neocortex accounts for interindividual and intergroup variations in both neural activity and endocrine secretions.

In the two studies just reviewed, narrative coherence is achieved by switching the neocortex on and off at strategic moments. Mason and his colleagues discovered that the cortisol levels of the PTSD group cluster in the "low normal" range. The clustering effect is statistically significant, but does it have explanatory or clinical significance? To argue that it does, one would have to explain first why the PTSD cortisol levels are "normal" (within the range established for physically healthy men) and then why they are "low."

The researchers' solution to the first question is to assert that the normal/abnormal distinction is meaningless in this context, since it was created for discerning endocrine pathology rather than the phenomenon that the hypothesis explains, differences in endocrine functioning. The important finding is, then, that the PTSD cortisol levels "cluster": they fall within relatively narrow values, and one would not expect to see this in a group of randomly selected healthy men. But why are the levels low, when the theory leads us to expect them to be high, preparing the individual for fight or flight? Mason et al. answer that a state of chronic arousal is maladaptive, and psychological defenses—denial, splitting, paranoid ideation—interpose against it, depressing secretions. In this way, the narrative is completed by switching the neocortex on at the last minute, during the write-up phase—just in time to reconcile observations with theory, but too late for a proper interrogation.

The article by Pitman and his collaborators is more ambitious, tapping locations all along the neural-hormonal circuit. There are two unexpected findings: hormonal secretions of men with PTSD are not elevated after exposure to the combat-simulation stressor, and a naloxone-reversible analgesia effect occurs anyway. Readers are given alternative narrative endings: the predicted hormonal changes *did* occur but were undetected, or predicted changes *did not* occur and endorphins were released through a neural circuit that future research has yet to identify.

The most striking contrast that emerges in the course of the experiment concerns the self-reported "emotion states." After watching the *Platoon* videotape, the men with PTSD reported *double* the amount of "disgust" and "sadness" reported by the control group and *five times* the amount of "guilt." (The figures compare naloxone groups.) While the experiment's biological findings are carefully deconstructed in this article, no attention is given to the particular meanings that "guilt," "disgust," and "sadness" held for the PTSD and non-PTSD groups. The decision to switch the neocortex off at this point makes good narrative sense, though, since the entire point of having biological stories about PTSD is to relocate the traumatic memory from the neocortex, where it is apprehended through words like "guilt" and "disgust," to more primitive regions of the brain—areas to which the researchers, rather than the veterans, have privileged access.

Epilogue

A Key to Post-Traumatic Stress Lies in Brain Chemistry, Scientists Find
New York Times, 12 June 1990

A single instance of overwhelming terror can alter the chemistry of the brain . . . scientists are finding. . . .

New studies in animals and humans suggest that specific sites in the brain undergo these changes. . . .

"Victims of a devastating trauma may never be the same biologically," said Dr. Dennis Charney, a psychiatrist at Yale and director of clinical neuroscience at the National Center for Post-Traumatic Stress Disorder. . . .

The next step, researchers say, is to develop drugs that counter the specific brain mechanisms underlying the disorder. . . .

The discovery of brain changes are finally putting to rest a dispute over whether there is such an entity as "post-traumatic stress." . . .

"The more scientific critics wanted clear evidence of a specific biological basis," said Dr. Krystal [Dr. Charney's colleague]. "Now we have it."

Conclusion

In 1994, the American Psychiatric Association published a fourth edition of its official nosology. The new manual, *DSM-IV*, perpetuates the Kraepelinian framework established by *DSM-III*. Disorders are generally represented as monothetic categories, each one bounded by a distinctive list of criterial features. The manual's most obvious departure from the previous editions is rhetorical and concerns the definition of its eponymic subject, "mental disorders." In *DSM-III* and *DSM-III-R*, the term is defined in a way that includes all of the factions and orientations that were then represented in the American Psychiatric Association:

> [E]ach of the mental disorders is conceptualized as a clinically significant behavioral or psychological syndrome or pattern that occurs in an individual and that is typically associated with either a painful symptom (distress) or impairment. . . . In addition, there is an inference that *there is a behavioral, psychological, or biological dysfunction.* (Amer. Psychia. Assoc. 1980:6; my emphasis; see also Amer. Psychia. Assoc. 1987:xxii)

The definition can be read as a kind of contract whose purpose is to reconcile the claims of rival parties—psychoanalysts, behavioralists, biological reductionists, eclectics—to the field of mental disorders. *DSM-IV* scraps this contract and advances a unitary orientation, an essentially biologized conception of "mental disorder":

> Although this volume is titled the *Diagnostic and Statistical Manual of Mental Disorders*, the term *mental disorder* unfortunately implies a distinction between "mental" disorders and "physical" disorders that is a reductionistic anachronism of mind/body dualism. A compelling literature documents that there is much "physical" in "mental" disorders and much "mental" in "physical" disorders. (Amer. Psychia. Assoc. 1994:xxi)

This kind of talk, criticizing an outdated dualism, is not new. It has been in the air since the 1950s, and *DSM-IV* simply makes explicit what many readers already knew in the days of *DSM-III*, namely, that "mental" disorders are deployed across domains of nature and (therefore) across domains of science. Mind-body dualism, because it insulates mental life and psychological processes from their biological substratum, torpedos the hierarchy of nature and science on which psychiatric fallibilism has constructed its distinctive epistemology (see chapter 8). In the contest between dualism (whose standard bearer is psychodynamic psychiatry) and fallibilism, dualism has lost out.

DSM-IV has also made several changes in the entry for PTSD. Two of these are relatively minor. First, the phenomenology of childhood PTSD is given closer attention. For example, readers are advised that symptomatic intrusions among children may take the form of "frightening dreams without recognizable content." When a child experiences recurrent distressful dreams, the clinician ought to consider the possibility of traumatic memory. Second, *DSM-IV* redraws the nosological boundaries of PTSD so as to distinguish it from a new short-term reactive phenomenon called "acute stress disorder," a syndrome that begins within a month of the etiological event and persists less than a month following onset.

The most important change to PTSD is in *DSM-IV*'s definition of the etiological event, i.e. the "stressor criterion." In chapter 4, I explained how this feature unifies the PTSD syndrome and distinguishes it from otherwise identical syndromes that are traced to non-traumatic etiologies and covered by other diagnoses—notably combinations of depression and anxiety disorders. In *DSM-III* and *DSM-III-R*, the event that triggers PTSD is associated with two features:

> *Feature 1*: A traumatic event is outside the range of usual human experience.
> *Feature 2*: Such an event is markedly distressful to almost anyone who experiences it.

In the early 1990s, the *DSM-IV* Task Force created a committee to assess the adequacy of the PTSD classification and to recommend changes. The committee's report recommended that these two features be revised, on the grounds that the first feature is "vague and unreliable," and the second ignores the fact that people may respond differently to outwardly similar events (Amer. Psychia. Assoc. 1991:H:15).

To call the first feature "vague" seems rather odd. If it is anything, the feature's core idea, that traumatic events are always outside the range of usual human experience, is overly precise. The definition was tailored to the exceptional experiences/memories of soldiers who had fought an exceptional kind of war in Vietnam. The problem with the definition is simply that it does not correspond with many cases that are routinely diagnosed as PTSD. The manual says one thing, and diagnosticians very often do something else. (See chapters 4 and 5 in this connection.) The committee's second point, where they call attention to subjective responses to events that might evoke fear or horror, identifies a similar lack of congruence between the official description of PTSD and widely accepted diagnostic practices. This problem has been obvious to diagnosticians from the earliest days of PTSD, but, until now, it was either ignored or explained by various auxiliary hypotheses.

The *DSM-IV* Task Force accepted the committee's criticisms and adopted their recommended changes:

Feature 1: The traumatized person experienced, witnessed, or was confronted with an event or events involving death (either actual or threatened) or serious injury (including threats to the physical integrity of oneself or others). To be "confronted" with traumatic events would include "learning about unexpected or violent death, serious harm, or threat of death or injury experienced by a family member or other close associates."

Feature 2: The traumatized person's response to these events involved intense fear, helplessness, or horror. (Amer. Psychia. Assoc. 1994:424, 427–428).

As in *DSM-III* and *DSM-III-R*, no time limit is set on the interval between the etiological event and the onset of symptoms (to be more exact, symptoms sufficient to cause "clinically significant distress or impairment in social, occupational, or other important areas of functioning" [Amer. Psychia. Assoc. 1994:429]).

The upshot of these changes is that *DSM-IV* enlarges the variety of experiences and memories that can be used to diagnose PTSD. According to the revised criteria, one must now accept not only that encounters with death and injury affect different people in different ways but *also* that different people can have profoundly different conceptions of what constitutes a realistic "threat." One must also accept that *accounts* of death or injury (in contrast to direct encounters) can be sufficient to constitute traumatic stressors. Finally, the use of the term "response" without any qualifying adjective (in the second revision) allows diagnosis to include cases in which individuals discover their distressful feelings (intense fear, etc.) long after the fact—a condition discussed in chapter 4 in connection with *DSM-III* and *DSM-III-R*. The new edition reminds readers that the inability of a trauma victim to recall important aspects of his or her trauma—aspects that include affect—is symptomatic of PTSD. A recent article in the *New York Times*, based on interviews with PTSD researchers, reports that it is precisely these people who are at risk for developing the disorder, even twenty years after the event:

> Those who react with an apparently unwarranted calm may be particularly prone to post-traumatic distress problems . . . which may not surface until months or even years later, experts say. . . .

> Those most at risk during such a crisis, Dr. [Charles] Marmar said, "are those people who try to cope with emotional difficulties in their lives by avoiding their feelings or keeping them to themselves. . . . The more you try to sweep your feelings under the rug, the more likely you are to dissociate during the trauma." (Goleman 1994:20)

The *practical* effect of these revisions is to make the stressor criterion consistent with ongoing practices. The revised classification excludes no one who has been diagnosed with PTSD under the previous rules. For example, veterans who were traumatized by the deaths and grievous injuries that they inflicted without remorse are still covered by the diagnosis. No one is being left out in the cold; no one will forfeit a service-connected pension because of these changes.

Despite these continuities between the revised PTSD entry and previous ones, the publication of *DSM-IV* is a signifying moment. It signals the repatriation of the traumatic memory, the act of bringing it back home from the jungles and highlands of Vietnam. The recent history of the traumatic memory is dominated by the experiences of Vietnam War veterans and by the resources and incentives the Veterans Administration have provided for PTSD research and specialized treatment. But this situation seem likely to change. The memories, welfare, and grievances of Vietnam War veterans are of diminishing interest to the majority of Americans. Their traumatogenic sufferings are eclipsed by new abominations and new victims, in Cambodia, Bosnia, Ruanda, and elsewhere. The collective memory of their war dims and gradually merges with memories of older, half-remembered wars fought in Korea, Europe, and the Pacific. As the veterans of Vietnam age and fade, and their patrons in government adopt new priorities, a chapter in the history of the traumatic memory draws to a close.

Notes

Introduction

1. This passage is found in Shakespeare's *Henry the Fourth, Part I*, act 2, scene 2, lines 35–41. In it, Lady Percy addresses her husband, Hotspur (Henry Percy). Later in the same speech (lines 42, 45–53):

> In thy faint slumbers I by thee have watched,
> . . . And thou hast talked
> Of sallies and retires, of trenches, tents,
> Of palisadoes, frontiers, parapets,
> Of basilisks, of cannon, culverin,
> Of prisoners' ransom, and of soldiers slain,
> And all the currents of a heady flight.
> *Thy spirit within thee has been so at war,*
> And thus hath bestirred thee in thy sleep,
> That beads of sweat have stood upon thy brow
> Like bubbles in a late-disturbed stream. . . .

Trimble's interpretation depends on the meaning given to the line I have italicized. Does Lady Percy refer to mental conflict and intrusive, traumatic memories? Alternatively, do these images mirror Hotspur's famous love of battle, including its bloody and frightful awfulness?

2. I am unable to devise a system of gender-free or gender-equal pronouns that will not also impose a burden on my reader's ability to follow my prose. The pronoun problem is compounded by the fact that, in some chapters, I refer to categories of people who are exclusively male. In the interest of simplicity, I generally employ masculine pronouns, confident that, in every case, the textual context will make it clear where these words signify males and where they are intended to include both women and men.

3. The author identifies himself as "Maurice Florence," but his editor suggests that this is a pseudonym and that his actual name is Michel Foucault (Gutting 1994:viii).

Chapter One

1. Charcot is not endorsing a psychological explanation of the woman's symptoms. According to him, hysteria has a physiological basis; therefore, hypnotic states must have a similar basis, since they are integral to hysteria. (He rejected etiologies based on organ pathology, because, he claimed, hysteria develops without visible lesions and its symptoms, while highly patterned, conform to no known anatomical pathways.) Charcot's physiological account of hyponosis was the basis of his dispute with Hippolyte Bernheim and other members of the Nancy school, who claimed that hypnotic states are products of suggestion, a mechanism that they explained in essentially psychological terms (Harris 1985; Kravis 1988:1201–1202; R. Smith 1992:125–129).

2. There is another difference that separates Spencer from both Crile and Cannon. Spencer started with a localizationist view of the brain, dividing the cerebral cortex into discrete regions, each the site of a distinctive mental faculty (Clarke and Jacyna 1987:220–234, 238–244; R. Young 1990:173, 180–181). In *Principles of Psychology* (1855), he shifted to an associationist position: simple ideas and perceptions are mentally connected into complex ideas and cognitive-affective structures through relations of resemblance (analogy), contiguity in time and place (including cause and effect), and sensations (notably pleasure and pain) (Richards 1992:158–159, 332–350). Spencer rejected the notion of a tabula rasa, however. In its place, he proposed a kind of evolutionary associationism that combined Lamarck's doctrine of acquired characteristics (traits acquired through use in one generation can be passed through heredity to following generations) with Haeckel's recapitulation theory (the evolutionary history of the species is recapitulated, in an accelerated and condensed form, in the biological development of its individual members). In Spencer's account, later adopted by Hughlings Jackson, the individual begins his mental life with a stock of phylogenetic memories, etched into his neural pathways and evidenced as reflexes, instincts, emotions, and other kinds of automatisms (Bowler 1988:84; Schacter 1989:116–147; C.U.M. Smith 1982a:76, 78–79; R. Young 1990: 178, 182–183, 186–187). While Spencer's version of evolutionary associationism was doomed by Mendelian genetics at the end of the century (Gould 1977:202–206), it lingered on for several decades. Pavlov continued to entertain the possibility of hereditary transmission of conditioned reflexes into the 1920s (Pavlov 1927:285), and Freud remained faithful to these principles, the foundations of the Oedipal complex, until his death in 1942 (Kitcher 1992:67–74, 104–109, 174–190; Sulloway 1983: chap. 4). Of course, evolutionary associationism did not disappear with the collapse of Spencer's Lamarckian version. It continued in the work of Crile and Cannon (among others), now mediated through Darwinian mechanisms: random variation, mutation, and natural selection.

3. The pain mechanism operates in higher life-forms through neural sensors ("noci-ceptors," in Crile's vocabulary), located on the body's surface. In the early years of the century, Crile conducted surgical experiments on turtles, armadillos, and skunks: animals that had evolved unique defenses, such as body armor and noxious chemical sprays, and, therefore, no longer relied on fleeing or fighting for survival. When these animals are confronted by predators, they are least vulnerable if they stay put. This arrangement ought to reduce the usefulness of pain and, if this is so, then the density of noci-ceptors in these creatures would be low in relation to animals like humans or dogs, whose survival depends on fighting and fleeing. According to Crile, experiments on a small number of animals tended to confirm his hypothesis (Crile 1910).

4. In *The Discovery of the Unconscious*, Henri Ellenberger remarks on similarities between Ribot's and Marcel Proust's conceptions of the self and its connections with the past:

[Proust] considered the human ego as being composed of many little egos, distinct though side by side, and more or less closely connected. Our personality thus changes from moment to moment, depending on the circumstances, the place, the people we are with. Events touch certain parts of our personality and leave others out. . . . The sum of our past egos is generally a closed realm, but certain past egos may suddenly reappear, bringing forth a revival of the past. It is then one of our past egos that is in the foreground, living for us. Among our many egos, there are also hereditary elements. Others (our social ego, for instance) are a creation of the thoughts and influences of other people upon us. This explains the continuous fluidity of mind, which is due to these metamorphoses of personality. Marcel Proust's work is of particular interest because its subtle analyses were not influenced by Freud and the other representatives of the new dynamic psychiatry. His academic sources went no further than Ribot and Bergson. It would be quite feasible to extract from his work a treatise on the mind, which would give a plausible picture of what the first dynamic psychiatry would have become had it followed its natural course. (Ellenberger 1970:167–168; see also Terdiman 1993:198–199, 202)

5. Ribot's conception of the mind includes two other aspects, underlying the conscious personality: organic consciousness and unconscious cerebration.

The organic consciousness is a sense of bodily unity or "coenaesthesis." It is "so vague that it is difficult to speak of it in precise terms." In physically and mentally healthy people, the organic consciousness is an unrecognized sense of well-being. Mental disorders begin when it is disturbed by physical causes, such as neurological lesions; normal feelings are replaced by sensations of either "melancholy, mental distress, and anxiety" or "undue joyousness, exuberant emotions, and extreme content," and there is discord among the elements composing the conscious personality. In cases of fully developed periodic amnesia (double consciousness), each coexisting personality rests on a sense of organic consciousness (Ribot 1883:108–109).

Unconscious cerebration consists of mental activity in its "organic phase." It underlies routine activities that are executed without consciousness and it also "sets obscure ideas in order." "Consciousness is the narrow gate through which a very small part of all this work is able to reach us" (Ribot 1883:37–40).

6. Léonie is not the first case in which one personality has access to the memories of all coexisting personalities. She is preceded in the literature by Félida X., introduced by E. Azam in 1876 and later mentioned by Janet (1901:122) and Ribot (1883:99–104).

7. Janet also practiced other techniques. Fixed ideas, he wrote, are often "incarnated in words," and it is these words that "call up the rest [of the symptoms]." A fixed idea can be deprived of pathogenic power by changing or decomposing the meaning given to its key words. Fixed ideas can also be treated through "substitution," a process in which a therapist "induce[s] hallucinations whereby the scenes imagined by the subject were transformed" (Janet 1925:676–677). Janet describes how he treated a patient named Justine with these techniques. Justine had once been employed to care for patients dying of cholera. As a result of these experiences, she now had recurrent visions of hideous corpses and would periodically fall into hysterical crises, during which she cried out that she was infected with cholera. Janet placed Justine in a hypnotic state and instructed her to dress a putrefying corpse (a recurring image) in the quaint clothes she had seen a Chinese general wearing at an exposition. In later hypnotic sessions, Janet decomposed the associations fixed in

the terrible word "cholera" and finished the process with the hypnotic suggestion that the general's name is "Cho Lé Ra" (Janet 1898:156–212; cited in Gauld 1992:375).

8. At the end of the century, according to Ruth Harris:

Psychiatric concepts were constructed around certain key dichotomies—normal and pathological, mind and body, higher and lower, right and left, equilibrium and destabilization, economy and excess, control and disinhibition. These polarities provided the boundaries of scientific debate, containing within them deeper cultural tensions. According to where the boundaries were drawn along the implied continuum, such designations generated intense controversy both within the medical community and from those outside it. (1989:19)

Physiology in general, and [Claude Bernard's] concept of the *milieu intérieur* in particular, provided a treasure trove of metaphorical expressions for talking about animal, human, and social organisms. Checks and balances, the division of labour, equilibrium and disequilibrium, reduction and recombination—these were the polarities which structured physiological explanation. They also provided a descriptive language in medical-legal discourse, which resonated with moral associations by describing the individual's "disequilibrated" and "disinhibited" propensities in a readily accessible fashion. (1989:33)

Chapter Two

1. For accounts of Rivers's career as an anthropologist, see Slobodin 1978 and Kucklick 1992.

2. Jacksonian ideas about neural evolution and functional hierarchy are continued today in Paul MacLean's notion of the "triune brain" (MacLean 1990).

3. On the differences between twilight states and fugues, see Enoch and Trethowan 1991:chap. 4.

4. Mott's readiness to explain shell shock in terms of organic pathology and degenerationist diathesis is consistent with his position and time. Before entering the RAMC, he served as a pathologist at several London asylums and is said to have had a special interest in elucidating the role of heredity in insanity. His eulogy credits him with establishing the connection between general paralysis of the insane and syphilis (Royal College of Physicians 1955).

5. Comparisons between shell shock and compensation neurosis are mentioned by Mott (1918:127) and Ross (1941:141).

6. Opinions concerning differential diagnosis of hysteria and neurasthenia were not specific to army physicians, of course. For works that put this subject into broader medical and social contexts, see Alam and Merskey 1992; Gosling 1987 and 1992; Micale 1990; Shorter 1992.

7. The encounter with A1 is more frightening in Yealland's uncut version. Yet the book received a favorable review in *The Lancet*, where Yealland was praised for his clinical sagacity. His eulogy reports that, through a "strong personality and kindly approach," Yealland reduced and sometimes eliminated his patients' epileptic seizures. He is also reported to have been "a witty companion, fond of harmless

practical jokes"—at least until later in life, after the war, when he became "very serious" and devoted much time to evangelical work among alcoholics (Royal College of Physicians 1955).

8. Rivers made fear the starting point for his account of the war neuroses and tied it to an instinctual response located, in a rather general way, within the nervous system. By the end of the war, other army doctors had been persuaded by the prewar work of Crile and Cannon to locate the fear response more concretely, within a clearly identified system of neuroendocrine connections. Like Rivers, Crile and Cannon began with the idea that fear stimulates a fight, flight, or freeze reaction. Spurts of kinetic energy are needed for fighting or fleeing, and the energy is made available by releasing adrenaline, which stimulates thyroid secretions and leads to a rise in blood pressure and the conversion of glycogen into sugar. When neither fight nor flight is possible but fear persists, these chemical products remain in the blood and eventually affect mind and body, producing symptoms characteristic of the war neuroses, such as the terrifying dreams, states of exhaustion, emotional lability, disordered action of the heart, and tremors (Bury 1918:98; Hurst 1918: chap. 5; McDougall 1926:245–246; Mott 1918b:127; Mott 1919:622; War Office Committee 1922:100). In contrast to previous biological accounts, it would have been relatively easy to measure these neural-hormonal effects. Systolic pressure might be correlated with pulse rate and tachycardia with fine tremors of the outstretched hands, for example. William Brown suggested that large numbers of soldiers be examined using these methods and the findings then analyzed statistically (Brown 1919:835). However, no systematic research of this sort was undertaken during the war.

Cannon likewise served with the Allied Expeditionary Force. In addition to working as a surgeon, he conducted research on the pathophysiology and prevention of shock among seriously wounded soldiers. During this period, he seems to have set aside his interest in specifically psychogenic shock (Benison et al. 1991).

Chapter Four

1. The "learning companion" that supplements *DSM-III-R* instructs its readers with case studies in which PTSD symptoms mirror their etiological events (Spitzer et al. 1989:88–90). Compare these trauma-inscribed symptoms with the ambiguous dream motifs adduced by Kardiner for his traumatized patients (see chapter 3 above).

2. The NVVRS and the Breslau data are even more divergent. Using the standard Diagnostic Interview Schedule for *DSM-III-R*, Breslau shows lifetime rates for men that are five times greater than the NVVRS current rates for men. Some of these inconsistencies can be traced to differences in study design. For instance, the VES and the NVVRS use different definitions of "current prevalence"; the ECA study and Breslau samples are similar socially and economically, but the Breslau sample is drawn from a high-risk cohort, people between the ages of twenty and thirty; the NVVRS sample includes both officers and enlisted men, while the VES sample includes only enlisted men. However, these differences can, at best, account for only a fraction of the large disparities among the prevalence rates.

Chapter Five

1. The Gulf of Tonkin Resolution was passed by Congress in August 1964, following alleged attacks on U.S. warships by North Vietnamese forces. The resolution delegated authority to President Johnson to take military action against the Republic of North Vietnam.

2. On the issue of the possible racial bias of these tests, see Pritchard and Rosenblatt 1980; Walters et al. 1983.

Chapter Six

1. Paul Kline identifies nineteen problems that are routinely encountered when researchers attempt to compare psychotherapy outcome studies. Given these obstacles, he concludes, "It is beyond comprehension how the enumeration of dozens of studies, the majority, if not all, with severe methodological inadequacies, can be held to demonstrate anything other than the inadequacy of meta-analysis" (Kline 1992:69). Given the profound differences among therapeutic doctrines concerning the meaning of the basic terms, such as "recovery" and "symptom substitution," it would seem that the more serious problem is epistemological rather than methodological.

2. According to the center's clinical director, these ideas about splitting originate with the account of instinctual drives provided by Freud in "Instincts and Their Vicissitudes" (1915). Psychoanalytic ideas about splitting are more commonly associated with object relations theory, notably in the work of Melanie Klein and Otto Kernberg. Questions about origins notwithstanding, the ideology of splitting is well represented in psychodynamic discourse:

> The lack of fusion of libidinal and aggressive drives, inherent in the mechanism of splitting, leads to dissociation of their corresponding affects. Thus "neutralization" of drives and admixture of libidinal and aggressive affects does not occur, and nascent, intense emotions are easily experienced. Anger, for instance, is felt only as rage; murderous and suicidal impulses readily surface in response to frustration. (Akhtar and Byrne 1983:1015; see also Brende 1983, Newberry 1985, and Parson 1986 on splitting in PTSD)

In the object relations literature, splitting is traced to a failure of normal development, and it is now often associated with borderline personality disorder (Kernberg 1985: 230, 234–238). More recently, object relations theorists have proposed that splitting produced by exposure to traumatic experiences is possible during any stage of life.

The idea that intrapsychic splitting can create a mirror image of itself in clinical settings, is also represented in the psychiatric literature:

> Splitting in the hospital has been well described in a number of papers on the intense countertransference evoked by . . . patients. . . . Staff members find themselves assuming polarized positions and defending those positions against one another with a vehemence that is out of proportion to the importance of the issue. The patient has represented one self-representation to one group of treaters and

another self-representation to another group of treaters. . . . [E]ach self-representation evokes a corresponding reaction in the treater that can be understood as an unconscious identification. (Gabbard 1989:446)

3. An "action script" is a stereotyped behavior pattern that is socially appropriate to one or more culturally recognizable, although not necessarily labeled, situations. A "narrative script" is a patterned way of recounting (describing, explaining, et cetera) situations and events. Action scripts and narrative scripts presuppose the cultural competence of actors and narrators (who encode meanings) and their audiences (who decode meanings). Because scripts are often played out in interactional and dialogic situations, actors may be obliged to simultaneously encode and decode meanings. Further, actual encounters tend to have a dynamic quality, and actors may find it advantageous to introduce additional scripts, in order to reframe or renegotiate the meaning of their words and actions.

Chapter Eight

1. For a general account of the biology of stress, see Rose 1984 and Watkins and Mayer 1982 and 1986; for a synoptic account of stress system disorders, see Chrousos and Gold 1992; for a review of research on psychobiologic mechanisms associated specifically with PTSD, see Charney et al. 1993.

2. Recent research, employing groups of healthy (diagnosis-free) human subjects in a laboratory setting, has added a cognitive component (imagistic, semantic) to the traumatogenic somatic memory, via a shared (memory-enhancing) mechanism: activation of the beta-adrenergic stress hormone system (Cahill et al. 1994).

3. "Neuroendocrine" is a more restrictive term than "neural-hormonal": it includes only the ductless glands, while the latter includes hormones released in the brain.

4. The opiate versus non-opiate distinction is based on evidence of different receptor sites for each group of endorphins. The neural versus hormonal distinction is based on evidence that (1) some cases of endorphin analgesia persist after the endocrine glands have been removed or deactivated, and (2) all cases of endorphin analgesia can be attenuated or ended by lesions in the nervous system. Therefore, when the release of endorphins cannot be traced to hormonal sources (case 1), it can be assumed that they are released neurally (Watkins and Mayer 1986; also Watkins and Mayer 1982).

Works Cited

Adrian, E. D., and Lewis R. Yealland. 1917. "The Treatment of Some Common War Neuroses." *Lancet* i:867–872.

Akhtar, Salman, and Jessica P. Byrne. 1983. "The Concept of Splitting and Its Clinical Relevance." *American Journal of Psychiatry* 140:1013–1016.

Alam, Chris N., and Harold Merskey. 1992. "The Development of the Hysterical Personality." *History of Psychiatry* 3:135–165.

Alexander, D. A., and A. Wells. 1991. "Reactions of Police-Officers to Body-Handling after a Major Disaster: A Before-and-After Comparison." *British Journal of Psychiatry* 159:547–555.

Alexander, Franz G., and Sheldon T. Selesnick. 1966. *The History of Psychiatry*. New York: Harper and Row.

American Psychiatric Association. 1952. *Diagnostic and Statistical Manual of Mental Disorders*. Washington, D.C.: American Psychiatric Association.

———. 1966. *Diagnostic and Statistical Manual of Mental Disorders*. 2d ed. Washington, D.C.: American Psychiatric Association.

———. 1980. *Diagnostic and Statistical Manual of Mental Disorders*. 3d ed. Washington, D.C.: American Psychiatric Association.

———. 1987. *Diagnostic and Statistical Manual of Mental Disorders*. 3d ed., revised. Washington, D.C.: American Psychiatric Association.

———. 1994. *Diagnostic and Statistical Manual of Mental Disorders*. 4th ed. Washington, D.C.: American Psychiatric Association.

Andraesen, Nancy C. 1980. "Post-Traumatic Stress Disorder." In *Comprehensive Textbook of Psychiatry /III*, 3d ed., ed. Harold Kaplan, Alfred Freedman, and Benjamin Sadock, 1517–1525. Baltimore: Williams and Wilkins.

Arbib, M., and M. Hesse. 1986. *The Construction of Reality*. Cambridge: Cambridge Univ. Press.

Armstrong, S. L., L. R. Gleitman, and H. G. Gleitman. 1983. "What Some Concepts Might Be." *Cognition* 13:263–308.

Atkinson, Ronald M., Robin G. Henderson, Landy F. Sparr, and Shirley Deale. 1982. "Assessment of Vietnam Veterans for Post-Traumatic Stress Disorder in Veterans Administration Disability Claims." *American Journal of Psychiatry* 139:1118–1121.

Atran, S. 1985. "The Nature of Folk-Botanical Life-Forms." *American Anthropologist* 87:298–315.

Azam, E. 1876. "Le dédoublement de la personnalité suite de l'histoire de Félida X." *Revue Scientifique*, 2d ser.: 265–269.

Babington, Anthony. 1983. *For the Sake of Example: Capital Courts-Martial 1914–1920*. New York: St. Martin's Press.

Babinski, Joseph F. F., and Jules Froment. 1918. *Hysteria or Pthiatism and Reflex Nervous Disorders in the Neurology of War*. London: London Univ. Press.

Bailar, J. C., and F. Mosteller. 1992. "Guidelines for Statistical Reporting in Arti-
  cles in Medical Journals." In *Medical Uses of Statistics*, ed. J. C. Bailar and
  F. Mosteller, 181–200. Boston: New England Journal of Medicine Books.
Baldessarnini, Ross J. 1985. "Drugs and the Treatment of Psychiatric Disorders." In
  The Pharmacological Basis of Therapeutics, ed. A. G. Gillman, L. S. Goodman,
  T. W. Rall, and F. Murad, 387–445. New York: Macmillan.
Barker, Pat. 1991. *Regeneration*. Harmondsworth: Penguin.
Barnes, Barry, and David Bloor. 1982. "Relativism, Rationalism and the Sociology
  of Knowledge.' In *Rationality and Relativism*, ed. Martin Hollis and Steven
  Lukes, 21–47. Cambridge, Mass.: MIT Press.
Barnes, Barry, and David Edge. 1982. "The Culture of Science." In *Science in Con-
  text*, ed. Barry Barnes and David Edge, 65–74. Cambridge, Mass.: MIT Press.
Bayer, Ronald, and Robert L. Spitzer. 1985. "Neurosis, Psychodynamics, and
  DSM-III." *Archives of General Psychiatry* 42:187–196.
Beard, George. 1880. *A Practical Treatise on Nervous Exhaustion (Neurasthenia),
  Its Symptoms, Nature, Sequences, Treatment*. New York: William Wood.
————. 1881. *American Nervousness, Its Causes and Consequences*. New York:
  G. P. Putnam.
Beck, Aaron. 1976. *Cognitive Therapy and the Emotional Disorders*. New York:
  International Universities Press.
Benison, Saul, A. Clifford Barger, and Elin L. Wolfe. 1991. "Walter B. Cannon and
  the Mystery of Shock: A Study of Anglo-American Co-Operation in World War
  I." *Medical History* 35:217–249.
Berios, German E. 1985. "Positive and Negative Symptoms and Jackson: A Con-
  ceptual History." *Archives of General Psychiatry* 42:95–97.
Beveridge, A. W., and E. B. Renvoize. 1988. "Electricity: A History of Its Use in
  the Treatment of Mental Illness in Britain during the Second Half of the 19th
  Century." *British Journal of Psychiatry* 153:157–162.
Blashfield, Roger K. 1984. *The Classification of Psychopathology: Neo-Kraepelin-
  ian and Quantitative Approaches*. New York: Plenum.
Blashfield, R., J. Sprock, D. Haymaker, and J. Hodgin. 1989. "The Family Resem-
  blance Hypothesis Applied to Psychiatric Classification." *Journal of Nervous and
  Mental Disease* 177:492–497.
Bloom, Floyd E. 1985. "Neurohormonal Transmission and the Central Nervous
  System." In *The Pharmacological Basis of Therapeutics*, ed. A. G. Gillman,
  L. S. Goodman, T. W. Rall, and F. Murad, 236–259. New York: Macmillan.
Bloor, David. 1976. *Knowledge and Social Imagery*. London: Routledge and Kegan
  Paul.
Boehnlein, J. K., and J. D. Kinzie. 1992. "Commentary. DSM Diagnosis of Post-
  Traumatic Stress Disorder and Cultural Sensitivity: A Response." *Journal of
  Nervous and Mental Disease* 180:597–599.
Bowler, Peter. 1988. *The Non-Darwinian Revolution: Reinterpreting a Historical
  Myth*. Baltimore: Johns Hopkins Press.
Brende, Joel O. 1983. "A Psychodynamic View of Character Pathology in Vietnam
  Combat Veterans." *Bulletin of the Menninger Clinic* 47:193–216.
Breslau, Naomi, and Glenn C. Davis. 1987. "Post-Traumatic Stress Disorder: The
  Stressor Criterion." *Journal of Nervous and Mental Disease* 175:255–264.

Breslau, Naomi, Glenn C. Davis, and Patricia Andreski. 1994a. "Risk Factors for PTSD Related Traumatic Events." Manuscript. *American Journal of Psychiatry* (in press).

Breslau, Naomi, Glenn C. Davis, Patricia Andreski, Belle Federman, and James C. Anthony. 1994b. "Epidemiologic Findings on PTSD and Comorbid Disorders in the General Population." Manuscript.

Breslau, Naomi, Glenn C. Davis, Patricia Andreski, and E. Peterson. 1991. "Traumatic Events and Post-Traumatic Stress Disorder in an Urban Population of Young Adults." *Archives of General Psychiatry* 48:216–222.

Brett, Elizabeth, Robert Spitzer, and Janet Williams. 1988. "*DSM III-R* Criteria for Post-Traumatic Stress Disorder." *American Journal of Psychiatry* 145:1232.

Breuer, Josef, and Sigmund Freud. 1955 [1893–1895]. *Studies on Hysteria*. Vol. 2 of *Standard Edition of the Complete Psychological Works of Sigmund Freud*. London: Hogarth Press.

Brill, Norman Q., and Gilbert Beebe. 1955. *A Follow-Up Study of War Neuroses*. Washington, D.C.: Government Printing Office.

British Medical Association. 1919. "Special Clinical Meeting of the Section on Medicine: War Neuroses." *Lancet* i:709–720.

Brown, William. 1919. "War Neurosis: A Comparison of Early Cases Seen in the Field with Those Seen at the Base." *Lancet* i:833–836.

Brown, William. 1920. "The Revival of Emotional Memories and Its Therapeutic Value." *British Medical Journal* i:16–19, 30–33.

Burns, David. 1980. *Feeling Good: The New Mood Therapy*. New York: New American Library.

Bury, Judson S. 1896. "Diagnosis of Functional from Organic Disease of the Nervous System." *British Medical Journal* ii:189–192.

———. 1918. "Pathology of the War Neuroses." *Lancet* ii:97–99.

Cahill, Larry, Bruce Prins, Michael Weber, and James L. McGaugh. 1994. "ß-Adrenergic Activation and Memory for Emotional Events." *Nature* 371:702–704.

Canguilhem, Georges. 1991. *The Normal and the Pathological*. New York: Zone Books.

Cannon, Walter B. 1914. "The Interrelations of Emotions as Suggested by Recent Physiological Research." *American Journal of Psychiatry* 25:256–281.

———. 1929. *Bodily Changes in Pain, Hunger, Fear and Pain*. Boston: Charles T. Branford.

———. 1942. "'Voodoo' Death." *American Anthropologist* 44:169–181.

Cantor, N., and N. Genero. 1986. "Psychiatric Diagnosis and Natural Categorization: A Close Analogy." In *Contemporary Directions in Psychopathology: Toward the DSM-IV*, ed. T. Millon and G. Klerman, 233–256. New York: Guilford.

Cazeneuve, Jean. 1972. *Lucien Lévy-Bruhl*. Oxford: Blackwell.

Centers for Disease Control. 1988. "Health Status of Vietnam Veterans: I. Psychosocial Characteristics." *Journal of the American Medical Association* 259:2701–2707.

Charcot, Jean-Marin. 1889. *Clinical Lectures on Diseases of the Nervous System Delivered at the Infirmary of la Salpêtrière*. London: New Sydenham Society.

Charlton, Bruce. 1990. "A Critique of Biological Psychology." *Psychological Medicine* 20:3–6.

Charney, D. S., A. Y. Deutch, J. H. Krystal, S. M. Southwick, and M. Davis. 1993. "Psychobiologic Mechanisms of Post-Traumatic Stress Disorder." *Archives of General Psychiatry* 50:294–305.

Chrousos, George P., and Philip W. Gold. 1992. "The Concept of Stress and Stress System Disorders: Overview of Physical and Behavioral Homeostasis." *Journal of the American Medical Association* 267:1244–1252.

Ciompi, Luc. 1991. "Affects as Central Organizing and Integrating Factors: A New Psychosocial/Biological Model of the Psyche." *British Journal of Psychology* 159:97–105.

Clark, Michael J. 1983. " 'A Plastic Power Ministering to Organisation': Interpretations of the Mind-Body Relation in Late Nineteenth-Century British Psychiatry." *Psychological Medicine* 13:487–497.

Clarke, Edwin and L. S. Jacyna. 1987. *Nineteenth-Century Origins of Neuroscientific Concepts*. Berkeley: Univ. of California Press.

Collie, John. 1917. *Malingering and Feigned Sickness, with Notes on the Workmen's Compensation Act, 1906, and Compensation for Injury, Including the Leading Causes Thereon*. London: Edward Arnold.

Collins, H. M. 1992. *Changing Order: Replication and Induction in Scientific Practice*. 2d ed. Chicago: Univ. of Chicago Press.

Colquhoun, John Campbell. 1836. *Isis Revelata: An Inquiry into the Origin, Progress and Present State of Animal Magnetism*. 2 vols. London: Longman, Brown, Green, and Longmans.

Connerton, Paul. 1989. *How Societies Remember*. Cambridge: Cambridge Univ. Press.

Copeland, J. 1981. "What Is a 'Case'? A Case for What?" In *What Is a Case?* ed. J. K. Wing, P. Bebbington, and L. Robins. Boston: Blackwell Scientific Publications.

Crabtree, Adam. 1993. *From Mesmer to Freud: Magnetic Sleep and the Roots of Psychological Healing*. New Haven: Yale Univ. Press.

Crile, George W. 1899. *An Experimental Research into Surgical Shock*. Philadelphia: Lippincott.

———. 1910. "Phylogenetic Association in Relation to Certain Medical Problems." *Boston Medical and Surgical Journal* 163:893–904.

———. 1915. *The Origin and Nature of the Emotions*. Philadelphia: W. B. Saunders.

Culpin, Millais. 1931. *Recent Advances in the Study of the Psychoneuroses*. London: J. and A. Churchill.

———. 1940. "Mode of Onset of the Neuroses in War." In *The Neuroses of War*, ed. Emanuel Miller, 33–54. London: Macmillan.

Daly, R. J. 1983. "Samuel Pepys and Post-Traumatic Stress Disorder." *British Journal of Psychiatry* 143:64–68.

Darwin, Charles. 1965 [1872]. *The Expression of the Emotions in Man and Animals*. Chicago: Univ. of Chicago Press.

Davidson, Jonathan., D. Hughes, D. G. Blazer, and L. K. George. 1991. "Post-Traumatic Stress Disorder in the Community: An Epidemiological Study." *Psychological Medicine* 21:713–721.

Davidson, Jonathan, Harold Kudler, William Saunders, and Rebecca Smith. 1990.

"Symptom and Comorbidity Patterns in World War II and Vietnam Veterans with Post-Traumatic Stress Disorder." *Comprehensive Psychiatry* 31:162–170.

Davidson, Jonathan., H. Kudler, R. Smith, S. L. Mahorney, S. Lipper, E. Hammett, W. B. Saunders, and J. O. Cavenar. 1990. "Treatment of Post-Traumatic Stress Disorder with Amitriptyline and Placebo." *Archives of General Psychiatry* 47:259–266.

de Mouillesaux, ———. 1789. "Rapport sur une somnambule magnétique." *Annales de la Société Harmonique des Amis Réunis de Strasbourg* 3:1–87.

Dean, Eric T. 1992. "The Myth of the Troubled and Scorned Vietnam Veteran." *Journal of American Studies* 26:59–74.

Dejerine, Jonathan, and E. Gauckler. 1915. *The Psychoneuroses and Their Treatment by Psychotherapy*. Philadelphia: J. B. Lippincott.

Duncan-Jones, P., D. A. Grayson, and P.A.P. Moran. 1986. "The Utility of Latent Trait Models in Psychiatric Epidemiology." *Psychological Medicine* 16:391–405.

Dworkin, Gerald. 1988. *The Theory and Practice of Autonomy*. Cambridge: Cambridge Univ. Press.

Eder, Montague D. 1916. "The Psycho-Pathology of the War Neuroses." *Lancet* ii:264–268.

Eder, Montague D. 1917. *War-Shock: The Psycho-Neuroses in War Psychology and Treatment*. Philadelphia: P. Blakiston's Son.

Eisenberg, Leon. 1986. "Mindlessness and Brainlessness in Psychiatry." *British Journal of Psychiatry* 148:497–508.

Ekman, Paul, Wallace V. Freisen, and Phoebe Ellsworth. 1972. "What Emotion Categories or Dimensions Can Observers Judge from Facial Behavior?" In *Emotion in the Human Face*, 2d ed., ed. Paul Ekman, 39–55. Cambridge: Cambridge Univ. Press.

Ellenberger, Henri F. 1970. *The Discovery of the Unconscious*. New York: Basic Books.

———. 1993 [1966]. "The Pathogenic Secret and Its Therapeutics." In *Beyond the Unconscious: Essays of Henri F. Ellenberger in the History of Psychiatry*, ed. Mark Micale, 341–359. Princeton: Princeton Univ. Press.

Ellis, Albert. 1977. *Handbook of Rational Emotive Therapy*. New York: Springer.

Engelhardt, H. Tristram. 1975. "John Hughlings Jackson and the Mind-Body Relation." *Bulletin of the History of Medicine* 49:137–151.

English, Peter C. 1980. *Shock, Physiological Surgery, and George Washington Crile: Medical Innovation in the Progressive Era*. Westport, Conn.: Greenwood Press.

Enoch, David, and William Trethowan. 1991. *Uncommon Psychiatric Syndromes*. 2d ed. Oxford: Butterworth, Heinemann.

Erdelyi, Matthew H. 1990. "Repression, Reconstruction, and Defense: History and Integration of the Psychoanalytic and Experimental Frameworks." In *Repression and Dissociation: Implications for Personality Theory, Psychopathology, and Health*, ed. Jerome L. Singer, 1–32. Chicago: Univ. of Chicago Press.

Erichsen, John E. 1859. *The Science and Art of Surgery*. Philadelphia: Blanchard and Lea.

———. 1866. *On Railway and Other Injuries of the Nervous System*. London: Walton and Maberly.

Erichsen, John E. 1872. *The Science and Art of Surgery*. 6th ed. London: Longmans, Green.

———. 1883. *On Concussion of the Spine, Nervous Shock and Other Obscure Injuries to the Nervous System, in Their Clinical and Medico-Legal Aspects*. William Wood: New York.

Eysenck, Hans, and S.B.G. Eysenck. 1975. *Manual of the Eysenck Personality Questionnaire*. Seven Oaks (Kent, U.K.): Hodder and Stoughton.

Faust, David, and Richard A. Miner. 1986. "The Empiricist and His New Clothes: *DSM-III* in Perspective." *American Journal of Psychiatry* 143:962–967.

Feighner, J. P., Eli Robins, Samuel Guze, R. A. Woodruff, G. Winokur, and R. Muñoz. 1972. "Diagnostic Criteria for Use in Psychiatric Research." *Archives of General Psychiatry* 26:57–63.

Fleck, Ludwik. 1979 [1935]. *Genesis and Development of a Scientific Fact*. Chicago: Univ. of Chicago Press.

Florence, Maurice. 1994. "Michel Foucault, 1926–." In *The Cambridge Companion to Foucault*, ed. Gary Gutting, 314–319. Cambridge: Cambridge Univ. Press.

Fontana, Alan, Robert Rosenhecht, and Elizabeth Brett. 1992. "War Zone Traumas and Post-Traumatic Stress Disorder Symptomatology." *Journal of Nervous and Mental Disease* 180:748–755.

Frances, Allen, and Arnold Cooper. 1981. "Descriptive and Dynamic Psychiatry: A Perspective on *DSM-III*." *American Journal of Psychiatry* 138:1198–1202.

Frances, Allen, H. A. Pincus, T. A. Widiger, W. W. Davis, and M. B. First. 1990. "*DSM-IV*: Work in Progress." *American Journal of Psychiatry* 147:1439–1448.

Frank, Jerome. 1961. *Persuasion and Healing*. Baltimore: Johns Hopkins Univ. Press.

———. 1973. *Persuasion and Healing*. 2d ed. Baltimore: Johns Hopkins Univ. Press.

———. 1991. *Persuasion and Healing*. 3d ed. Baltimore: Johns Hopkins Univ. Press.

Frankel, Fred H. 1994. "The Concept of Flashbacks in Historical Perspective." *International Journal of Clinical and Experimental Hypnosis* 42:321–336.

Freud, Sigmund. 1953a [1900]. *The Interpretation of Dreams*. Vols. 4 and 5 of *Standard Edition of the Complete Psychological Works of Sigmund Freud*. London: Hogarth Press.

———. 1953b [1904]. "Freud's Psychoanalytic Procedure." *Standard Edition of the Complete Psychological Works of Sigmund Freud*, 7:97–108. London: Hogarth Press.

———. 1955a [1919]. "Introduction to Psycho-Analysis and the War Neuroses." In *Standard Edition of the Complete Psychological Works of Sigmund Freud*, 17:205–210. London: Hogarth Press.

———. 1955b [1920]. "Memorandum on the Electrical Treatment of War Neurotics." In *Standard Edition of the Complete Psychological Works of Sigmund Freud*, 17:211–215. London: Hogarth Press.

———. 1955c [1920]. *Beyond the Pleasure Principle*. In *Standard Edition of the Complete Psychological Works of Sigmund Freud*, 18:3–143. London: Hogarth Press.

———. 1955d [1923]. "Two Encyclopedia Articles: (A) Psycho-Analysis." In

Standard Edition of the Complete Psychological Works of Sigmund Freud, 18:235–254. London: Hogarth Press.

———. 1957 [1915]. "Instincts and Their Vicissitudes." In *Standard Edition of the Complete Psychological Works of Sigmund Freud*, 14:109–140. London: Hogarth Press.

———. 1958 [1914]. "Remembering, Repeating and Working-through." In *Standard Edition of the Complete Psychological Works of Sigmund Freud*, 12:147–159. London: Hogarth Press.

———. 1962 [1896]. "On the Etiology of Hysteria." In *Standard Edition of the Complete Psychological Works of Sigmund Freud*, 3:62–68. London: Hogarth Press.

———. 1964a [1923]. "Splitting of the Ego in the Process of Defense." In *Standard Edition of the Complete Psychological Works of Sigmund Freud*, 23:271–278. London: Hogarth Press.

———. 1964b [1938]. *Moses and Monotheism: Three Essays*. In *Standard Edition of the Complete Psychological Works of Sigmund Freud*, 23:3–137. London: Hogarth Press.

———. 1966a [1888]. "Hysteria." In *Standard Edition of the Complete Psychological Works of Sigmund Freud*, 1:39–57. London: Hogarth Press.

———. 1966b [1892]. "On the Theory of Hysterical Attacks." In *Standard Edition of the Complete Psychological Works of Sigmund Freud*, 1:151–154. London: Hogarth Press.

Fuller, Richard B. 1985. "War Veterans' Post-Traumatic Stress Disorder and the U.S. Congress." In *Post-Traumatic Stress Disorder and the War Veteran Patient*, ed. William Kelley, 3–11. New York: Brunner/Mazel.

Fussell, Paul. 1975. *The Great War and Modern Memory*. New York: Oxford Univ. Press.

Gabbard, Glen O. 1989. "Splitting in Hospital Treatment." *American Journal of Psychiatry* 146:444–451.

Gaines, Atwood D. 1992. "Medical/Psychiatric Knowledge in France and the United States: Culture and Sickness in History and Biology." In *Ethnopsychiatry: The Cultural Construction of Professional and Folk Psychiatries*, ed. Atwood D. Gaines, 171–201. Albany: State Univ. of New York Press.

Gasser, J. 1988. "La notion de mémoire organique dans l'oeuvre de T. Ribot." *History and Philosophy of the Life Sciences* 10:293–313.

Gauld, Alan. 1992. *A History of Hypnotism*. Cambridge: Cambridge Univ. Press.

Gersons, Berthold, and Ingrid Carlier. 1992. "Post-Traumatic Stress Disorder: The History of a Recent Concept." *British Journal of Psychiatry* 161:742–748.

Geuss, Raymond. 1981. *The Idea of Critical Theory: Habermas and the Frankfurt School*. Cambridge: Cambridge Univ. Press.

Gigerenzer, Gerd, Zeno Swijtink, Theodore Porter, Lorraine Daston, John Beatty, and Lorenz Krüger. 1989. *The Empire of Chance: How Probability Changed Science and Everyday Life*. Cambridge: Cambridge Univ. Press.

Gilbert, G. Nigel, and Michael Mulkay. 1984. *Opening Pandora's Box*. Cambridge: Cambridge Univ. Press.

Gillespie, R. D. 1942. *Psychological Effects of War on Citizen and Soldier*. New York: W. W. Norton.

Goldstein, Jan. 1987. *Console and Classify: The French Psychiatric Profession in the Nineteenth Century*. Cambridge: Cambridge Univ. Press.

Goleman, Daniel. 1994. "Those Who Stay Calm in Disasters Face Psychological Risks, Studies Say." *New York Times*, 17 April.

Goodwin, Donald W., and Samuel B. Guze. 1984. *Psychiatric Diagnosis*. 3d ed. New York: Oxford Univ. Press.

Gosling, F. G. 1987. *Before Freud: Neurasthenia and the American Medical Community, 1870–1910*. Urbana: Univ. of Illinois Press.

Gould, Stephen Jay. 1977. *Ontogeny and Phylogeny*. Cambridge: Harvard Univ. Press.

———. 1981. *The Mismeasure of Man*. New York: W.W. Norton.

Gowers, W. R. 1903. *Diseases of the Brain and Cranial Nerves, General and Functional Diseases of the Nervous System*. Vol. 2 of *A Manual of Diseases of the Nervous System*. Philadelphia: P. Blakiston's Son.

Graham, J. 1984. *Handbook of Psychological Assessment*. Oxford: Oxford Univ. Press.

Green, Bonnie L., J. B. Lindy, M. C. Grace, and G. C. Gleser. 1989. "Multiple Diagnosis in Post-Traumatic Stress Disorder, the Role of War Stressors." *Journal of Nervous and Mental Disease* 177:329–335.

Grinker, R. R. and J. P. Spiegel. 1945. *War Neuroses*. Philadelphia: Blakiston.

Grob, G. N. 1991. "Origins of *DSM-I*: A Study in Appearance and Reality." *American Journal of Psychiatry* 148:421–431.

Grünbaum, Adolph. 1984. *The Foundations of Psychoanalysis: A Philosophical Critique*. Berkeley: Univ. of California.

———. 1993. *Validation in the Clinical Theory of Psychoanalysis: A Study in the Philosophy of Psychoanalysis*. Madison, Conn.: International Universities Press.

Gutting, Gary. 1994. Preface to *The Cambridge Companion to Foucault*, ed. Gary Gutting, vii–viii. Cambridge: Cambridge Univ. Press.

Guze, Samuel B. 1989. "Biological Psychiatry: Is There Any Other Kind?" *Psychological Medicine* 19:315–323.

Hacking, Ian. 1982. "Language, Truth and Reason." In *Rationality and Relativism*, ed. Martin Hollis and Steven Lukes, 48–66. Cambridge, Mass.: MIT Press.

———. 1984. "Five Parables." In *Philosophy in History: Essays in the Historiography of Philosophy*, ed. Richard Rorty, J. B. Schneewind, and Quention Skinner, 103–124. Cambridge: Cambridge Univ. Press.

———. 1985. *Representing and Intervening: Introductory Topics in the Philosophy of Natural Science*. Cambridge: Cambridge Univ. Press.

———. 1986. "Making Up People." In *Reconstructing Individualism: Autonomy, Individuality, and the Self in Western Thought*, ed. Thomas Heller, Morton Sosna, and David Wellberry, 222–236. Stanford: Stanford Univ. Press.

———. 1992a. "'Style' for Historians and Philosophers." *Studies in the History and Philosophy of Science* 32:1–20.

———. 1992b. "The Self-Vindication of the Laboratory Sciences." In *Science and Practice and Culture*, ed. A. Pickering, 29–64. Chicago: University of Chicago Press.

———. 1992c. "Statistical Language, Statistical Truth and Statistical Reason: The Self-Authentication of a Style of Scientific Reasoning." In *The Social Dimen-*

sions of Science, ed. Earnan McMullin, 130–157. Notre Dame, Ind.: Univ. of Notre Dame Press.

Halbwachs, Maurice. 1980. *The Collective Memory*. New York: Harper and Row.

Haley, Sarah. 1974. "When the Patient Reports Atrocities: Specific Considerations of the Vietnam Veteran." *Archives of General Psychiatry* 30:191–196.

Hargreaves, G. Ronald, Eric Wittkower, and A.T.M. Wilson. 1940. "Psychiatric Organisation in the Services." In *The Neuroses of War*, ed. Emanuel Miller, 163–179. London: Macmillan.

Harris, Ruth. 1985. "Murder under Hypnosis in the Case of Gabrielle Bompard: Psychiatry in the Courtroom in Belle Époque Paris." In *The Anatomy of Madness: Essays in the History of Psychiatry*, ed. W. F. Bynum, Roy Porter, and Michael Shepherd, 2:97–241. London: Tavistock.

———. 1989. *Murders and Madness: Medicine, Law, and Society in the Fin de Siécle*. Oxford: Clarendon Press.

Head, Henry. 1922. "The Diagnosis of Hysteria." *British Medical Journal* i:827–829.

Helzer, John, and Lee Robins. 1988. "The Prevalence of Post-Traumatic Stress Disorder." *New England Journal of Medicine* 318:1692.

Helzer, John, Lee Robins, and L. McEvoy. 1987. "Post-Traumatic Stress Disorder in the General Population." *New England Journal of Medicine* 317:578–583.

Hendin, H., and A. P. Haas. 1991. "Suicide and Guilt as Manifestations of PTSD in Vietnam Combat Veterans." *American Journal of Psychiatry* 148:586–591.

Herman, Judith L. 1992. *Trauma and Recovery*. New York: Basic Books.

Hesse, Mary. 1966. *Models and Analogies in Science*. Notre Dame, Ind.: Univ. of Notre Dame Press.

———. 1988. "Theories, Family Resemblances and Analogy." In *Analogical Reasoning: Perspectives of Artificial Intelligence, Cognitive Science, and Philosophy*, ed. David H. Helman, 317–340. Dordrecht, Netherlands: Kluwer.

Holmes, David S. 1990. "The Evidence for Repression: An Examination of Sixty Years of Research." In *Repression and Dissociation: Implications for Personality Theory, Psychopathology, and Health*, ed. Jerome L. Singer, 85–102. Chicago: Univ. of Chicago Press.

Horowitz, Leonard M., D. L. Post, R. De Sales French, K. D. Wallis, and E. Y. Siegelman. 1981. "The Prototype as a Construct in Abnormal Psychology: 2. Clarifying Disagreements in Psychiatric Judgements." *Journal of Abnormal Psychology* 90:575–585.

Horowitz, Mardi. 1976. *Stress Response Reactions*. New York: Basic Books.

———. 1986. *Stress Response Reactions*. 2d ed. New York: Basic Books.

———. 1993. "Stress-Response Syndromes: A Review of Post-Traumatic Stress and Adjustment Disorders." In *International Handbook of Traumatic Stress Syndromes*, ed. John P. Wilson and Beverley Raphael, 49–60. New York: Plenum.

Horowitz, Mardi, Henry C. Markman, Charles H. Stinson, Bram Fridhandler, and Jess. H. Ghannam. 1990. "A Classification Theory of Defense." In *Repression and Dissociation: Implications for Personality Theory, Psychopathology, and Health*, ed. Jerome L. Singer, 61–84. Chicago: Univ. of Chicago Press.

Horowitz, Mardi, N. Wilner, and W. Alvarez. 1979. "Impact of Event Scale: A Measure of Subject Stress." *Psychosomatic Medicine* 41:209–218.

Hurst, Arthur F. 1918. *Medical Diseases of the War*. London: Edward Arnold.

———, ed. 1941. *Medical Diseases of the War*. 2d ed. London: Edward Arnold.

Israëls, Han, and Morton Schatzman. 1993. "The Seduction Theory." *History of Psychiatry* 4:23–59.

Jackson, Bruce. 1990. "The Perfect Informant." *Journal of American Folklore* 103:400–416.

Jackson, John Hughlings. 1931a. *Selected Writings of John Hughlings Jackson*. Vol. 1. London: Hodder and Stoughton.

———. 1931b. *Selected Writings of John Hughlings Jackson*. Vol. 2. London: Hodder and Stoughton.

James, William. 1896. *Principles of Psychology*. Vol. 2. New York: Holt.

Janet, Pierre. 1889. *L'automatisme psychologique: Essai de psychologie expérimentale sur les formes inférieures de l'activité humaine*. Paris: Alcan.

———. 1898. *Névroses et idées fixes*. Paris: Alcan.

———. 1901. *The Mental State of Hystericals: A Study of Mental Stigmata and Mental Accidents*. New York: G.P. Putnam.

———. 1925. *Psychological Healing*. New York: Macmillan.

Johnson, Mark. 1993. *Moral Imagination: Implications of Cognitive Science for Ethics*. Chicago: Univ. of Chicago.

Jones, A. Bassett, and Llwellyn J. Llwellyn. 1917. *Malingering or the Simulation of Disease*. London: Heinemann.

Jordan, B. Kathleen, William Schlenger, Richard Hough, Richard Kulka, Daniel Weiss, John Fairbank, and Charles Marmar. 1991. "Lifetime and Current Prevalence of Specific Psychiatric Disorders among Vietnam Veterans and Controls." *Archives of General Psychiatry* 48:207–215.

Jordan, Edward Furneaux. 1880. *Surgical Inquiries, Including the Hastings Essay on Shock, the Treatment of Surgical Inflammations, and Clinical Lectures*. London: J. and A. Churchill.

Karasu, Toksoz. 1986. "The Specificity versus Nonspecificity Dilemma: Toward Identifying Therapeutic Change Agents." *American Journal of Psychiatry* 143:687–695.

Kardiner, Abram. 1941. *The Traumatic Neuroses of War*. Washington, D.C.: National Research Council.

———. 1959. "Traumatic Neuroses of War." In *American Handbook of Psychiatry*, ed. Silvano Arieti, 245–257. New York: Basic Books.

Kardiner, Abram, and Herbert Spiegel. 1947. *War Stress and Neurotic Illness*. New York: Paul B. Hoeber.

Keane, Terrence M., Anne Marie Albano, and Dudley Blake. 1993. "Current Trends in the Treatment of Post-Traumatic Stress Disorder." In *Torture and Its Consequences: Current Treatment Approaches*, ed. Metin Basoglu, 363–401. Cambridge: Cambridge Univ. Press.

Keane, Terrence M., J. M. Caddell, and K. Taylor. 1988. "Mississippi Scale for Combat-Related Post-Traumatic Stress Disorder: Three Studies of Reliability and Validity." *Journal of Consulting and Clinical Psychology* 56:85–90.

Keane, Terrence M., John A. Fairbank, J. M. Caddell, R. T. Zimering, and M. E. Bender. 1985. "A Behavioral Approach to Assessing and Treating Post-

Traumatic Disorder in Viet Nam Veterans." In *Trauma and Its Wake*, ed. Charles Figley, 257–294. New York: Brunner/Mazel.

Keane, Terrence M., and D. G. Kaloupek. 1982. "Imaginal Flooding in the Treatment of Post-Traumatic Stress Disorder." *Journal of Consulting and Clinical Psychology* 50:138–140.

Keane, Terrence M., P. Malloy, and J. Fairbank. 1984. "Empirical Development of an MMPI Subscale for the Assessment of Combat-Related Post-Traumatic Stress Disorder." *Journal of Clinical and Consulting Psychology* 52:888–891.

Keane, Terrence M., and W. E. Penk. 1988. "The Prevalence of Post-Traumatic Stress Disorder." *New England Journal of Medicine* 318:1690–1691.

Kendell, R. E. 1988. "What Is a Case?," *Archives of General Psychiatry* 45:374–376.

Kernberg, Otto. 1986. *Severe Personality Disorders*. New Haven: Yale Univ. Press.

Kirk, Stuart A., and Herb Kuchins. 1992. *The Selling of DSM: The Rhetoric of Science in Psychiatry*. New York: Aldine de Gruyter.

Kitcher, Patricia. 1992. *Freud's Dream: A Complete Interdisciplinary Theory of Mind*. Cambridge, Mass.: MIT Press.

Klein, Donald, and Judith K. Rabkin. 1984. "Specificity and Strategy in Therapy Research and Practice." In *Psychotherapy Research*, ed. Janet Williams and Robert L. Spitzer, 306–329. New York: Guilford Press.

Klerman, Gerald L. 1984. "The Advantages of *DSM-III*." *American Journal of Psychiatry* 141:539–545.

———. 1986. "Historical Perspectives on Contemporary Schools of Psychopathology." In *Contemporary Directions in Psychopathology: Towards the DSM-IV*, ed. Theodore Millon and Gerald Klerman, 3–28. New York: Guilford Press.

———. 1989. "Psychiatric Diagnostic Categories: Issues of Validity and Measurement." *Journal of Health and Social Behavior* 30:11–25.

———. 1990. "Approaches to the Phenomenon of Comorbidity." In *Comorbidity of Mood and Anxiety Disorders*, ed. J. D. Maser and C. R. Cloninger, 13–37. Washington, D.C.: American Psychiatric Press.

———. 1991. "The *Osheroff* Debate: Finale." *American Journal of Psychiatry* 148:387–388.

Kline, Paul. 1992. "Problems of Methodology in Studies of Psychotherapy." In *Psychotherapy and Its Discontents*, ed. W. Dryden and C. Feltham, 64–86. Buckingham (U.K.): Open Univ. Press.

Kolb, Lawrence C. 1985. "The Place of Narcosynthesis in the Treatment of Chronic and Delayed Stress Reactions of War." In *The Trauma of War: Stress and Recovery in Vietnam Veterans*, ed. S. M. Sonnenberg, A. S. Blank, and T. A. Talbott, 211–226. Washington, D.C.: American Psychiatric Press.

———. 1987. "A Neuropsychological Hypothesis Explaining Post-Traumatic Stress Disorder." *American Journal of Psychiatry* 144:989–995.

———. 1989. "Heterogeneity of PTSD." *American Journal of Psychiatry* 146:811–812.

Kolb, Lawrence C., B. C. Burris, and S. Griffiths. 1984. "Propranolol and Clonidine in Treatment of the Chronic Post-Traumatic Stress Disorders of War." In *Post-Traumatic Stress Disorder: Psychological and Biological Sequelae*, ed. Bessel A. van der Kolk, 97–105. Washington, D.C.: American Psychiatric Press.

Kosten, T. R., J. W. Mason, E. L. Giller, R. B. Ostroff, and L. Harkness. 1987.

"Sustained Urinary Norepinephrine and Epinephrine Elevation in Post-Traumatic Stress Disorder." *Neuroendocrinology* 12:13–20.

Kraepelin, Emil. 1902. *Clinical Psychiatry: A Textbook for Students and Physicians*. New York: Macmillan.

———. 1973 [1920]. "Comparative Psychiatry." In *Themes and Variations in European Psychiatry*, ed. S. R. Hirsch and M. Shepherd, 7–30. Charlottesville: Univ. of Virginia Press.

Kramer, Milton. 1985. "Historical Roots and Structural Bases of the International Classification of Diseases." In *International Classification in Psychiatry: Unity and Diversity*, ed. Juan Mezzich and Michael von Cranach, 3–29. Cambridge: Cambridge Univ. Press.

Kravis, Nathan Mark. 1988. "James Braid's Psychophysiology: A Turning Point in the History of Dynamic Psychiatry." *American Journal of Psychiatry* 145:1191–1206.

Kubey, Craig. 1986. *The Vietnam Vet Survival Guide*. New York: Facts on File.

Kucklick, Henrika. 1991. *The Savage Within: The Social History of British Anthropology, 1885–1945*. Cambridge: Cambridge Univ. Press.

Kulka, R. A., W. E. Schlenger, J. A. Fairbank, R. L. Hough, B. K. Jordan, C. R. Marmar, and D. S. Weiss. 1990a. *Trauma and the Vietnam War Generation: Report of Findings from the National Vietnam Veterans Readjustment Study*. New York: Brunner/Mazel.

———. 1990b. *Trauma and the Vietnam War Generation: Tables of Findings and Technical Appendices*. New York: Brunner/Mazel.

Lachar, D. 1974. *The MMPI: Clinical Assessment and Automated Interpretation*. Los Angeles: Western Psychological Services.

Lakin, Martin. 1988. *Ethical Issues in the Psychotherapies*. New York: Oxford Univ. Press.

Lakoff, George. 1990. "The Invariance Hypothesis: Is Abstract Reasoning Based on Image-Schemas?" *Cognitive Linguistics* 1:39–74.

Lakoff, George, and Zoltan Kovecses. 1985. "The Cognitive Model of Anger in American English." In *Cultural Models in Language and Thought*, ed. Dorothy Holland and Naomi Quinn, 195–211. Cambridge: Cambridge Univ. Press.

Lambert, Michael J., David A. Shapiro, Allen E. Bergin. 1986. "The Effectiveness of Therapies." In *Handbook of Psychotherapy and Behavior*, 3d ed., ed. Sol L. Garfield and Allen E. Bergin, 157–211. New York: John Wiley.

Latour, Bruno. 1987. *Science in Action: How To Follow Scientists and Engineers through Society*. Cambridge: Harvard Univ. Press.

———. 1993. *We Have Never Been Modern*. Cambridge: Harvard Univ. Press.

Laufer, Robert S., E. Brett, and M. S. Gallops. 1985a. "Symptom Patterns Associated with Post-Traumatic Stress Disorder among Vietnam Veterans Exposed to War Trauma." *American Journal of Psychiatry* 142:1304–1311.

Laufer, Robert S., E. Brett, and M. S. Gallops. 1985b. "Dimensions of Post-Traumatic Stress Disorder among Vietnam Veterans." *Journal of Nervous and Mental Disease* 173:538–545.

Laufer, Robert S., M. S. Gallops, and E. Frey-Wouters. 1984. "War Stress and Trauma: The Vietnam Veteran Experience." *Journal of Health and Social Behavior* 25:65–85.

Laufer, Robert S., T. Yager, E. Frey-Wouters, and J. Donnellan. 1981. "Postwar Trauma: Social and Psychological Problems of Vietnam Veterans in the Aftermath of the Vietnam War." In *Legacies of Vietnam: Comparative Adjustment of Veterans and Their Peers*, ed. A. Egendorf, C. Kadushin, R. S. Laufer, G. Rothbart, and L. Sloan, vol. 3. Washington D.C.: Government Printing Office.

Leed, Eric J. 1979. *No Man's Land: Combat and Identity in World War I*. Cambridge: Cambridge Univ. Press.

Leese, Peter J. 1989. *A Social and Cultural History of Shellshock, with Particular Reference to the Experience of British Soldiers during and after the Great War*. Ph.D. diss. Milton Keynes: Open University.

Lepore, Randall F. 1986. *Post-Traumatic Stress Disorder: V.A. Disability Claims and Military Review*. Boston: Dominus Vobiscum Publications.

Lévy-Bruhl, Lucien. 1985 [1905]. *How Natives Think*. Princeton: Princeton Univ. Press.

Leys, Ruth. 1994. "Traumatic Cures: Shell Shock, Janet, and the Question of Memory." *Critical Inquiry* 20:623–662.

Lewis, Thomas. 1917. *Report upon Soldiers Returned as Cases of "Disordered Action of the Heart" (D.A.H) or "Valvular Disease of the Heart" (V.D.H.)*. London: His Majesty's Stationery Office.

———. 1920. *The Soldier's Heart and the Effort Syndrome*. New York: Paul B. Hoeber.

Lifton, Robert. 1967. *Death in Life: Survivors of Hiroshima*. New York: Random House.

———. 1973. *Home from the War: Vietnam Veterans, Neither Victims Nor Executioners*. New York: Simon and Schuster.

Lindy, J., B. L. Green, and M. C. Grace. 1987. "The Stressor Criterion and Post-Traumatic Stress Disorder." *Journal of Nervous and Mental Disease* 175:269–272.

Livesley, W. J. 1985. "The Classification of Personality Disorder: I. The Choice of Category Concept." *Canadian Journal of Psychiatry* 30:353–358.

Loftus, Elizabeth, and Katherine Ketcham. 1994. *The Myth of the Repressed Memory: False Memories and Allegations of Sexual Abuse*. New York: St.Martin's Press.

Luria, A. R. 1968. *The Mind of a Mnemonist*. Cambridge: Harvard Univ. Press.

Lutz, Catherine. 1988. *Unnatural Emotions: Everyday Sentiments on a Micronesian Atoll and Their Challenge to Western Theory*. Chicago: Univ. of Chicago Press.

McDougall, William. 1920a. "The Revival of Emotional Memories and Its Therapeutic Value." *British Medical Journal* i:23–29.

———. 1920b. "Four Cases of 'Regression' in Soldiers." *Journal of Abnormal Psychology* 15:136–156.

———. 1926. *An Outline of Abnormal Psychology*. London: Methuen.

McFarlane, Alexander C. 1986. "Post-Traumatic Morbidity of a Disaster: A Study of Cases Presenting for Psychiatric Treatment." *Journal of Nervous and Mental Disease* 174:4–14.

———. 1988. "Relationship between Psychiatric Impairment and a Natural Disaster: The Role of Distress." *Psychological Medicine* 18:129–139.

McDougall, William. 1989. "The Aetiology of Post-Traumatic Morbidity: Predisposing, Precipitating and Perpetuating Factors." *British Journal of Psychiatry* 154:221–228.

———. 1993. "Synthesis of Research and Clinical Studies: The Australia Bushfire Disaster." In *International Handbook of Traumatic Stress Syndromes*, ed. John P. Wilson and Beverley Raphael, 421–429. New York: Plenum Press.

MacLean, Paul D. 1990. *The Triune Brain in Evolution.* New York: Plenum.

MacRae, K. D. 1992. "Statistics in Psychiatric Research." In *The Scientific Basis of Psychiatry*, ed. M. Weller and M. Eysenck, 75–110. London: W. B. Saunders.

Marsden, C. D. 1986. "Hysteria—A Neurologist's View." *Psychological Medicine* 16:277–288.

Mason, John W., Earl L. Giller, T. R. Kosten, R. Ostroff, and L. Podd. 1986. "Urinary Free-Cortisol Levels in Post-Traumatic Stress Disorder Patients." *Journal of Nervous and Mental Disease* 174:145–149.

Mason, John W., Earl L. Giller, T. R. Kosten, and Rachel Yehuda. 1990. "Psychoendocrine Approaches to the Diagnosis and Pathogenesis of PTSD." In *Biological Assessment and Treatment of PTSD*, ed. Earl L. Giller, 65–86. Washington, D.C.: American Psychiatric Press.

Mellman, T. A., C. A. Randolph, O. Brawman-Mintzer, L. P. Flores, and F. J. Milanes. 1992. "Phenomenology and Course of Psychiatric Disorders Associated with Combat-Related Post-Traumatic Stress Disorder." *American Journal of Psychiatry* 149:1568–1574.

Menninger, W. C. 1948. *Psychiatry in a Troubled World: Yesterday's War and Tomorrow's Challenge.* New York: Macmillan.

Merskey, Harold. 1991. "Shell-Shock." In *150 Years of British Psychiatry*, ed. German Berios and Hugh Freeman, 245–267. London: Gaskell, Royal College of Psychiatrists.

Meyer-Gross, W., E. Slater, and M. Roth. 1954. *Clinical Psychiatry.* Baltimore: Williams and Wilkins.

Mezzich, Juan. 1989. "An Empirical Approach to the Definition of Psychiatric Illness." *British Journal of Psychiatry* 154 (suppl. 4):42–46.

Micale, Mark S. 1990. "Charcot and the Idea of Hysteria in the Male: A Study of Gender, Mental Science, and Medical Diagnosis in Late Nineteenth-Century France." *Medical History* 34:363–411.

———. 1993. "On the 'Disappearance' of Hysteria: A Study in the Clinical Deconstruction of a Diagnosis." *Isis* 84:496–536.

———. 1994. "Charcot and *Les Névroses Traumatiques*: Historical and Scientific Reflections." *Revue Neurologique* 150:498–505.

Michels, Robert. 1984. "First Rebuttal." *American Journal of Psychiatry* 141:548–551.

Miller, Emanuel. 1940. "Psychopathological Theories of Neuroses in War-Time." In *The Neuroses of War*, ed. Emanuel Miller, 105–118. London: Macmillan.

Millon, Theodore. 1986. "On the Past and the Future of the DSM-IV." In *Contemporary Directions in Psychopathology: Toward the DSM-IV*, ed. T. Millon and G. L. Klerman, 29–70. New York: Guilford Press.

Morris, Edwin. 1867. *A Practical Treatise on Shock after Surgical Operations and*

Injuries, with Especial Reference to Shock Caused by Railway Accidents. London: Robert Hardwicke.

Mott, Fredrick W. 1916. "The Lettsomian Lectures on the Effects of High Explosives upon the Central Nervous System." *Lancet* i:331–338, 441–449, and 545–553.

———. 1917. "The Microscopic Examination of the Brains of Two Men Dead of Commotio Cerebri (Shell Shock) without Visible Injury." *British Medical Journal* ii:612–615.

———. 1918a. "Neurasthenia: The Disorders and Disabilities of Fear." *Lancet* i:127–129.

———. 1918b. "The Psychology of Soldiers' Dreams." *Lancet* i:169–172.

———. 1919. *War Neuroses and Shell Shock.* London: Henry Frowde, Hodder and Stoughton.

Mulkay, Michael, and G. Nigel Gilbert. 1981. "Putting Philosophy to Work: Karl Popper's Influence on Scientific Practice." *Philosophy of Social Science* 11:389–467.

Musil, Robert. 1953. *The Man Without Qualities.* New York: Perigree Books.

Myers, Charles S. 1940. *Shellshock in France 1914–18: Based on a War Diary.* Cambridge: Cambridge Univ. Press.

Myers, Fred R. 1988. "The Logic and Meaning of Anger among Pintupi Aborigines." *Man* 23:589–610.

Myers, Greg. 1990. *Writing Biology: Texts in the Social Construction of Scientific Behavior.* Madison: Univ. of Wisconsin Press.

Needham, Rodney. 1980. *Reconnaissances.* Toronto: University of Toronto Press.

Newbury, Thomas. 1985. "Levels of Countertransference toward Vietnam Veterans with Post-Traumatic Stress Disorder." *Bulletin of the Menninger Clinic* 49:151–160.

Ofshe, Richard, and Ethan Watters. 1994. *Making Monsters: False Memories, Psychotherapy, and Sexual Hysteria.* New York: Scribner's.

Oppenheim, Hermann. 1888. *Die traumatischen Neurosen.* Berlin.

———. 1894. *Lehrbuch der Nervenkrankheiten für Arzte und Studirende.* Berlin: S. Karger.

———. 1911. *Text-Book of Nervous Diseases for Physicians and Students.* Vol. 2. London: T. N. Foulis.

Ortony, Andrew, and T. J. Turner. 1990. "What's Basic about Basic Emotions?" *Psychological Review* 97:315–331.

Otis, Laura. 1993. "Organic Memory and Psychoanalysis." *History of Psychiatry* 4:349–372.

Ourousoff, Alexandra. 1993. "Illusions of Rationality: False Premises of the Liberal Tradition." *Man* 28:281–298.

Page, Herbert W. 1883. *Injuries of the Spine and Spinal Cord without Apparent Mechanical Lesion, and Nervous Shock, in Their Surgical and Medico-Legal Aspects.* London: J. and A. Churchill.

Parfit, Derek. 1984. *Reasons and Persons.* Oxford: Clarendon Press.

Parry-Jones, Brenda, and William L.L. Parry-Jones. 1994. "Post-Traumatic Stress Disorder: Supportive Evidence from an Eighteenth Century Natural Disaster." *Psychological Medicine* 24:15–27.

Parson, Edwin R. 1986. "Transference and Post-Traumatic Stress Disorder: Combat Veterans' Transference to the Veterans Administration Center." *Journal of the American Academy of Psychoanalysis* 14:349–375.

Pavlov, Ivan P. 1927. *Conditioned Reflexes: An Account of the Physiological Activity of the Cerebral Cortex.* London: Oxford Univ. Press.

Penk, W., R. Robinowitz, J. Black, M. Dolan, W. Bell, W. Roberts, and J. Skinner. 1989. "Co-Morbidity: Lessons Learned about Post-Traumatic Stress Disorder (PTSD) from Developing PTSD Scales for the MMPI." *Journal of Clinical Psychology* 45:709–728.

Perconte, Stephen, and Anthony J. Gorenczy. 1990. "Failure to Detect Fabricated Post-Traumatic Stress Disorder with the Use of the MMPI in a Clinical Population." *American Journal of Psychiatry* 147:1057–1060.

Pick, Daniel. 1989. *Faces of Degeneration: A European Disorder, c. 1848–1918.* Cambridge: Cambridge Univ. Press.

Pitman, Roger K. and S. P. Orr. 1990. "The Black Hole of Trauma." *Biological Psychiatry* 27:469–471.

Pitman, Roger K., S. P. Orr, D. F. Forgue, J. B. deJong, and J. M. Claiborn. 1987. "Psychophysiologic Assessment of Post-Traumatic Stress Disorder Imagery in Vietnam Combat Veterans." *Archives of General Psychology* 44:970–975.

Pitman, Roger K., B. A. van der Kolk, S. P. Orr, and M. S. Greenberg. 1990. "Naloxone-Reversible Analgesic Response to Combat-Related Stimuli in Post-Traumatic Stress Disorder." *Archives of General Psychiatry* 47:541–544.

Popper, Karl. 1972. *Objective Knowledge: An Evolutionary Approach.* Oxford: Clarendon Press.

Porter, T. M. 1992. "Quantification and the Accounting Ideal in Science." *Social Studies in Science* 22:633–652.

Pritchard, D. A., and A. Rosenblatt. 1980. "Racial Bias in the MMPI: A Methodological Review." *Journal of Consulting and Clinical Psychology* 48:263–267.

Pruyser, P. 1975. "What Splits in Splitting? A Scrutiny of the Concepts of Splitting in Psychoanalysis and Psychiatry." *Bulletin of the Menninger Clinic* 39:1–46.

Reiser, Morton F. 1988. "Are Psychiatric Educators 'Losing the Mind'?" *American Journal of Psychiatry* 145:148–153.

Renan, E. 1923 [1890]. *L'avenir de la science: Pensées de 1848.* Paris: Calmann-Lévy.

Ribot, Théodule A. 1883. *Diseases of Memory: An Essay in the Positive Psychology.* London: Kegan Paul, Trench.

Richards, Graham. 1992. *Mental Machinery: The Origins and Consequences of Psychological Ideas, Part One: 1600–1850.* Baltimore: Johns Hopkins Press.

Ricoeur, P. 1981. "Narrative Time." In *On Narrative*, ed. W.J.T. Mitchell, 165–186. Chicago: Univ. of Chicago Press.

Riegelman, R. K., and R. P. Hirsch. 1989. *Studying a Test and Testing a Study: How to Read the Medical Literature.* Boston: Little, Brown.

Rivers, W.H.R. 1906. *The Todas.* London: Macmillan.

———. 1916. "The FitzPatrick Lectures on Medicine, Magic, and Religion." *Lancet* i:59–65, 117–123.

———. 1917. "Freud's Theory of the Unconscious." *Lancet* i:912–914.

————. 1918. *Dreams and Primitive Culture: A Lecture*. Manchester: Manchester Univ. Press.

————. 1919. "Inaugural Address." *Lancet* i:889–892.

————. 1920. *Instinct and the Unconscious: A Contribution to a Biological Theory of the Psycho-Neuroses*. Cambridge: Cambridge Univ. Press.

————. 1923. *Conflict and Dreams*. London: Kegan Paul, Trench, Trubner.

Rivers, W.H.R., and Henry Head. 1908. "A Human Experiment in Nerve Division." *Brain* 31:323–450.

Robins, Eli, and Samuel B. Guze. 1970. "Establishment of Diagnostic Validity in Psychiatric Illness: Its Application to Schizophrenia." *American Journal of Psychiatry* 126:983–987.

Robins, Lee N., and John E. Helzer. 1986. "Diagnostic and Clinical Assessment: The Current State of Psychiatric Diagnosis." *Annual Review of Psychology* 37:409–432.

Robins, Lee N., John E. Helzer, J. Croughhan, and K. S. Ratcliff. 1981. "National Institute of Mental Health Diagnostic Interview Schedule: Its History, Characteristics, and Validity." *Archives of General Psychiatry* 38:381–389.

Rorty, Amélie O. 1985. "Self-Deception, *Akrasia* and Irrationality." In *The Multiple Self*, ed. Jon Elster, 115–131. Cambridge: Cambridge Univ. Press.

Rosaldo, Michelle Z. 1984. "Toward an Anthropology of Self and Feeling." In *Culture Theory: Essays on Mind, Self, and Emotion*, ed. Richard Shweder and Robert LeVine, 137–157. Cambridge: Cambridge Univ. Press.

Rosch, E. 1977. "Human Categorization." In *Studies in Cross-Cultural Psychology*, ed. N. Warren, 1:1–49. New York: John Wiley.

Rose, R. M. 1984. "Overview of Endocrinology and Stress." In *Textbook of Endocrinology*, 6th ed., ed. R. H. Williams, 645–671. Philadelphia: W. B. Saunders.

Ross, T. A. 1941. "Anxiety Neuroses of War." In *Medical Diseases of the War*, 2d ed., ed. A. Hurst, 135–160. London: Edward Arnold.

Roth, Michael S. 1989. "Remembering Forgetting: *Maladies de la Mémoire* in Nineteenth-Century France." *Representations* 26:49–68.

Royal College of Physicians of London. 1955. "Mott, Sir Frederick Walker, K.B.E." In *Munk's Roll: Lives of the Fellows of the Royal College of Physicians of London, 1826–1925*, 358–359. London: Royal College of Physicians of London.

————. 1955. "Yealland, Lewis Ralph." In *Munk's Roll: Lives of the Fellows of the Royal College of Physicians of London, 1826–1925*, 465. London: Royal College of Physicians of London.

Royal Society of Medicine. 1916. "Special Discussion on Shell Shock without Visible Signs of Injury, January 25, 1916." *Proceedings of the Royal Society of Medicine, Section on Psychiatry* 9 (part 3):i–xliv.

————. 1919. "Report." *Lancet* i:437–438.

Rycroft, Charles. 1968. *A Critical Dictionary of Psychoanalysis*. Harmondsworth: Penguin.

Sabshin, Melvin. 1990. "Turning Points in Twentieth-Century American Psychiatry." *American Journal of Psychiatry* 147:1267–1274.

Salmon, Thomas W. 1917. *The Care and Treatment of Mental Diseases and War*

Neuroses ("Shell Shock") in the British Army. New York: War Work Committee of the National Committee for Mental Hygiene.

Sassoon, Siegfried. 1936. *Sherston's Progress.* London: Faber and Faber.

Scadding, J. G. 1990. "The Semantic Problems of Psychiatry." *Psychological Medicine* 20:243–248.

Schacter, Daniel L. 1982. *Stranger behind the Engram: Theories of Memory and the Psychology of Science.* Hillsdale, N.J.: Lawrence Erlbaum.

Scott, Wilbur. 1990. "PTSD in DSM-III: A Case in the Politics of Diagnosis and Disease." *Social Problems* 37:294–310.

Selye, Hans. 1950. *The Physiology and Pathology of Exposure to Stress: A Treatise Based on the Concepts of the General-Adaptation-Syndrome.* Montreal: Acta.

Sharfstein, Steven S., and Howard Goldman. 1989. "Financing the Medical Management of Mental Disorders." *American Journal of Psychiatry* 146:345–349.

Shepherd, Michael. 1994. "Neurolepsis and the Psychopharmacological Revolution: Myth and Reality." *History of Psychiatry* 5:89–96.

Shorter, Edward. 1992. *From Paralysis to Fatigue: A History of Psychosomatic Illness in the Modern Era.* New York: Free Press.

Showalter, Elaine. 1985. *The Female Malady.* New York: Pantheon.

Sierles, Frederick, Jang-June Chen, Robert McFarland, and Michael Taylor. 1983. "Post-Traumatic Stress Disorder and Concurrent Psychiatric Illness: A Preliminary Report." *American Journal of Psychiatry* 140:1177–1179.

Silver, Jonathan M., Diane Sandberg, and Robert Hales. 1990. "New Approaches in the Pharmacotherapy of Post-Traumatic Stress Disorder." *Journal of Clinical Psychiatry* 51 (suppl. 10): 33–38.

Slobodin, Richard. 1978. *W.H.R. Rivers.* New York: Columbia Univ. Press.

Smith, C.U.M. 1982a. "Evolution and the Problem of Mind: Part I. Herbert Spencer." *Journal of the History of Biology* 15:55–88.

———. 1982b. "Evolution and the Problem of Mind: Part II.John Hughlings Jackson." *Journal of the History of Biology* 15:241–262.

Smith, G. Elliot. 1916. "Shock and the Soldier." *Lancet* i:813–817.

———. 1922. "The Late Dr. W.H.R. Rivers." *Lancet* i:1222.

Smith, Roger. 1992. *Inhibition: History and Meaning in the Sciences of Mind and Brain.* Berkeley: Univ. of California Press.

Snyder, Solomon. 1986. *Drugs and the Brain.* New York: Scientific American Library.

Solomon, Susan D., Ellen Gerrity, and Alyson Muff. 1992. "Efficacy of Treatments for Post-Traumatic Stress Disorder." *Journal of the Americal Medical Association* 268:633–638.

Sparr, Landy, and Loren D. Pankratz. 1983. "Factitious Post-Traumatic Stress Disorder." *American Journal of Psychiatry* 140:1016–1019.

Spence, Donald. 1982. *Narrative Truth and Historical Truth.* New York: W.W. Norton.

Spencer, Herbert. 1855. *Principles of Psychology.* London: Longman, Brown, Green, and Longmans.

Spiegel, David, and Etzel Cardena. 1990. "New Uses of Hypnosis in the Treatment of Post-Traumatic Stress Disorder." *Journal of Clinical Psychiatry* 51 (suppl. 10):39–43.

Spitzer, Robert L., Jacob Cohen, Joseph Fleiss, and Jean Endicott. 1967. "Quantification of Agreement in Psychiatric Diagnosis." *Archives of General Psychiatry* 17:83–87.

Spitzer, Robert L., Jean Endicott, and Eli Robins. 1978. "Research Diagnostic Criteria." *Archives of General Psychiatry* 35:773–782.

Spitzer, Robert L., M. B. First, J.B.W. Williams, K. Kendler, H. A. Pincus, and G. Tucker. 1992. "Dr. Spitzer and Associates Reply." *American Journal of Psychiatry* 149:1619–1620.

Spitzer, Robert L., Miriam Gibbon, Andrew Skodol, Janet Williams, and Michael First. 1989. *DSM-III-R Casebook: A Learning Companion to the Diagnostic and Statistical Manual of Mental Disorders (Third Edition, Revised).* Washington, D.C.: American Psychiatric Press.

Spitzer, Robert L., and Janet Williams. 1980. "Classification in Psychiatry." In *Comprehensive Textbook in Psychiatry/III*, 3d ed., ed. H. I. Kaplan, A. Freedman, and B. J. Sadock, 1035–1072. Baltimore: Williams and Wilkins.

Star, Susan Leigh. 1992. "The Skin, the Skull, and the Self: Toward a Sociology of the Brain." In *So Human a Brain: Knowledge and Values in the Neurosciences*, ed. Anne Harrington, 204–228. Boston: Birkhäuser.

Stone, Martin. 1988. "Shellshock and the Psychologists." In *The Anatomy of Madness: Essays in the History of Psychiatry*, ed. W. F. Bynum, R. Porter, and M. Shepherd, 2:242–271. London: Tavistock.

Strachey, James. 1962. "The Emergence of Freud's Fundamental Hypotheses." In *Standard Edition of the Complete Psychological Works of Sigmund Freud*, ed. James Strachey, 3:62–68. London: Hogarth Press.

Sulloway, Frank. 1983. *Freud, Biologist of the Mind.* New York: Basic Books.

Taine, Hippolyte. 1870. *De l'intelligence.* Paris: Hachette.

Talbot, John, and Robert L. Spitzer. 1980. "An In-Depth Look at *DSM-III*: An Interview with Robert Spitzer." *Hospital and Community Psychiatry* 31:25–32.

Tambiah, Stanley J. 1990. *Magic, Science, Religion, and the Scope of Rationality.* Cambridge: Cambridge University Press.

Taylor, Gabrielle. 1985. *Pride, Shame, and Guilt: Emotions of Self-Assessment.* Oxford: Clarendon Press.

Terdiman, Richard. 1993. *Present Past: Modernity and the Memory Crisis.* Ithaca: Cornell Univ. Press.

Terr, Lenore. 1994. *Unchained Memories: True Stories of Traumatic Memories.* New York: Basic Books.

Thompson, John B. 1984. *Studies in the Theory of Ideology.* Berkeley: Univ. of California Press.

Trimble, Michael R. 1985. "Post-Traumatic Stress Disorder: History of a Concept." In *Trauma and Its Wake*, ed. Charles R. Figley, 5–14. New York: Brunner/Mazel.

Tuke, D. Hack, ed. 1892. *A Dictionary of Psychological Medicine.* Philadelphia: Blakiston.

Turner, William A. 1916. "Arrangements for the Care of Cases of Nervous and Mental Shock Coming from Overseas." *Lancet* i:1073–1075.

Tversky, Amos, and Daniel Kahneman. 1981. "The Framing of Decisions and the Psychology of Choice." *Science* 211:453–458.

Ursano, R., and J. E. McCarroll. 1990. "The Nature of a Traumatic Stressor: Handling Dead Bodies." *Journal of Nervous and Mental Disease* 178:396–398.

Vaillant, George E. 1984. "The Disadvantages of *DSM-III* Outweigh Its Advantages." *American Journal of Psychiatry* 141:542–545.

Vaillant, George E., and Paula Schnurr. 1988. "What Is a Case?: A 45-Year Study of Psychiatric Impairment within a College Sample Selected for Mental Health." *Archives of General Psychiatry* 45:313–319.

van der Kolk, Bessel A., M. Greenberg, H. Boyd, and J. Krystal. 1985. "Inescapable Shock, Neurotransmitters, and Addiction to Trauma: Toward a Psychobiology of Post-Traumatic Stress." *Biological Psychiatry* 20:314–325.

van der Kolk, Bessel A., and Onno van der Hart. 1989. "Pierre Janet and the Breakdown of Adaptation in Psychological Trauma." *American Journal of Psychiatry* 146:1530–1540.

Wallace, Edwin. 1988. "What is 'Truth'?: Some Philosophical Contributions to Psychiatric Issues." *American Journal of Psychiatry* 145:137–147.

Walters, G. D., R. L. Green, T. B. Jeffrey, D. J. Kruzich, and J. J. Haskin. 1983. "Racial Variations in the MacAndrew Alcoholism Scale of the MMPI." *Journal of Clinical and Consulting Psychology* 51:947–948.

War Department. 1946. "Psychiatric Nomenclature." *Journal of Nervous and Mental Disease* 104:108–199.

War Office Committee. 1922. *Report of the War Office Committee of Enquiry into "Shell Shock."* London: His Majesty's Stationery Office.

Ware, J. H., F. Mosteller, F. Delgado, C. Donnelly, and J. A. Inglefinger. 1992. "P Values." In *Medical Uses of Statistics*, ed. J. C. Bailar and F. Mosteller, 181–200. Boston: New England Journal of Medicine Books.

Warnock, Mary. 1987. *Memory*. London: Faber and Faber.

Watkins, L. R. and D. J. Mayer. 1986. "Multiple Endogenous Opiate and Non-Opiate Analgesia Systems: Evidence of Their Existence and Clinical Implications." *Annals of the New York Academy of Science* 467:273–297.

———. 1988. "Organization of Endogenous Opiate and Nonopiate Pain Control Systems." *Science* 216:1185–1192.

Watson, I.P.B., L. Hoffman, and G. V. Wilson. 1988. "The Neurophysiology of Post-Traumatic Stress Disorder." *British Journal of Psychiatry* 152:164–173.

Weiner, Norman, and Palmer Taylor. 1985. "Neurohormonal Transmission: The Autonomic and Somatic Nervous Systems." In *The Pharmacological Basis of Therapeutics*, ed. A. G. Gilman, L. S. Goodman, T. W. Rall, and F. Murad, 66–99. New York: Macmillan.

Weissman, Myrna, and Gerald L. Klerman. 1978. "Epidemiology of Mental Disorders." *Archives of General Psychiatry* 35:705–712.

Widiger, T. and A. Frances. 1985. "The *DSM-III* Personality Disorders." *Archives of General Psychiatry* 42:615–623.

Williams, Janet, Robert Spitzer, and Andrew Skodol. 1985. "*DSM-III* in Residency Training: Results of a National Survey." *American Journal of Psychiatry* 142:755–758.

Wilson, Mitchell. 1993. "DSM-III and the Transformation of American Psychiatry: A History." *American Journal of Psychiatry* 150:399–410.

Wittkower, Eric, and J. P. Spillane. 1940a. "Neuroses in War." *British Medical Journal* i:223–225, 265–267.

———. 1940b. "A Survey of the Literature of Neuroses in War." In *The Neuroses of War*, ed. Emanuel Miller, 1–32. London: Macmillan.

Wolfsohn, Julian M. 1918. "The Predisposing Factors of War Psycho-Neuroses." *Lancet* i:177–180.

Wright, Lawrence. 1994. *Remembering Satan*. New York: Knopf.

Yager, Thomas, Robert Laufer, and M. S. Gallops. 1984. "Some Problems Associated with War Experiences in Men of the Vietnam Generation." *Archives of General Psychiatry* 41:327–333.

Yapko, Michael D. 1994. *Suggestions of Abuse: True and False Memories of Childhood Sexual Trauma*. New York: Simon and Schuster.

Yealland, Lewis R. 1918. *Hysterical Disorders of Warfare*. London: Macmillan.

Yehuda, Rachel, Steven M. Southwick, and Earl L. Giller. 1992. "Exposure to Atrocities and Severity of Chronic Post-Traumatic Stress Disorder in Vietnam Combat Veterans." *American Journal of Psychiatry* 149:333–336.

Young, Robert M. 1990. *Mind, Brain, and Adaptation in the Nineteenth Century: Cerebral Location and Its Biological Context from Gall to Ferrier*. New York: Oxford Univ. Press.

Index

Abraham, Karl, 78
abreactive therapy, 36–37, 73–75, 81, 92, 224–225
Adrian, E. D., 51, 55, 56, 70–72, 74
affect logic, 6, 124–125, 217–223, 285, 293n.5
Akhtar, Salman, 262n.2
Alam, Chris, 294n.6
Alexander, D. A., 137–138
Alexander, Franz, 96
alternating personalities, 30, 33
amnesia (*see also* memory; consciousness), 26ff., 75–77
analogical reasoning (*see also* classification), 122–123, 283–285; and meaning variance, 123–124, 128, 131, 136
Andraesen, Nancy, 110–112, 114, 122
anxiety, 61, 66, 77–81; anxiety neurosis, 64, 66
Arbib, M., 10
Armstrong, S. L., 119
Atkinson, Ronald, 112–114
Atran, Scott, 119
autognosis, 67, 74–77, 83
Azam, E., 293n.6

Babington, Anthony, 63
Babinski, Joseph, 73, 136
Bailar, J. C., 266
Baldessarini, Ross, 276
Barker, Pat, 68
Barnes, Barry, 122
Bayer, Ronald, 99, 101
Beard, George, 52
Beaunis, H. E., 32
Beck, Aaron, 178, 180, 206
Benison, Saul, 85, 295n.8
Bergson, Henri, 293n.4
Bernard, Claude, 294n.8
Bernheim, Hippolyte, 32, 46, 291n.1
Beveridge, A. W., 72
Blashfield, R., 95
Bloom, Floyd, 276
Bloor, David, 122
Boehnlein, J. K., 3

Bowler, Peter, 292n.2
Brende, Joel, 296n.2
Breslau, Naomi, 116, 129–130, 138–139, 295n.2
Brett, Elizabeth, 146
Breuer, Josef, 6, 36–37, 125, 224–225
Brill, Norman, 92–93
Brown, William, 62, 63, 73–75, 295n.8
Burns, David, 178
Bury, Judson, 53, 54, 295n.8

Cahill, Larry, 297n.2
Canguilhem, Georges, 39–40
Cannon, Walter B., 21–27, 40, 41, 82, 85, 292n.2, 295n.8
Cantor, N., 120
Cazeneuve, Jean, 45
Charlton, Bruce, 270
Charcot, Jean-Martin, 5, 19–21, 38, 39, 46, 57, 73, 119, 224, 291n.1
Charney, Dennis, 284, 286, 297n.1
Chouros, George, 297n.1
Ciompi, Luc, 219
Clark, Michael, 47
Clarke, Edwin, 292n.2
classification (*see also* analogical reasoning; diagnosis of mental disorders): Aristotelian, 94; contrast between exemplary and typical cases, 120, 122; family resemblances, 118; monothetic classification, 118–119, 121, 128; polythetic classification, 118–119, 128; validity, 104–105, 121, 134
Cleland, Max, 114
collective memory (*see also* memory), 128–129
Collie, John, 57–58
Collins, H. M., 10
Colquhoun, John C., 31
compensation paid for traumatogenic disorders: a motive for dissimulating, 7, 17, 20; a motive for opposing psychiatric recognition of PTSD, 113; policies of the Veterans Administration, 113, 213–214, 290; comparison with shell-shock, 294n.5

About the Author

ALLAN YOUNG is Professor of Anthropology at McGill University, in the Departments of Social Studies of Medicine, Anthropology, and Psychiatry.